IT STARTED WITH *Copernicus*

ALSO BY KEITH PARSONS

The Science Wars

Rational Episodes

God and the Burden of Proof

IT STARTED WITH *Copernicus*

VITAL QUESTIONS ABOUT SCIENCE

KEITH PARSONS

59 John Glenn Drive
Amherst, New York 14228

Published 2014 by Prometheus Books

Cover image © Lawrence Manning/Media Bakery
Cover design by Jacqueline Nasso Cooke

Portions of this book were previously published by McGraw-Hill (2006) as *Copernican Questions: A Concise Invitation to the Philosophy of Science*.

Prometheus Books recognizes the following registered trademarks mentioned within the text: Hula Hoop®, Kindle®, Nook®, Pixar®.

Inquiries should be addressed to
Prometheus Books
59 John Glenn Drive
Amherst, New York 14228
VOICE: 716–691–0133
FAX: 716–691–0137
WWW.PROMETHEUSBOOKS.COM

18 17 16 15 14 5 4 3 2 1

Library of Congress Cataloging-in-Publication Data

Parsons, Keith M., 1952–
[Copernican questions]
It started with Copernicus : vital questions about science / by Keith Parsons.
pages cm
Originally published as: Copernican questions (Boston : McGraw-Hill, 2005).
Includes bibliographical references and index.
ISBN 978-1-61614-929-1 (pbk.) — ISBN 978-1-61614-930-7 (ebook)
1. Science—Philosophy—History. 2. Science—Methodology. 3. Constructive realism. I. Title.

Q174.8.P37 2014
501—dc23

2014006791

Printed in the United States of America

CONTENTS

PREFACE

It Started with Copernicus is an expansion and revision of the book *Copernican Questions: A Concise Invitation to the Philosophy of Science*, published by McGraw-Hill in 2006. That earlier book was intended as an introductory text in the philosophy of science, and I am happy to hear that it has been used successfully by a number of instructors. However, *Copernican Questions* is now out of print, and Prometheus Books decided that in bringing out this new book, it would be so different that it should have a new concept and a new title. This is a trade book meant for any reader interested in being immersed in some of the philosophical issues and controversies raised by modern science.

Three entirely new chapters have been added, nearly doubling the size of the book. The last two of these new chapters deal with topics not normally covered in standard philosophy-of-science texts—the "naturalization" of the theory of knowledge and ethics, and the impact of science on the human self-image. The first topic is normally found in texts on epistemology rather than the philosophy of science. The latter topic is sometimes addressed in books on cognitive science and its implications such as Paul Thagard's *The Brain and the Meaning of Life* (Princeton: Princeton University Press, 2010) and Owen Flanagan's *The Problem of the Soul* (New York: Perseus Book Group, 2002). I think that these topics should be included here, in a text that focuses on the philosophy of science. In his book *Darwin's Dangerous Idea* (New York: Simon& Schuster, 1995) Daniel Dennett says that Darwinism is "universal acid" that corrodes all ideologies and dogmas. Actually, all of science is acidic in that sense—eating away at complacent assumptions and staid beliefs. Which of our comfortable old ideas are dissolved by scientific knowledge and methods, and which can be retained? This is a compelling question.

While I still think that this book is quite appropriate for an introductory course, it is not a typical introduction to the philosophy of science. A number of such introductory texts exist, and quite a few of them are excellent. This book differs from such textbooks in aim, content, and tone. My aim is not to introduce the whole field of philosophy of science, and not even to comprehensively survey the topics covered. Instead, I want to invite the reader to jump into some of the biggest, deepest, and sometimes nastiest controversies that have roiled around the nature, authority, role, extent, and implications of natural science. The academic nature of these debates masks—sometimes very poorly—the intensely passionate nature of the disagreements. It is not a detached discussion, but is often the intellectual equivalent of a barroom brawl. The deep disagreements about science that surfaced in the "science wars" of the 1990s generated debates that were animated and at times rancorous.

Another way that this book differs from a textbook is that I cover fewer topics but go into much greater depth on the ones I consider. My experience is that people learn more from in-depth discussions than from surveys, however skillfully these are done. For instance, this book goes into some detail with Thomas Kuhn's *The Structure of Scientific Revolutions* because Kuhn's book has had a major impact on academic and even popular discussions of science ever since its publication over fifty years ago. The entire second chapter deals with Kuhn's various doctrines of incommensurability, which, in my view, constitute his most significant challenge to traditional notions of scientific rationality.

In the so-called science wars of the 1990s, much more radical critiques of scientific rationality were subjects of lively and often-acrimonious debate. Chapter 3 examines a few of the most radical critiques of scientific objectivity from the "academic left"—social constructivists, postmodernists, and gender feminists—as well as those arising from the antievolutionary right, as represented by Phillip E. Johnson.

While the "science wars" involved academics from many fields and specialties, from physics to comparative literature, professional philosophers of science were more concerned with the issue of scientific realism, that is, whether the aim of science should be to discover the deep structure of the universe, or whether, more modestly, it should aim merely at predictive accuracy. Here also a particular book prompted much of the discussion, Bas van Fraassen's *The Scientific Image*. Van Fraassen's book and other arguments over realism and antirealism will be the subject of chapters 4 and 5. Standard introductory topics such as explanation and confirmation are examined *only* to the extent that they bear on the issues of rationality and realism.

Another way that this book departs from the norm is in tone. Textbooks generally maintain a guardedly neutral or reserved tone. Here I will not hesitate to make pointed critiques and frank assessments. I do not conceal my own views on controversial issues. It will be clear to any reader precisely what I think on the issues I discuss here. I do not try to hide these views behind a façade of neutrality because I think that readers have every right to know where the author stands on disputed points and why.

I think that texts that present arguments in a didactic manner, with all viewpoints getting equal and neutral treatment, create a false impression—one all too common in today's "dumbed-down" milieu—that in controversies it all comes down to "just someone's opinion." By drawing definite conclusions and offering arguments to support them, I am driving home the point that some philosophical conclusions *are* better than others. However, I certainly do not consider any of my arguments to be the last word on any topic. One reason that I take such definite stands is, as I say above, to provide a clear focus for readers to agree or disagree with me as they see fit. By stating my position and arguing for it vigorously, I am inviting readers to take up the debate.

The referees who read the manuscript for *Copernican Questions* had far more complaints about chapter 3 than any other part of the book. One charge was that I take a sarcastic or dismissive tone toward the writers I examine there. I have made every effort to remove or reword any passages that might give that impression. Another complaint of several referees of the original text was that the particular writers I examine in the third chapter are not the best representatives of social constructivism, postmodernism, or feminism. Rather, they charged that I had selected the most extreme and least plausible representatives of those viewpoints. First, I disagree that the particular authors I chose are unrepresentative of their respective viewpoints. They may state their views more immoderately, with less subtlety, or with fewer qualifications than others do, but I do not see that their positions are far outside the "mainstream" writers in those camps. As I see it, many of the "science critics" intentionally took controversial and confrontational positions that moved "extreme" views close to the norm.

Second, let me emphasize again that the aim of this book is not to offer a survey or overview of any sort. When I mention, for instance, the social constructivists, postmodernists, or feminist philosophers of science, I examine only a very few of these writers and I make no attempt to present a comprehensive survey of their fields. Major figures in these fields, with the exception of one or two individuals, are not mentioned at all. For instance, any survey of feminist philosophy of science would have to deal with Evelyn Fox Keller, Helen Longino, and Kathleen Okruhlik. I mention none of these simply because my aim was not to write a conspectus but to get into the issues that, in my experience, are most effective in engaging and motivating readers. Sandra Harding filled the bill very well here. She makes strong claims and backs them with clever arguments, so I focus on her as the representative of feminist philosophy of science.

Actually, some of the topics in this book will seem somewhat out of date for professional philosophers. The realism/antirealism issue is presently deadlocked and the issues about Kuhn, incommensurability, and the rationality of science are well-trodden ground. Professional philosophers of science might regard both issues with a bit of ennui. But this book is not primarily aimed at professional philosophers, though I hope that any who read it will find points of interest here. Chiefly, I want to engage readers who are not specialists in the philosophy of science because I think that these issues are far too interesting and important to leave just to the professionals. Further, the best way to begin to understand the current state of the philosophy of science is to understand its recent past, and that is what I try to do in chapters 2 through 5.

I also go into much greater detail in describing theories, controversies, and incidents from current science and the history of science than is usual for an elementary philosophy-of-science book. These narratives often go into considerable detail since I do not assume that readers will have a broad background in the natural sciences or the history of science. The reason for including so many such examples of actual scientific practice is to show that the problems of the philosophy of science are *real* problems that arise from real science and are not just armchair amusements for philosophers. Also, it is important to show how real science bears on the evaluation of philosophers' claims about science. I think that one very valuable contribution of Thomas Kuhn was to show that legitimate philosophy of science must be true to how science really has been done.

A detailed, chapter-by-chapter summary of this book's contents is found at the end of chapter 1. The reason it is placed there rather than in this preface is that the information given in the first chapter is really necessary for readers to understand what the rest of the book is about.

As I say, in this book I have not attempted to maintain the impersonal tone of studious solemnity typical of much philosophical writing. As is already obvious, I do not eschew the use of the first person, singular pronoun. The book's tone becomes more personal and polemical in the final chapter, which is only appropriate for the topics covered there.

I also write in a more conversational and informal style than is usual for textbooks. My reason for writing in this more personal manner is simply that I find that this works best in my teaching. Students respond much more readily to a looser conversational style than to a stiff, formal "lecture" mode, and I think that others will also. Throughout I have tried to follow Einstein's dictum that when you attempt to convey specialized knowledge to a nonspecialist audience, you should explain things as simply as possible—but no simpler. So I have not shied away from or watered down some rather-technical points. Though I have striven for clarity, I do not patronize readers by talking down to them. I expect them to be willing to think hard and to discover for themselves the deep pleasure of philosophical debate.

ACKNOWLEDGMENTS

COPERNICAN QUESTIONS, 2006

Much of this book was written while I was on faculty-development leave during the fall term of 2004. I would like to thank the faculty committee; my dean, Bruce Palmer; and the University Provost, Jim Hayes, for recommending and approving the leave time, as well as for their many encouraging remarks and expressions of support. I have always felt that my research and writing were strongly supported by administrators and colleagues, and I am grateful for that.

Jon-David Hague, Rona Allison, and the staff at McGraw-Hill were unfailingly polite and helpful during the process of producing this book. Five referees who read the (very raw) draft of some of the manuscript chapters offered many helpful criticisms and comments, for which I am greatly appreciative. I made a special effort to rewrite sections where, as they pointed out, my rhetoric got a bit carried away. Others of their suggestions, such as favoring the addition of extra topics, while eminently reasonable, could not be accommodated without fundamentally changing the nature of the text. Some criticisms of some referees struck me as wrongheaded, and I chose to ignore these. Naturally, I have nobody but myself to blame for any faults in the text resulting from this decision.

Others who read some or all of the manuscript and made many helpful suggestions include philosopher Cory Juhl of the University of Texas at Austin, philosopher Robert Almeder of Georgia State University, and two of my friends here at University of Houston–Clear Lake, physicist David Garrison and biochemist Ron Mills. Everyone these days is exceedingly busy, and it means a lot to me when people take the time to read things and give you thoughtful feedback.

I would like to thank my professors from the Department of History and Philosophy of Science and the Department of Philosophy at the University of Pittsburgh for creating an outstanding environment for the study of the philosophy of science. I especially would like to thank Professors John Earman and Clark Glymour, who jointly conducted a seminar on the realism/antirealism issue that I attended as a grad student at Pitt. This was a terrific seminar that showed how good graduate education can (all too rarely) be. Also, I would like to take a moment to honor the memory of Professor Wesley Salmon, whose tragic death in an automobile accident saddened us all and deprived the philosophy of science of one of its most penetrating and lucid intellects.

Naturally, while writing a book you impose on your family more than anyone else. I would like to thank my wife, Carol, for putting up with the weekends spent at the office writing and revising, and for cheerfully tolerating my general air of distraction when some problem would preoccupy me. We did take one week of the faculty-development leave time for a wonderful vacation in Paris. So, I lovingly dedicate this book to Carol, and I hope that whenever she sees it, she will think of Paris.

IT STARTED WITH COPERNICUS, 2014

When you commit to two major publishing projects with short due dates, as—probably foolishly—I did, you are bound to inconvenience some people. I would like first and foremost to thank Dr. Robert Zaballa, my coauthor on an immediately upcoming project, for his patience in waiting the extra months it took me to finish this book. I was confronted with two projects, neither of which I felt I could forego, and I greatly appreciate his forbearance.

As always, I would like to thank my wife, Carol, for her patience and understanding when I am absorbed with a major project. I would also like to thank a number of my colleagues who in the goodness of their hearts continue to be on speaking terms with me though my publishing deadlines kept me from doing my fair share of some important work, forcing them to take up the slack. I hope to make it up to them by doing extra work once these big projects are done.

A number of philosophers read parts of the new material and gave me very valuable feedback. John Beversluis read each of the new chapters and had many helpful comments and suggestions that improved each of those chapters. Charles Echelbarger also read and commented on a number of passages. Andrew Melnyk read the portions of chapter 8 that referred to his argument and returned valuable comments to me. When I posted online some of my critiques of their "argument from reason," Professors Charles Taliafero and Stewart Goetz kindly sent me a response by private e-mail. Unfortunately, the present work did not have room for their response and my reply, so I hope to further address these points in a later publication. Of course, any remaining errors of fact or interpretation are my own responsibility.

Special thanks must go to our faculty-suite secretary Jacque Darragh for her help in generating the manuscript for this book. "Support" staff always deserve more appreciation than they get. Pretty much nothing would get done without them.

Finally, as always, Steven L. Mitchell and his staff at Prometheus Books did their usual highly professional job in producing this book. The high production values of Prometheus and the dedication of its staff always make it a pleasure to work with them.

CHAPTER ONE

COPERNICAN QUESTIONS

On May 24, 1543, Europe's foremost astronomer lay dying. The story goes that Nicholas Copernicus was on his deathbed when he received from the printer the first copies of his great work *De Revolutionibus Orbium Caelestium—On the Revolutions of the Heavenly Spheres*. This work proposed nothing less than a radical revision of the established view of the universe. Copernicus argued that the earth is not the immovable center of the universe but rather is part of a solar system. He proposed that the earth, along with its sister planets Mercury, Venus, Mars, Jupiter, and Saturn, are arrayed in a series of concentric circular orbits about the sun (or, to be precise, a point very close to the sun). Copernicus was not the first to make this startling suggestion; several thinkers in ancient Greece had entertained the notion of a heliocentric (sun-centered) cosmology. But by Copernicus's day, the geocentric (earth-centered) cosmology had gained the full support of science, philosophy, and, most important of all in those days, theology. So deeply entrenched was the geocentric view that Copernicus's system was inevitably regarded as shocking, absurd, or perhaps even heretical. Yet despite the opposition of scientists, philosophers, and the church—which famously condemned Galileo for defending the Copernican view—the heliocentric theory had won by about 1650.

The Copernican Revolution, like the Darwinian revolution of the nineteenth century, impacted not just science but something deep within the human psyche. Anyone who has followed the recent debates over "scientific creationism" or "intelligent design theory" knows that the Darwinism continues to elicit passionate feelings. In a sense, we are also still coming to terms with Copernicus.

WHAT WAS COPERNICUS'S REVOLUTION?

Just what was so radical about Copernicus's theory, and why did it shock so many of his contemporaries? Why did others find it so inspiring that it is fair to say that the whole Scientific Revolution began with Copernicus? To answer these questions we have to get deeper into the history. Merely to say that with Copernicus science moved from an earth-centered to a sun-centered cosmology is hardly adequate to understand the depth and breadth of a transformation so profound that it really marks the beginning of the modern world and the demise of the medieval one.

New theories do not arise in a vacuum. They are proposed in the face of established theories that have served long and honorably and have faced down many previous challengers. So, when new theories win, old theories lose. The fate of discarded theories is not pretty. They become the objects of mirth or pity as later generations find it hard to imagine how the universe could ever have been conceived in such terms.

Such condescension is inappropriate. As the saying goes, the past is another country, and when we disdain past views merely because they are old, we are behaving like someone who laughs at the customs and beliefs of other cultures. This does not mean that we must regard past theories—or even the customs and beliefs of other cultures—as equal to our own. The point is that theories accepted by people in previous centuries, including theories long since recognized as false, were in their day eminently reasonable views that explained the universe in ways that were rigorously logical, coherent, beautiful, satisfying, and comprehensive. The intellectual pillars of the late medieval worldview included the teachings of the church, of course, but also the doctrines of two eminent thinkers of pagan antiquity, the philosopher/scientist Aristotle and the astronomer Claudius Ptolemy.

Aristotle (384–322 BCE) is always ranked with Plato as one of the two greatest philosophical intellects of the ancient world. Aristotle studied under Plato, but he developed a distinct philosophy of extraordinary scope and power. He thought about everything and had important things to say about almost every conceivable topic. As opposed to Plato, whose mind ascended to the transcendent realm of eternal essences, Aristotle's intellect was focused on the physical world. The painting *The School of Athens* by the Renaissance master Raphael, depicts Plato and Aristotle as its two central figures. As they walk, engaged in animated debate, Plato is pointing heavenward while Aristotle gestures with his palm down and his arm forward and parallel to the ground, as if to emphasize his concern with the earthly. It may seem odd that so this-worldly a thinker as Aristotle could have become the intellectual paragon of deeply religious intellectuals of Christian Europe. How this happened is one of the most remarkable stories in intellectual history.

By the late centuries of the first millennium, during the so-called Dark Ages of Europe, Aristotle's writings, except for a few of his works on logic and method, had been lost to the Latin-speaking scholars of the West. By contrast, in the East, in the intellectual centers of Muslim culture, the works of Aristotle were very well known and were the focus of much brilliant scholarship. However, during the twelfth century and the first half of the thirteenth, much philosophical literature, including the lost works of Aristotle, became available to scholars in the Latin West. These scholars were stunned to find in the works of a pagan philosopher a system of knowledge more comprehensive and sophisticated than anything they possessed. To get some idea of the effect, imagine what it would be like if today archaeologists unearthed tablets of ancient hieroglyphics which, when translated, contained science and philosophy more advanced than our own. Aristotle made so powerful an impression that Thomas Aquinas (1224–1274), perhaps the leading Christian thinker of the Middle Ages, referred to Aristotle simply as "The Philosopher." In *The Divine Comedy*, Dante called Aristotle "The Master of Those Who Know."

However, many of the more conservative elements in the church were alarmed by the influx of Aristotelian ideas. In the year 1277, the Bishop of Paris issued an edict condemning 219 distinct propositions derived from or inspired by the philosophy of Aristotle. It is true that there were some glaring inconsistencies between what Aristotle taught and what the church believed. For instance, Aristotle held that the world had existed forever; the church taught, as it says in the biblical book of Genesis, that the earth was created in six literal days. It took the genius of Aquinas and others to synthesize Aristotelian philosophy and Christian orthodoxy.

Aristotle wrote a treatise on astronomy, which was known to medieval scholars by its Latin name *De Caelo* (*On the Heavens*). Like all educated people of his day, Aristotle recognized that the earth is a sphere (that Christopher Columbus set out to show that the world is round is a silly but oddly persistent notion). He held that the earth stood forever as the immobile center of the universe, the hub around which everything else revolved. He also held that the cosmos is spherical, or rather a set of nested spheres centered on the earth. Aristotle endorsed the standard view that our world is composed of the four elements: earth, air, fire, and water. Further, each of these elements has a natural motion. He held that each element has weight or lightness as an innate property. Innately heavy elements like earth will tend to move in a straight line toward the center of the universe; light elements, like fire, will tend to move directly away from the center. In Aristotle's universe, up and down had absolute meanings, referring to directions toward or away from the world's center. Objects composed of the four elements can deteriorate or waste away, and the elements themselves can be changed into one another. So, the physical objects we encounter daily are unstable and impermanent; they are created and destroyed.

For Aristotle, the heavens are very different from the earth, being composed of entirely different material and obeying different laws. Each heavenly body is situated on its own sphere, and each such sphere is part of a rather-complicated system of perfectly transparent crystalline spheres, each having its center at the center of the earth. The system of spheres has to be rather complicated since the movements of the heavenly bodies are complex. For instance, the sun not only makes its daily journey across the sky, but once a year it makes a complete circuit of the ecliptic, the path of the sun through the zodiacal constellations. The various rotations of the heavenly spheres, each rotating at a different uniform speed around its own axis, account for these complex motions of the heavenly bodies. The heavenly spheres are composed of a sublime substance called "aether," an indestructible fifth element. The natural motion of the aether is not up or down but to move eternally in a perfect circle. The moon occupies the lowest level of the heavenly spheres, followed by Mercury, Venus, the sun, Mars, Jupiter, Saturn, and the realm of the fixed stars (Uranus and Neptune were unknown to the ancients).

As you can see, there is nothing naïve or primitive about Aristotle's cosmology. It is a highly sophisticated, if premature, attempt to make systematic sense of our bewildering universe. Aristotle's system is also very beautiful, with its radiant heavenly bodies carried along on aethereal crystalline spheres. Remarkably, it is a system that is also highly congruent with common sense. Making the earth the immovable, solid center of everything certainly feels right. We still naturally speak of the sun rising and setting even though we have known for centuries that this is an optical illusion caused by the rotation of the earth on its axis. It still *looks* like we are standing still and the heavenly bodies are moving around us. For most practical purposes, it is fine to think this way. Finally, and perhaps most significantly, Aristotle's cosmology fits well with Christian theology. The central place of earth, the humanly abode, is perfectly congruent with the theological view that the cosmos is the great stage where the drama of human redemption is acted out. Also, the distinction Aristotle made between the imperfect sublunary (below the moon) realm, where things decay and die, and the perfect realm of the heavens made good sense for Christians. After all, human sin has corrupted the earth, whereas the visible heavens border on the realm of God and his angels and so approach the perfection of the divine.

Claudius Ptolemy's dates are not well known, but he seems to have done most of his work around the middle of the second century of the Common Era. Ptolemy's astronomical writings represent the culmination and synthesis of the great tradition of Greek astronomy. Yet he was an original and creative thinker who did far more than merely compile or summarize the work of his predecessors. It is fitting that Claudius lived in Alexandria, which for several centuries had been the intellectual center of the Hellenistic and Roman world. Ptolemy's writings, like most of Aristotle's, were long lost to the West. Once again, Arabian scholars came to the rescue, compiling Ptolemy's astronomical works into a book that they simply called *Almagest*, which is Arabic for "The Greatest."

For Ptolemy, the aim of astronomical theory was to provide a geometrical model of the heavens that would "save the appearances." An astronomical model "saves the appearances" when its geometrical representation of the heavenly bodies accurately predicts their positions and movements. For instance, when our model accurately tells us that at a given time Mars will be in a certain place and will be observed to move in a certain way, then our model succeeds to that extent. Further, for Ptolemy and the other ancient astronomers, the mathematical model had to meet certain stringent requirements. The model had to represent the motions of planets in terms of uniform motion around perfectly circular paths. Why perfect circles? There were aesthetic and even religious reasons for insisting on perfect circles. After all, the heavens were divine, and surely the divine must move in the most beautiful way. A more "scientific" motivation for preferring circles might be that circles seemed the simplest figures. Scientists try to provide explanations that are no more complicated than they have to be. Unfortunately, as we shall see, the insistence on modeling the heavens in terms of uniform motion along perfect circles eventually led to insufferable complexity.

For Ptolemy, as for all ancient astronomers, the biggest astronomical problem was to construct a model that could account for the complex movements of the planets. *Planet* did not mean quite the same thing for Ptolemy that it does for us. *Planet* comes from a Greek word meaning "wanderer." Most of the lights you see in the night sky are the "fixed" stars, the stars that remain in the same constellations year after year. The movements of these stars are simple and regular, and they have hardly changed their relative positions over human history. Could we be transported back to the time of Ptolemy, or even to ancient Egypt or Babylon, the night sky would look very familiar to us. But the ancients recognized seven bodies—Mercury, Venus, the moon, the sun, Mars, Jupiter, and Saturn—that did not follow the simple rules of the fixed stars. Note that Ptolemy classified the sun and the moon as planets because, like the other "wanderers," they move through the constellations. Generally, they move from west to east against the background of the constellations. For instance, we might see Mars in the western part of the constellation Taurus. If we look again a week or two later, we will see that it has moved considerably toward the more eastern part of the constellation. Sometimes, however, a planet will slow in its eastward journey, stop, and then move for a while in "retrograde" fashion back toward the west. Eventually it will stop its retrograde motion and resume its normal eastward course. Planets moving in perfect circles around the earth should not show such retrograde motion. Also, the planets do not move across the sky at a uniform speed; sometimes they move faster across the constellations than at other times. How can we explain such odd and complicated movements if our model must stick to uniform motions around perfect circles?

Ptolemy's devices for solving these problems were quite ingenious. One trick was the epicycle and the deferent. Imagine a planet in a perfectly circular orbit. However, the center of the planet's orbit is not the earth or the sun but just a point in space. Suppose further that this point in space is itself moving along a circular path, a circle much bigger than the little circle of the planet's orbit. The big circle is called the "deferent" and the little circle is the "epicycle." If you are located at the center of the deferent circle, the motion of the planet that you observe will be a compound of two different motions—the movement of the planet around the epicycle and the movement of the epicycle around the deferent. Obviously, using such devices you can make things as complicated as you like, with smaller epicycles moving around bigger epicycles and the whole system moving around a deferent. The flexibility of the epicycle-on-deferent system permits it to model very complex motions.

Two other models used by Ptolemy were the eccentric and the equant. When we say that a body is moving with uniform motion about a circle, we mean that its motion is uniform with respect to the center of the circle, that is, as viewed from the circle's center, the body will sweep out equal angles in equal times. But, as noted above, the motion of planets around the earth often does not appear uniform. Ptolemy therefore placed the earth at the "eccentric" point, a point away from the center of the planet's orbit. In this way, he preserved the uniform motion of the planet—it *is* uniform around the center of its orbit—but when viewed from the eccentric point (where the earth is) it will *appear* nonuniform.

Finally, Ptolemy considered a model in which the earth and a location in space called the "equant" point were located at equal distances on opposite sides of the center of a planet's orbit. In the equant model, the planet's motion is uniform, not with respect to the earth or to the orbit's center, but with respect to the equant point. Actually, this is cheating. For the planet's motion to *appear* uniform from the equant point, the planet would *actually* have to move faster when farther away from the equant point and slower when closer to it, so the equant model required Ptolemy to fudge. However, to save the appearances accurately, that is, to predict the observed motions of the planets with an adequate degree of accuracy, Ptolemy needed all three models.

Though Ptolemy's models worked fine, his system obviously got very complicated. There is a story that Alphonso X "The Wise," the king of Castile asked his court astronomers to explain the Ptolemaic system to him. Overawed by the complexity of epicycles, deferents, eccentrics, and equants, Alphonso was supposed to have muttered, "If the Lord had asked me for advice at the Creation, I would have suggested something simpler." Copernicus also yearned for something simpler. He hated Ptolemy's equant model as a violation of the rules, which, in fact, it was, and yearned for a system that would eliminate the equants. The result of his labors was the Copernican system with the sun in the center and the planets orbiting the sun in the now-familiar order: Mercury, Venus, Earth (with the Moon in orbit), Mars, Jupiter, and Saturn.

Despite the radical step of putting the sun at the center, the Copernican system was conservative in many respects. Copernicus preserved the circular orbits and the uniform motions of the geocentric astronomers. Copernicus even retained the Ptolemaic devices of the epicycle-on-deferent and the eccentric, though he did get rid of the despised equants. Still, the Copernican system in its original form was not really any simpler than the Ptolemaic system, and it could not predict the motions of the planets any better. However, Copernicus did not carry out the Copernican Revolution by himself. He had the help of later astronomers who brilliantly defended and developed his system, particularly Galileo Galilei and Johannes Kepler. Kepler took the biggest step when he broke with the tradition of ages and postulated that the planets followed an elliptical rather than circular orbit. This move resulted in the great simplification of the Copernican system—no more bizarre epicycles or eccentrics—and a very considerable improvement in the accuracy of planetary predictions. The improvements in the Copernican system by Kepler, announced in his *New Astronomy* of 1609, and the brilliantly pugnacious defense of the Copernican system by Galileo in his 1632 classic *Dialogues concerning the Two Chief World Systems*, meant that the Copernican system had its full impact only in the seventeenth century.

The acceptance of the Copernican system meant that the familiar old cosmology of Aristotle and Ptolemy had to be completely rejected. There was no place in the new cosmology for Aristotle's crystalline spheres. Worse, the telescopic observations of Galileo, published in 1610 in his *Sidereus Nuncius* (*The Starry Messenger*), showed that the moon was not a perfectly smooth sphere as Aristotle had said all celestial bodies must be. Rather, Galileo could see that it had plains, valleys, and mountains. He could even calculate the height of some of the mountains. Old-fashioned astronomers reacted furiously, but Galileo's evidence, and his aggressive, polemical style, carried the day. Kepler's elliptical orbits spelled doom for the Aristotelian notion that the movements of the heavenly bodies were "natural" and required no force to operate on them.

The final blow against Aristotle's view fell well after the Copernican view was generally accepted. In 1687, Isaac Newton published his epochal *Philosophiae Naturalis Principia Mathematica* (*Mathematical Principles of Natural Philosophy*). Newton added a theory of universal gravitation to the Copernican cosmology, thus showing that the same forces that govern the fall of an apple also explain the movements of the heavenly bodies. Gone forever was Aristotle's vision of a heavenly realm of absolute perfection moving in perfect spheres around the corrupt, polluted earth. Gone also was Ptolemy's geocentrism with its concept, so reassuring to theology and common sense, of a central, immobile earth. The poet John Donne (1572–1631) expressed the woe of a conservative and deeply religious intellectual for whom the "new philosophy"—the Copernican system—seemed disorienting and disturbing:

> And the new Philosophy calls all in doubt;
> The element of Fire is quite put out;
> The sun is lost, and th' earth, and no man's wit
> Can well direct him where to look for it.

Donne's perplexity is understandable. Really, the Copernican Revolution had implications so profound that we are still dealing with them today. For one thing, after Copernicus, people had to get used to a vastly larger universe. It is not that the universe was exactly cozy prior to Copernicus. Ptolemy had regarded the earth as merely a point in relation to the whole of space. Yet the universe had to be incomprehensibly vaster for the heliocentric theory to be true. The reason is this: We can observe no stellar parallax with the naked eye. Parallax is the way objects in the foreground shift with respect to the background when we observe those nearer objects from different viewpoints. Close one eye and point to a distant object. Close that eye and open the other one while not moving your finger. The tip of your finger will no longer appear lined up with the distant object because the distance between your two eyes means that each eye has a slightly different perspective. The apparent shift of your fingertip with respect to the background object is parallax. If, as Copernicus claimed, the earth revolves around the sun, the stars closer to earth should shift their apparent positions with respect to more distant stars as the earth moves along its orbit. Yet the naked eye observes no such parallax in the stars (it was not observed even by telescope until the nineteenth century). The only possible explanation for the lack of visible parallax is that the stars, even the closest ones, are inconceivably far away. Copernicus was right; the stars are inconceivably far away (Alpha Centauri, the closest bright star to the sun, is 4.3 light-years or about 26,000,000,000,000—twenty six million million—miles away).

It is not just the inconceivable distances that make us uncomfortable, it is the awareness, growing since the time of Copernicus, that our earth is not in any sense at the center of things; it is not the body around which everything else revolves. A famous poster shows a giant spiral galaxy with an arrow pointing to a minute speck in one of the outer arms. The caption reads "You Are Here." As the Voyager spacecraft was leaving the solar system, its cameras looked back to take a final glimpse at the planets. Our earth was a tiny blue dot lost in the immensity of space. As Carl Sagan pointed out, it is humbling to think that everyone we have ever known or heard of called that tiny blue dot home. How important can we be in the whole scheme of things when our home is just one stray speck in that inconceivable immensity?

WHAT HAPPENS WHEN YOUR WORLD CHANGES?

Yet there is an even deeper sense in which the issues raised by the Copernican Revolution are still with us. What happens when the effects of a scientific change are so profound that our whole worldview is altered? The Copernican Revolution succeeded so completely that it is hard for us now to realize how differently the cosmos appears to us now than it did to even the most educated and sophisticated persons in the Middle Ages. The fact of the matter is that we do not—we cannot—see the world in the same terms that our medieval ancestors did. Noted philosopher John Searle mentions a charming example. A Gothic church in Venice is called the Madonna del Orto (Madonna of the Orchard). It has that name because while it was being built, a statue of the Madonna was found in an adjacent orchard. Everyone assumed that the statue had come from heaven and signified that the church should be named after the Madonna. As Searle notes, if the church were being built today and somebody found a statue of the Madonna in the orchard, no one, however devout, would assume that the statue had a heavenly origin. Everyone would just take it for granted that it had fallen there by accident. Why is this? It is not that we are necessarily less religiously inclined now than we were then. Rather, we see the world in fundamentally different terms. We no longer automatically give a supernatural meaning to mysterious events. As Searle puts it: "We no longer think of odd occurrences as cases of God performing speech acts in the language of miracles. Odd occurrences are just occurrences we do not understand" (Searle, 1998, 35). Further, unless the occurrence is *really* odd, we just assume that it has an ordinary explanation.

Unquestionably, the main reason that we see things so differently from medieval people is that we live on the far side of the Scientific Revolution from them. The Scientific Revolution, of which the Copernican Revolution was just the opening act, gave us a whole new set of concepts for understanding the world and imbued us with a new set of assumptions about how the world works. More profoundly, when we reject old theories and accept very different new ones, we change much more than just our beliefs. We change the way that we spontaneously act, think, and feel. It is not just that we now regard it as improbable that a Madonna statuette found in the orchard fell there from heaven. It does not even *occur* to us to think that it did, and the idea now seems quaint or superstitious to us. Put another way, when a change in outlook is sweeping enough, the transformation we experience is not merely intellectual but visceral.

The purely intellectual aspects of scientific change long preoccupied historians and philosophers of science. The historians sought to reconstruct the reasoning behind the big discoveries and to delineate the evidence from experiment and observation that convinced scientists to accept new theories. Philosophers offered formal models of scientific explanation and theory confirmation. Of course, even the most severe rationalists were aware that science was more than just an intellectual exercise—that scientists were merely human, and that scientific theories impinge on many aspects of human life. But the truly revolutionary effects of scientific change were regarded as irrelevant to the historian's or the philosopher's job, which was to focus on the rational (or, as philosophers say, the "epistemic") factors. This all changed in 1962 with the publication of Thomas Kuhn's *The Structure of Scientific Revolutions* (hereafter referred to as "SSR"), a work surely on anyone's short list of the most influential books of the twentieth century.

Kuhn had a doctorate in physics, but he turned his attention to the history of science. In 1957, he published his book *The Copernican Revolution*, which tells the story with great clarity and insight. SSR is Kuhn's attempt to generalize the lessons he had learned by studying Scientific Revolutions. No summary could possibly convey the richness and subtlety of Kuhn's thought, and here I can give only a rather brief summary. As Kuhn says at the very beginning of the book, he aims to show that a proper understanding of the history of science has profound implications for the image of science in our culture. Those educated in the natural sciences are taught as if their disciplines hardly had a history. The textbooks they learn from present theory and fact as though it came from no one in particular, and even if historical figures like Newton or Darwin are mentioned, only meager information is offered about them and their discoveries. By contrast, no one gets a doctorate in philosophy without a thorough grounding in the thought of Plato, Aristotle, Descartes, Hume, Kant, and the other major figures in the history of philosophy. A scientist generally will not care what past scientists said except insofar as their views have been absorbed, often greatly altered in form or content, into *current* theory or practice. Physics students get their Newton from current textbooks, not from reading the *Principia*. Worse, few biology students have read *On the Origin of Species*.

For Kuhn, the history of science offers a number of deep lessons. First, the image of scientific change as continuous and cumulative is an illusion. True, there will be periods of what Kuhn calls "normal science" when scientists successfully solve their problems by developing and applying the guiding theories of their disciplines. During these periods scientific growth will appear stable as increments of knowledge steadily accumulate. Under the regime of normal science, a scientific discipline is committed to a core theory or set of theories that is so deeply entrenched that it largely *defines* the nature of the particular discipline. Kuhn calls this core theory or set of theories the "paradigm" that rules, and defines, a scientific discipline at any given time.

The role of the paradigm is crucial for several reasons. First, it defines the sorts of problems that a scientific field should study and imposes limits on the kinds of experiments or observations that will be relevant to their solution. For instance, when behaviorism was the predominant paradigm in psychology, it stipulated that behavior was what psychologists should attempt to explain, not feelings, perceptions, or other sorts of purely internal experiences. Also, the data behaviorists sought were measurements of observed behavior; data from introspection were ruled out as irrelevant. Second, the paradigm delimits the kinds of answers that scientists are allowed to give when they solve puzzles. For instance, paleontologists have long been puzzled by the mass extinction that occurred at the end of the Cretaceous Period, wiping out the dinosaurs and many other forms of life. No one knew what caused this mass extinction, but it was assumed that the answer, when and if it came, would involve earthly processes such as climate change or a drop in sea levels. Such earth-bound causes were expected because they were consistent with the reigning paradigm in the earth sciences, which required that changes in the earth be explained in terms of the established types of the slow, gradual, generally low-intensity causes that geologists had so far observed. Thus, when in 1980 a group of scientists proposed that a cataclysmic impact by an extraterrestrial body had wiped out the dinosaurs, such a hypothesis conflicted starkly with the reigning paradigm. Predictably, an enormous controversy erupted over the claim.

Things are relatively peaceful so long as a paradigm reigns unchallenged, and scientists are happily busy solving puzzles on the basis of that paradigm. Eventually, though, the paradigm runs into anomalies, facts that are difficult to accommodate on the basis of the reigning theories. Scientists will first make every effort to explain away the anomalies or show that they are, after all, compatible with the ruling paradigm. However, anomalies can pile up to the point that a critical mass is eventually reached and, says Kuhn, science enters a "crisis" phase. In this phase, the old paradigm is effectively dead, but nothing has yet been devised to replace it. Eventually, brilliant (usually young) scientists will devise a new paradigm that resolves the anomalies that stumped the old paradigm. The new view is rapidly accepted, over the die-hard resistance of (usually elderly) scientists who cling to the old paradigm. Once the new paradigm is in place, and scientists are once more busy at puzzle solving, a new era of normal science emerges.

But what exactly happens when a new paradigm replaces an old one? The traditional view is that theory change in science is a rather pedestrian affair. New evidence, in the form of neutral observations, data, and experimental results, gradually build up in favor of the new theory until the scientific community, calmly and rationally, decides that theory change is in order. What makes this an orderly and rational transition, on the traditional view, is that observation, data, and evidence are seen as entirely independent of the theories in dispute. In other words, the relation between theory and evidence is one-directional: theories depend on evidence, but evidence does not depend on theory. Therefore, evidence can serve as the objective grounds for a neutral and impartial assessment of the rival claims of competing theories.

Kuhn notes that things are not really quite so simple. A paradigm sets the standard for good science for its discipline. That is, for practitioners of that field, good science will be science like the paradigm, and anything that departs too radically from the paradigm will be regarded as bad science. It follows that each paradigm comes with its own methods and standards, which are largely incompatible with those of other paradigms, and so such methods and standards cannot serve as neutral criteria for deciding *between* competing paradigms. Not even the data can be considered neutral. Each paradigm will specify what kinds of data are relevant and which can simply be dismissed. As we noted above, where strict behaviorism ruled in psychology, data about subjective feelings or inner experience were simply ruled out as irrelevant. The consequence is that scientific judgments about the value of any piece of evidence will largely be determined by the scientists' *prior* theoretical commitments. Scientists who favor paradigm A will find the types of evidence that goes with A convincing; proponents of paradigm B will, on the contrary, find the types of evidence that go with B convincing. Evidence that looks good to the proponent of one theory will therefore look bad or irrelevant to the proponent of another.

In fact, a new paradigm is so different that Kuhn says that the new and the old paradigm are often "incommensurable" with each other. *Incommensurability* is a term borrowed from mathematics. For instance, the length of a side of a square is incommensurable with the length of the square's diagonal. To say that the side and the diagonal are incommensurable means that there are *no* possible units of *any* size such that whole numbers of those units will give you the length of both the side and the diagonal. Thus, there are no units that provide a "common measure" of both the side and the diagonal. Kuhn's use of the term is ambiguous; as we shall see in the next chapter, he means different things by *incommensurable* at different times. However, in general, he means that proponents of rival paradigms have such different viewpoints, assumptions, and even vocabularies that they often fail to make contact when they try to communicate—they just talk past each other. When communication fails to this extent, people do not even succeed in disagreeing with each other in any meaningful way; they just glare at each other across a chasm of mutual incomprehension.

So, clearly theory change in science is not the simple, straightforward process it was once thought to be. What *does* happen when a scientist changes his or her mind and switches from one paradigm to another? Kuhn suggests that when scientists switch allegiance from an old paradigm to a new one, this is like a religious conversion—a wholesale shift in one's view of reality. The religious convert "sees the light" or is "born again" into a whole new way of experiencing the world. Prominent religious converts such as St. Augustine or Leo Tolstoy testify that after their conversions things that had seemed vitally important before now seemed trivial, and things that previously had seemed silly or pointless were now full of meaning. Likewise, Kuhn suggests, the scientific convert gets a whole new perspective on reality; it is as though the convert is living in a different world than the one occupied by his or her old self. In fact, Kuhn says that the convert to a new scientific paradigm *is* living in a new world. Kuhn admits that he is not entirely sure what he means when he says, for instance, that Aristotle and Galileo lived in different worlds, but he thinks that it is true in some deep sense.

Kuhn may have been unsure just what he meant with his talk about "world changes," but many of his contemporaries thought they knew quite well what he meant. Kuhn was taken as advocating a form of what philosophers call "relativism," or more precisely, "epistemological relativism." Epistemological relativism is the position that there is no absolute, objective truth—no truth "out there" waiting to be discovered—but only "truth" relative to diverse cultures, conceptual frameworks, or worldviews. In other words, truth is parochial, as are all of the standards for judging what is true, plausible, or rational. Different societies, or different groups within a society, may have very different principles for evaluating knowledge-claims. For instance, when it comes to addressing some questions, one group might appeal to science and another group to the Bible. According to epistemological relativism, each group is merely articulating what is reasonable from its perspective, and neither can claim absolute authority for its standards. All we can say is that appealing to science is "right" for one group and appealing to the Bible is "right" for the other group.

Relativism is a very controversial doctrine. Some find it exhilarating. For them, relativism liberates us from an oppressive, narrow rationalism that privileges a monolithic standard of truth and rationality and instead opens our minds to the many diverse "voices of humankind." Others think that relativism is an egregiously wrongheaded doctrine that subverts reason itself. In particular, they think that a relativistic view of science means that scientific change is irrational.

For much of his career, Kuhn complained that both enemies and would-be friends had badly misconstrued his arguments in SSR. He has especially emphasized that his conclusions were not as radical as many have thought, and he went to considerable effort to correct what he regarded as overstatements of his views by other people. Sadly, as every author soon comes to realize, once a book is published and launched into the world, it is no longer the author's possession. People will take it as they will, however loudly the author protests and reasserts his or her original intentions. All authors can do is express their views so clearly and candidly that they can hope that readers of good will are not going to badly misconstrue them (and even then, one of Murphy's laws applies: "When you speak so clearly that no one can misunderstand you, someone will misunderstand you."). However, crucial passages in SSR are not terribly clear, and Kuhn's talk about "incommensurability," "world changes," and "conversions" makes it look like, when read literally, an assertion of some rather-radical claims. So, in all fairness to Kuhn's interpreters, it is not unreasonable to read SSR as making some very bold claims.

Another problem with interpreting Kuhn is that there is hardly a claim in SSR that later writings did not qualify, moderate, or retract. So, a criticism of the early Kuhn may be deemed unfair if it does not take into account the later revisions. However, it was Kuhn the young radical whom people heard, not Kuhn the middle-aged moderate, and my aim here is chiefly to present the *issues* SSR raised for its readers. Therefore, in this book I shall take the early Kuhn as canonical, especially the Kuhn of SSR. So, I now turn to what people, both friends and foes, *heard* Kuhn saying at the time.

It is plausible to take Kuhn as asserting that truth is relative to paradigms (and this is precisely how many have understood him). If this is Kuhn's claim, it means that judgments about the truth of any claim can only be made from *within* a paradigm, not across or between paradigms. After all, it makes no sense to say that one opinion is truer than another unless the two rival claims can be compared, and to be compared they have to be expressible in the same language. Kuhn's doctrine of incommensurability was taken as asserting that the claims of rival paradigms cannot even be expressed in a common language. That is, advocates of one paradigm do not even possess the terms that would allow them to state the claims of another paradigm. Yet if the claims made by rival paradigms are incommensurable in this strong sense—that is, they are not even expressible in the same language—then it makes no sense to say that the claims of one paradigm are truer, in any absolute sense, than the claims of another. Neither can there be any shared truths between paradigms since it makes no sense to say that one truth claim is the same as another if the two cannot even be stated in the same language. For instance, if we English speakers hear a German speaker say "*Schnee ist weiss*," and we find out that this means the same thing as "Snow is white" in English, then we can say that the German speaker is saying something we also think to be true. On the other hand, if we heard someone utter "*Boojum glebt farkle*," and no one could tell us how to express this in any language we know, we could not say whether or not this utterance affirmed or denied anything we think to be true. The upshot is that each paradigm incorporates its own "truth," which can neither contradict nor agree with the "truth" found in other paradigms.

The picture that emerges is a very disturbing one for anyone imbued with traditional views of science as rational and progressive. It looks like the history of science is not a march of more or less steady progress toward truth but the successive unfolding of radical changes of perspective—a succession of paradigms, each bearing its own "truths." This means that Newton's view cannot have been truer, in any absolute sense, than Aristotle's, nor can Einstein's theories really be any truer than Newton's. Each paradigm comes with its own comprehensive set of "truths," that is, its own "world."

COPERNICAN QUESTIONS: RATIONALITY AND REALISM

Paradoxically, therefore, it looks like the deepest questions we can ask about incidents like the Copernican Revolution, which for many would be the very touchstone of rational progress in science, are whether scientific change is really rational or progressive at all. Those revolutionary episodes in the history of science—when one whole way of looking at the cosmos is replaced by another, what Kuhn calls a change of paradigms—therefore raise two distinct but related questions: (a) Is it possible to make a rational comparison between rival paradigms so that the scientific decision to switch from one to the other can be based on objective reasoning and impartial evidence? Let us call this the Rationality Question. (b) Is it reasonable to say that one paradigm is closer to truth than another, so that science progresses toward truth as new paradigms replace old ones? Let us call this the Realism Question. If the answer to (a) is no, then the traditional image of science as the very model of human rationality will have to be discarded. Instead, changes in science will look much more like changes in religious or political persuasion and will be subject to the same sorts of psychological or sociological explanation. If a negative answer is given to (b), then we can no longer look upon truth as the goal of science. Put bluntly, we cannot say that science over the last four hundred years has progressed any closer to truth.

In later writings, starting with the "postscript" Kuhn added to the second edition of SSR, Kuhn denied that it was ever his intention to portray scientific change as irrational. He has affirmed that he holds that scientists rationally choose between theories on the basis of the usually recognized criteria—such as accuracy, fruitfulness, scope, consistency, and simplicity. He says that all he ever meant to argue is that there can be no automatic, cut-and-dried methodology, decision procedure, or technique for determining which of two theories best exemplifies these sorts of virtues. Likewise, there is no way to compel agreement between proponents of different theories, no absolutely knock-down arguments or overwhelming evidence that will show one side to be absolutely right and the other one just plain silly. Further, Kuhn has denied that his doctrine of incommensurability rules out all or even most meaningful debate between advocates of different paradigms. He says he only claimed *partial* incommensurability between rival paradigms (we shall examine in the next chapter some of the senses that Kuhn says theories can be incommensurable). Still, I think it is undeniable that for most readers, SSR raised both what I call the Rationality Question and the Realism Question. That is, it made them query whether there really can be a rational basis for switching from one comprehensive outlook to another one. Also, it got them to ask whether the history of science really is a history of ever-closer approximations of the true picture of the universe.

It is possible to give an affirmative answer to the Rationality Question and a negative one to the Realism Question. The reason is simply that the fact that there is some basis for rational comparison between two theories is not enough to show that either theory is closer to being true. Two equally false theories can differ a great deal in how they stand in regard to the evidence. For instance, suppose that Johnson committed a murder, but suspicion falls on two completely innocent persons, Smith and Jones. Upon investigation, we find that Smith was in another country when the murder occurred but that Jones was seen within a mile of the murder scene close to the time of the murder. Clearly, Smith and Jones are equally innocent of the murder, but the evidence so far clearly favors the hypothesis that Jones did it.

In fact, a number of prominent philosophers of science, such as Larry Laudan and Bas van Fraassen, reject the relativist view that rival paradigms are incommensurable, yet they do not regard successive scientific theories as moving closer to truth. That is, they give an affirmative answer to the Rationality Question and a negative one to the

Realism Question. They hold that science is certainly rational, in the sense that rival theories can be compared on an objective basis. They also hold that science is progressive in certain senses. For instance, Laudan, following Kuhn's lead, takes the pragmatic view that theories progress by being better at solving problems than their predecessors. That is, new theories are better if they can solve conceptual problems and accommodate anomalies better than the old ones. "Realism" is the position in the philosophy of science that the goal of science is to discover the truth about the world, not only the truth about the observable parts of the world, but truth about the deep, hidden structure of the cosmos, like what it is ultimately made of. Further, realists hold that science has historically progressed toward truer views of the universe, culminating in our current theories, which we justifiably regard as approximately true. Antirealists such as Laudan and van Fraassen think that there are neither persuasive reasons to think that successive scientific theories converge toward truth, nor that current theories, however successful, are approximately true. So, antirealists hold that science, which they still view as an eminently worthwhile and reasonable activity, should not have the attainment of *theoretical* truth as its goal (learning the truth about *observable things* is fine for antirealists).

Interestingly, when Copernicus's theory first appeared in print, there was a dispute about whether the heliocentric system should be taken as claiming to be a true representation of the cosmos. In 1542, the year before his death, Copernicus entrusted the manuscript of his great work to his friend, the mathematician Georg Rheticus, who was to oversee its printing. However, Rheticus had just taken a new job in another city and had to leave the uncompleted task in the hands of his friend, one Andreas Osiander. Osiander finished the job, but he also added to the text an unsigned and unauthorized preface that he himself had written. This unsigned preface, which readers would naturally think had been written by Copernicus himself, basically said that contents of the book should not be taken as true. Osiander's preface stated that the astronomer's job is solely to save the appearances by making whatever hypothetical suppositions are necessary to permit the accurate calculations of the movements of the heavenly bodies. In other words, the sole job of astronomical hypotheses is to permit astronomers to use the principles of geometry to describe the celestial motions correctly and predict them accurately. We need not regard such hypotheses as true or even probable. However, Copernicus *did* think his system was true, and not merely a handy, practical calculating device. When Rheticus saw the unauthorized preface, he was so incensed that he sued to get the printer to remove it. The lawsuit failed and *De Revolutionibus* went into the world with its unauthorized addendum.

MORE QUESTIONS: METHOD, NATURALISM, AND MEANING

The Scientific Revolution started by Copernicus prompted other deep questions. First was the issue of method. How should science be done? What procedures are best for building the edifice of scientific knowledge? What, indeed, is the aim of science? Does it seek certainty, or will something less do? Once we have identified the goal of science, how do we get there? What methods, techniques, or procedures should we follow? What kind of reasoning should we use? Perhaps the most important problem is this: Science seeks to know about things in general. It seeks universal laws that apply to all physical phenomena. Of course, some particular things, the moon, for instance, are of great scientific interest, but science understands particular things in terms of general types of causes, which, in turn, are ultimately understood in terms of basic physical entities, laws, and processes.

The first methodologist was Aristotle. He held that the goal of science is to achieve certain knowledge, that is, to demonstrate the truth of conclusions by deriving them in an infallible way from premises known with certainty. For Aristotle, scientific knowledge derives from certain knowledge of the true essences of things. He postulated a mental faculty for extracting the essential nature of things from the observation of many things of that type. If I have examined, say, reptiles copiously and in detail, I can have infallible insight into the essential nature of reptiles. I then can apply an infallible form of deductive reasoning, the syllogism, to establish other, equally certain truths.

So powerful and persuasive was Aristotle's account of scientific knowledge and how it is acquired that his method still reigned two thousand years later. However, the rejection of all things Aristotelian in the seventeenth century extended to methodology as well as cosmology. Francis Bacon, using his very considerable rhetorical powers, attacked the Aristotelian view of science as obscurantist and blamed the lack of scientific progress on blind devotion to Aristotle. He recommended a brand-new start that would overhaul the scientific method. As he saw it, Aristotle relied far too much on the syllogism and insufficiently on inductive methods, that is, basing conclusions on observation and experiment. He proposed an approach to science that put critical observation and experiment—inductive method—front and center.

With an emphasis on experiment, observation, and inductive inference, science seemed to be on the methodological high road. Perhaps the certainty sought by Aristotle was unobtainable, but, surely, science could confidently assert the probability of its conclusions. Such confidence ran head-on into the skepticism of David Hume (1711–1776). Hume, with a simple but apparently devastating argument, seemingly showed that no amount of observation or experiment can establish a universal claim. Maybe we have observed *A*s on a million occasions and have seen that all *A*s have been *B*s. However, Hume says that we can have confidence that the million-and-first *A* will be *B* only if we assume that nature is uniform, that is, that what happened in the past will continue in that same manner in the future. However, there is no way to establish the uniformity of nature because (a) it cannot be proven and (b) to try to support it with inductive methods of observation and experiment is to argue in a circle since inductive methods only work if we *assume* nature to be uniform! The consequence, for Hume, was that there is no *rational* basis for projecting past regularities into the future, but only a *psychological* propensity to do so. Apparently, then, inductive methodology, far from improving on Aristotle, provided no rational basis for science at all.

Despite Hume's apparently devastating critique, the standard method advocated by philosophers of science in the twentieth century was what is called the "hypothetico-deductive" method that combined both deductive and inductive reasoning. Yet this method was attacked vigorously by Sir Karl Popper (1902–1994). Popper affirmed that Hume's critique was sound and unanswerable. He therefore advocated a scientific method that, he claimed, would appeal only to deductive reasoning. His method of "conjectures and refutations" proposed bold theories and subjected them to rigorous test. The survivors of such testing were not confirmed, he said, but only not falsified, that is, not shown to be false. The goal of science, said Popper, is not to confirm theories—nothing can do that—but only to certify the best theories as (not yet) falsified. Concomitantly, falsifiability, not verifiability, is the hallmark of a true scientific theory.

Yet Popper in his turn has been severely criticized. Inductive methodology has many current defenders, and others advocate a third form of reasoning—inference to the best explanation—as a model of scientific method.

There are problems that arise from the very success of science. First, if we have science and its various methods, can there be other legitimate forms of reasoning? What are the limits of scientific cognition? Can there be "other ways of knowing" that are just as legitimate as science but that stand outside scientific ways of thinking? For instance, can there be, as some philosophers have maintained, a form of *a priori* intuition that gives us direct and immediate knowledge of deep metaphysical or moral truth? Is there a realm of "first-person experience" that is more certain than the third-person accounts we have to give in science? In particular, can philosophy offer insights that science cannot, or, perhaps, can we simply dispense with philosophy once we have developed science sufficiently?

The latter possibility has been raised by some eminent philosophers. In particular, W. V. O. Quine, one of the best-known philosophers of the twentieth century, recommended that epistemology, a core area of philosophy, be "naturalized." That is, he argued that the philosophical project of "grounding" knowledge had failed and would always fail, and that the philosophical approach to knowledge should simply be replaced by scientific fields, like cognitive psychology, that take knowledge as their subject. Quine's disturbing "replacement" recommendation is highly controversial, and not just because philosophers fear that science will push them out of a job. The big question is how a naturalized epistemology can intelligibly speak of norms—the standards that should guide our intellectual practices by telling us what we ought or ought not to do in our cognitive endeavors. After all, it seems that science can only study what is, not what ought to be, and norms seem to be an indispensable part of epistemology.

The same question about norms arises in attempts to naturalize ethics. Ethical naturalism is not a new philosophical development but one that goes back to Aristotle, whose *Nicomachean Ethics* proposed that morality be grounded in human biology. In the Enlightenment of the eighteenth century, the leading ethical thinkers rejected such naturalism. Immanuel Kant (1724–1804) argued that ethical obligations must be certain, necessary, and universal, applying to all rational creatures as such. Ethical imperatives must therefore be grounded in pure reason and not anything merely empirical, like biology. Ethical naturalism also faces Hume's famous challenge that you cannot derive an "ought" from an "is," that is, ethical obligations cannot be derived from any statement of fact. Science deals with fact and therefore has nothing to say about the basis of moral obligation. For Hume, it is the collective feeling of humanity that ultimately underlies moral judgment. We deem wrong those things we find repugnant and right those things that engender positive feelings. Recent decades, however, have seen a strong revival of a neo-Aristotelian form of ethical naturalism.

A second kind of question made urgent by the success of science is this: If we think of everything in scientific terms, not exempting ourselves, how should we then view the nature or meaning of being human? This problem already disturbed René Descartes (1596–1650) in the seventeenth century. If science explains *everything*, including us, how do we continue to see ourselves as having some of the remarkable attributes we traditionally claim? How, for instance, can we have free will? What happens to free choice if we are merely physical, like a machine? Are we not machines ourselves, with no more free will than an automated dishwasher? Such a prospect so disturbed Descartes that he argued for the existence of a spiritual, nonphysical soul that, he says, constitutes the human person. That is, Descartes reacted to the perceived threat of an all-consuming science by exempting the human essence from the physical universe.

Descartes's worries still concern many philosophers today. In particular, a number of theistic philosophers argue that if humans are conceived in purely physical terms, it is impossible to account for the fact that we know firsthand by considering our own experience that we have free choice and decide questions rationally, that is, on the basis of reasons or evidence. Their solution is to follow Descartes in exempting the human soul from the physical universe. Other philosophers, though convinced atheists, agree with the theists that a consistent physicalism is incompatible with free will and other aspects of alleged human uniqueness. However, they accept physicalism and reject free will and all of those other cozy features of our comfortable self-image. We are machines. Deal with it.

THE PLAN OF THE BOOK

So the Copernican Revolution has left us with some big questions, and we are still grappling with them. First, how do we understand large-scale conceptual change, as occurs when science undergoes a revolution? That is, how do we understand a shift in perspective so deep that we do not merely change our answers to certain questions but also change the entire conceptual framework in which those questions were posed? Can there be a basis for rational comparison between alternative worldviews, or are rival paradigms incommensurable? Second, even if we concede that there is some basis for rational comparison between different paradigms, is it reasonable to see science as progressing toward truth? However much better our current theories might "work," in some sense, than earlier theories, is there any reason to think that we have gotten any closer to the inner essence of reality? Third, can we describe a general "scientific method"? In particular, what kind of methods, techniques, or procedures can give us confidence in universal truths when our experience is always of particulars? Fourth, are there nonscientific forms of reasoning that are just as legitimate as scientific sorts and that can lead us to truths that science cannot attain? Specifically, is philosophy a distinct form of reasoning that can do some things that science cannot, such as, perhaps, providing us with norms to guide our cognitive and ethical decisions? Finally, what are the consequences for the human self-image of adopting a comprehensively scientific perspective, that is, one that views humans as just as much a part of the physical universe as anything else? What happens to our notions that we possess free will, inherent dignity, and rationality of a sort that no animal or machine can have?

The following chapters will explore these big questions. Chapters 2 and 3 focus on the Rationality Question, and chapters 4 and 5 on the Realism Question. Chapter 2 focuses on the question of incommensurability as posed in various senses by Kuhn. Because many have seen incommensurability as implying that theories cannot be rationally compared, this seems to be Kuhn's biggest challenge to traditional scientific ideals. Chapter 3 will examine arguments by two sociologists of science, two authors representing the "postmodernist" style of science critique, and feminist theorist Sandra Harding. These works, representing (but not necessarily representative of) the recent science critique of the "academic left," challenge standard views of scientific objectivity. These critics argue that science is merely a "social construct" like any other cultural artifact, and that science is not a value-neutral mode of inquiry but is (and should be!) drenched with ideological and political agendas. Chapter 3 also examines a challenge to scientific objectivity coming from the "academic right." Phillip E.

Johnson, professor of law and well-known antievolutionary activist, argues that science has abandoned objectivity in favor of a dogmatic commitment to naturalism.

Chapter 4 deals with the question of progress in science, which is an aspect of the bigger question of how we should view the history of science. Some historians, strongly influenced by the social-constructivist program, argue that even prototypical instances of scientific progress, like the development of the experimental method in the seventeenth century, are products of social and political agendas. Chapter 4 examines and rebuts this claim. However, even if we admit that science does progress in some sense, it is a deep question whether it progresses toward *truth*. Chapter 4 continues with an examination of Larry Laudan's claim that the history of science does not support the idea that science progresses toward truth. Although it is not as famous as Kuhn's SSR, Bas van Fraassen's book *The Scientific Image* (1980) has certainly had a profound effect on the philosophy of science. Van Fraassen argues that truth should not be the goal of science. He advocates an antirealist position he calls "constructive empiricism" that promotes "empirical adequacy"—saving the appearances—as the goal of scientific theory. Chapter 5 examines van Fraassen's case for constructive empiricism and focuses on the important question of the goal of scientific inquiry.

Chapter 6 addresses the issue of scientific methodology. It begins with Aristotle, the first scientific methodologist, whose concept of science dominated the intellectual world for centuries. Aristotle's methods were stridently rejected by Francis Bacon, the prophet, or perhaps the propagandist, of the Scientific Revolution. We look at his views in some detail. This topic segues into a presentation of the hypothetico-deductive method that was standard for the logical, empiricist philosophers of science who dominated the field for much of the twentieth century. David Hume raised deep problems with inductive reasoning, which standard accounts of scientific methodology had highlighted since Bacon. In this chapter I will present Hume's arguments against induction and Karl Popper's vigorous and sophisticated defense of those arguments. Finally, the chapter briefly considers whether another form of reasoning, inference to the best explanation, might have crucial advantages over induction.

The seventh chapter asks whether we still need philosophy, conceived as a distinct intellectual enterprise not reducible to science, or whether we can "naturalize" such fields as epistemology and ethics, that is, essentially replace philosophy with science or absorb philosophy in the scientific enterprise. The chapter considers Quine's "replacement thesis" and takes up the question of how norms are to be justified on a scientific basis. Hilary Kornblith's project for naturalizing epistemology by relating it to studies of animal cognition is examined in detail. Also, Aristotelian and neo-Aristotelian defenses of naturalized epistemology will be presented. The question of the place of philosophy in a scientific worldview concludes the chapter.

The eighth chapter considers the impact of science on the human self-image. It begins with a debate between a physicalist philosopher, Andrew Melnyk, and two Christian philosophers, Stewart Goetz and Charles Taliafero. Melnyk defends the claim, which seems to be implied by current neuroscience, that all mental activities are fully realized in the operations of the brain. That is, the "mental" is a set of operations performed by the brain. Goetz and Taliafero respond that such a claim is incompatible with the data of first-person consciousness, namely our awareness that our choices are free and that we form beliefs on the basis of reasons. As they see it, if all causation is physical, we can neither choose nor reason but have only the illusion that we do. Much of the chapter looks at and evaluates the recent book *The Atheist's Guide to Reality* by philosopher Alex Rosenberg (2011). Rosenberg agrees with Taliafero and Goetz that a scientific view of humanity is incompatible with the traditional self-image that introspection seems to support. His argument, however, is that we should recognize that our first-person self-image is a massive illusion and therefore accept the fact that there is no free will or morality. The chapter asks whether there is a middle course that accepts neither a repudiation of physicalism nor a rejection of free choice or morality.

The topics of the last two chapters are seldom included in books on the philosophy of science, but I think that they should be. The extent to which the results and methods of science should impact other aspects of our lives—philosophical, personal, and religious—is a big and important question. Recent philosophers have had some very interesting things to say about these points, and this book takes us deep into those discussions.

FURTHER READINGS FOR CHAPTER ONE

The story of the Copernican Revolution never has been and probably never will be better told than in Thomas Kuhn's *The Copernican Revolution: Planetary Astronomy in the Development of Western Thought* (New York: MJF Books, 1985). Kuhn provides a thorough review of the history of astronomy before Copernicus and a very clear and insightful exposition of Copernicus's innovations and how they were advanced by Galileo, Kepler, and others. Kuhn's knowledge is astonishing, and he is eager to share it with the reader. Though debate continues on how to interpret and evaluate Kuhn's philosophical views, there is no question about his excellence as a historian of science.

Icon Books has a fine series called "Revolutions in Science." These works are succinct, highly readable, and authoritative. The series includes John Henry's *Moving Heaven and Earth: Copernicus and the Solar System* (Duxford, Cambridge: Icon Books, 2001). Henry's book can be read in an afternoon, and, while not as detailed as Kuhn's classic, it tells the story with verve and lucidity.

One of the best introductions to Aristotle's thought, both as a scientist and as a philosopher, is still G. E. R. Lloyd's *Aristotle: The Growth and Structure of His Thought* (Cambridge: Cambridge University Press, 1968). Lloyd also wrote two books that perhaps still constitute the best overall introduction to the science of the ancient Greeks: *Early Greek Science: Thales to Aristotle* (New York: W. W. Norton, 1970) and *Greek Science after Aristotle* (New York: W. W. Norton, 1973). Like Kuhn and Henry, Lloyd communicates deep learning without being boring or pedantic. Aristotle is a difficult thinker and a writer of technically proficient but flat and uninspired prose. The subtlety and beauty of his ideas are hard to communicate to modern audiences. Aristotle is also pretty deep; students often find his system a tough nut to crack. Lloyd's enthusiasm and skill as an expositor helps the reader to both understand and appreciate the scope and power of Aristotle's mind.

For a clear exposition of the Ptolemaic system, see Kuhn's book cited above. A very easy-to-understand introduction is in David C. Lindberg's *The Beginnings of Western Science* (Chicago: University of Chicago Press, 1992). Geocentric cosmology seems so "obviously" wrong to us today that it is tempting to view it with pity or amusement. Kuhn and Lindberg represent Ptolemy as the first-class scientist that he was and present his system as the great intellectual achievement that it was.

John Searle is a particularly clearheaded philosopher and his book quoted in the text, *Mind, Language, and Society* (New York: Basic Books, 1998), is an especially accessible introduction to his thought. Searle states very clearly just how our worldview has changed in modern times compared to the ancient and medieval worlds.

Anyone with an interest in the philosophy of science, or the intellectual history of the twentieth century, should read Kuhn's *The Structure of Scientific Revolutions*, third edition (Chicago: University of Chicago Press, 1996). One measure of a book's importance is the quality of the opposition lined up against it. *Criticism and the Growth of Knowledge*, edited by Imre Lakatos and Alan Musgrave (Chicago: University of Chicago Press, 1970), is a compilation of some of the early critical reactions to SSR. Among the critics were Karl Popper, Imre Lakatos, and Stephen Toulmin—certainly three of the leading lights in the philosophy of science in the twentieth century. Another eminent early critic was Israel Scheffler, whose book-length critique of Kuhn, *Science and Subjectivity*, was first published in 1966. The second edition is available from Hackett Publishing Company, Indianapolis, Indiana (1982). These critics clearly took Kuhn to be saying that scientific change was not rational but was "mob psychology," as one of them put it. That is, they interpreted Kuhn as saying that there is no basis for objective comparison between competing paradigms and that only an emotional "conversion" could motivate a scientist to switch from one paradigm to another. In their view, Kuhn had debased science by injecting subjectivism and relativism into his analysis of scientific change. Their criticisms were therefore often mordant.

More recent examinations of Kuhn's work have been more balanced and less polemical. Perhaps the best monograph introducing Kuhn's work is Alexander Bird's *Thomas Kuhn* (Princeton: Princeton University Press, 2000). Two good collections of essays on Kuhn are *World Changes: Thomas Kuhn and the Nature of Science*, edited by Paul Horwich (Cambridge: MIT Press, 1993), and *Thomas Kuhn*, edited by Thomas Nickles (Cambridge: Cambridge University Press, 2003).

CHAPTER TWO

IS SCIENCE REALLY RATIONAL?

THE PROBLEM OF INCOMMENSURABILITY

We saw in the last chapter that Thomas Kuhn's groundbreaking study of the history of Scientific Revolutions raised many deep questions about the traditional view of scientific rationality. The challenge is this: When scientific change is so deep that it involves a fundamental alteration of our worldview, how can such change be rational? After all, when we say that our worldview has changed, doesn't this mean that our view of the *whole* world has changed—our understanding of the facts themselves and not just of our theories for explaining the facts? But if *everything* looks different from two different paradigms, and if our thinking is always governed by paradigms, where do find any neutral ground to stand on so that we can judge between paradigms? How can there be any neutral body of evidence, or, indeed, any shared body of methods, standards, criteria, or values to guide our choice between competing theories? Aren't we just sealed in our own worldviews until, perhaps, an emotionally charged conversion experience knocks us into a different one?

Nevertheless, scientists often do change their theories, and they think that they are acting reasonably when they do so, even when the new theory involves a revolutionary change of perspective. When a promising new theory is proposed in science, it has to prove its mettle by taking on all rival theories. If it wins over all challengers, then it achieves the (almost) universal approval of the qualified experts. When consensus emerges and a scientific community crowns the winning theory, scientists working in that community are convinced that their theory-choice decision has been a rational one, made in the light of impartial evidence by a process of objective reasoning. Are scientists right about the reasonableness of their decisions, or are they just deluded or engaging in self-justifying rhetoric?

What, precisely, *is* the challenge that Kuhn poses to ideals of scientific rationality? As we have noted, there has been much controversy about this. For many of Kuhn's critics, the problem is that his conclusions about the incommensurability of rival paradigms mean that scientists have no logical basis for comparing opposing paradigms and making a rational decision between them. If paradigms cannot be compared in the light of neutral evidence and impartial reasoning, relativism seems to be the consequence. That is, we must conclude that there are no objective criteria for judging one paradigm as rationally preferable or closer to truth than another. Each paradigm will have its own criteria for defining rationality, and each will be, by definition, "true" by its own criteria. But what exactly is incommensurability and how precisely is it supposed to make impossible the rational evaluation of rival paradigms?

The problem of incommensurability allegedly arises when two parties have such radically different views that their ability to communicate breaks down, at least to some degree. But to be a philosophically interesting idea, incommensurability has to mean more than this. Communication can break down for all sorts of reasons. Prior to the Civil War, Southerners and Northerners could no longer have meaningful debates about the issue of slavery, or much of anything else. They hurled insults and epithets back and forth, delivered sonorous diatribes, and employed all the devices of the florid oratory of the day. But long before the first shot was fired at Fort Sumter, they had ceased any meaningful exchange of ideas. However, the problem certainly did not seem to be that North and South spoke a different language or could find no common terms to express their disagreements. The problem was that feeling ran so high and opinion had become so polarized that hardly anyone was willing or even able to listen to reason anymore. Tragically, in a situation like this, when people can no longer settle their disagreements by rational, peaceful means, violence is almost inevitable.

Since scientists are merely human, and are prone to rancorous disputes the same as everybody else, it is not surprising that they too sometimes do not listen to their opponents. But it is no detriment to the ideals of scientific rationality that scientists sometimes fail to live up to them. Really to threaten ideals of scientific rationality, incommensurability must mean more than that people have simply gotten too angry, stubborn, or indifferent to listen to each other. It must mean that people who have very different views—even if each is intelligent, has the best intentions, and is motivated by a sincere desire to understand the other side—simply *cannot* find logical, rational ways to settle all of their differences. Despite their best intentions, at certain points in their debates they fail even to disagree with each other in a meaningful way and they wind up just talking past one another.

But just how strong a thesis is incommensurability? How severe is the impairment of communication when proponents of rival theories are trying to have a rational debate but unavoidably fail to make contact? Kuhn himself steadfastly maintains that incommensurability does not imply incomparability. In an essay titled "Commensurability, Comparability, Communicability" (CCC) written some years after SSR, Kuhn insists that incommensurability between theories is only partial and local and that many terms retain their meaning across theory change. Further, "the terms that preserve their meaning across theory change provide a sufficient basis for the discussion of differences and for comparisons relevant to theory choice" (CCC, 36).

Still, in many passages Kuhn argues that incommensurability means that theories might not be comparable in ways that everybody, from the seventeenth century on, has assumed that they are. So, it does sound like Kuhn is making a strong claim about the comparability of theories.

To get clear on just what Kuhn's thesis is, we need to consider how philosophers of science prior to Kuhn viewed the process of theory choice. The reigning model of theory confirmation before Kuhn was called the "hypothetico-deductive method" (abbreviated here as "H-D method"; we will examine this method in greater detail in chapter 6). According to this model, scientists use a very straightforward and simple method for choosing between rival theories: A hypothesis (or theory; I'll use the terms interchangeably) is proposed. From that hypothesis, and certain other additional statements, an "observational consequence" is deduced. That is, if the hypothesis and the additional statements are all true, we can be sure that another statement must be true, a statement predicting that at a given time and place, a particular observation will be made. If we look at that time and place and we observe what is predicted, we say the hypothesis is "confirmed"; if we observe something else, we say the hypothesis is "disconfirmed." When two rival hypotheses are being compared, we deduce contradictory predictions from each hypothesis. If we look and see that one prediction came true and the other did not, the hypothesis making the true prediction is confirmed and the other is disconfirmed.

For example, in the early nineteenth century, there were two rival theories about the nature of light. One theory said that light is a stream of tiny particles; the other theory said that light is a wave. English scientist Thomas Young realized that these two opposing theories made different predictions. Wave phenomena can produce interference effects; particles cannot. Suppose that you have two identical waves that are exactly in phase, that is, the crest and trough of one wave is lined up with the crest and trough of the other. If these two waves intersect to form one wave, the crests and the troughs of the component waves will combine constructively to produce a new wave of double intensity. This is *constructive interference*. If two identical intersecting waves are exactly one half wavelength out of phase, so that the crest of one lines up with the trough of the other, the two waves will cancel out each other. This is *destructive interference*. Particles, naturally, would not interfere in this way but would just pile up. So Young deduced that in a certain experimental situation, intersecting beams of light should produce interference patterns—alternating lines of constructive and destructive interference—if light is a wave, but they should only produce intensified light if light is a stream of particles. He set up such an experiment and found that the overlapping beams definitely produced an interference pattern and not just intensified light. This experiment confirmed the wave theory and disconfirmed the particle theory.

The H-D method therefore permits rival theories to be compared on a point-by-point basis. The theories can be compared side by side because each theory implies observational consequences that are precisely denied by the observational consequences of the other. One theory predicts "*p*," and the other one predicts "not-*p*." Kuhn notes that for two theories to be comparable in this straightforward way—for their predictions to clash head-to-head—they must be able to express those contradictory predictions in the same terms. For instance, to conduct his experiment, Young had to be able to express intelligibly what he expected to see if light is a wave and the contrary expectation if light is made of particles. However, Kuhn asserts that with some rival theories it is not always possible to make them clash in this way. Sometimes the meanings of the terms used to express the observational consequences of one theory will mean something so different from those expressing the observational consequences of its rival—even if each theory uses the same words—that one theory no longer denies what the other asserts. It is like the situation where two people are eating bowls of chili. One takes a bite, breaks into a sweat, takes a quick swig of cold beer, and exclaims, "Man, is that chili hot!" The other person says, "Mine is cold. I'm going to have to heat it in the microwave." One says the chili is hot and the other says it is cold, but they are not really disagreeing because one means that it is extremely spicy and the other means that its temperature is too low.

What Kuhn wants to deny, therefore, is that theories are always point-by-point comparable by a straightforward examination of their observational consequences. So, for Kuhn, rival theories are comparable in some ways but sometimes not in others. The problem is that Kuhn uses the term *incommensurable* in several different senses. Following W. H. Newton-Smith in his fine book *The Rationality of Science* (1981), I identify three senses of incommensurability in Kuhn's work: incommensurability of standards, incommensurability of values, and radical meaning variance. Two rival theories may each claim to explain a given set of natural phenomena, but differ considerably on the issue of what counts as a legitimate explanation. Proponents of these theories may therefore talk past each other because each is assuming a very different standard about what counts as a good explanation. One thinks that he has explained things well and the other just does not get it. Theorists might also assume very different values about what a good theory is supposed to accomplish. One thinks that her theory is better because it is far simpler than its rivals; her opponent favors a theory that is less simple but broader in scope (i.e., applies to more kinds of problems). The former scientist values simplicity over scope, and just the opposite for the latter. Because their underlying disagreement is about values and not the contents of the theories *per se*, they might simply fail to communicate. Each will think the other is just being pigheaded. Finally, two different theories may use terms that sound the same, but their proponents fail to recognize that the meanings of the terms differ radically between the two theories. In this case, scientists may think that they are disagreeing when they argue with a proponent of the opposing theory, but they are really just talking past each other.

Since the purpose of this chapter is to elucidate the rational resources available to scientists in choosing between theories, we shall consider incommensurability in each of these senses and determine what, if any, problems are raised about the comparability of theories. We shall see, in agreement with Kuhn, that even theories he regards as incommensurable are rationally comparable in many ways. However, we shall also see that even radically different theories can clash head-to-head in ways that Kuhn questions.

INCOMMENSURABILITY OF STANDARDS

One way that theories can be incommensurable, says Kuhn, is that they have different standards about what constitutes good science:

> We have already seen several reasons why proponents of competing paradigms must fail to make complete contact with each other's viewpoints. Collectively these reasons have been described as the incommensurability of the pre- and postrevolutionary normal-science traditions, and we need only recapitulate them briefly here. In the first place, the proponents of competing paradigms will often disagree about the list of problems that any candidate for paradigm must resolve. Their standards or their definitions of science are not the same. Must a theory of motion explain the cause of the attractive forces between particles of matter, or may it simply note the existence of such forces? Newton's dynamics was widely rejected because, unlike both Aristotle's and Descartes's theories, it implied the latter answer to the question. When Newton's theory had been accepted, a question was therefore banished from science. That question, however, was one that general relativity may proudly claim to have solved (SSR, 148).

When you hold a big rock in your hand, it feels heavy. The sensation of heaviness you feel is due to the force that pulls the rock toward the center of the earth. We call that force *gravity*. Why does gravity exist? As we saw in the first chapter, Aristotle offered an answer for this: Rocks tend to fall down because they are composed of elements that possess an intrinsic natural motive force that impels them toward their natural place, which, for things composed mostly of earth (like rocks) is the center of the cosmos. Isaac Newton, however, when asked why gravitational attraction exists, like between a rock and the earth, famously said, "*Hypotheses non fingo*" ("I frame no hypotheses"). Newton offered a mathematical law, the inverse-square law of gravitation, that describes *how* rocks fall (and which encompasses all other gravitational phenomena) with great accuracy, but he declined to speculate on *why* gravitational force exists. Newton just dismissed the question of why gravitational force exists. He simply accepted its existence and set about giving a precise mathematical description of *how* gravity works. He held that his answer to the "how" question was enough for physics. You still sometimes hear people assert (wrongly) that science explains "how" but not "why." However, when Einstein devised his theory of general relativity, one of the remarkable things about that theory is that it explained, in terms of the curvature of space in the presence of massive bodies, just why gravitational attraction exists.

So the question of why rocks fall down was at one time a meaningful scientific question, at another time dismissed as pointless speculation, and then, once again, considered an important scientific question to which Einstein gave a compelling answer. The lesson Kuhn draws is that Aristotelians and Newtonians on the one hand, and Newtonians and Einsteinians on the other, would have problems communicating because they have very different assumptions about what is a meaningful scientific question, much less what answers should be given to the questions.

Kuhn's concern is certainly legitimate. There have been times in the history of science when conflicting assumptions—over what sorts of questions can be asked or what sorts of answers can be given—have impeded communication and scientific progress. Let's look at what may, for many people, be a clearer example from another branch of science (physics is abstract and obscure for many people; the earth sciences are for them more accessible). Consider the tremendous imbroglio I mentioned in the previous chapter, the one that erupted in the 1980s when some scientists proposed a radical theory about the extinction of the dinosaurs. First a bit of background: In 1980, paleontologists, like other earth scientists, continued to accept many of the standards of good geological science laid down by Charles Lyell's 1830 classic *Principles of Geology*. Lyell opposed "catastrophism," geological theories that explained large-scale features of the earth's surface in terms of sudden catastrophes—massive floods and the like—of a kind or degree never witnessed by humans. Lyell held that for geology to be a genuine science it had to explain the earth in terms of the gradual, steady operation of those geological forces we actually see operating around us. We can invoke floods to explain geological features, but they cannot be floods, like the biblical deluge of Noah, of a degree vastly greater than any floods geologists had witnessed.

By 1980, earth scientists had already come to accept that some geological processes might have occurred with a degree of intensity that humans, in the very limited time we have had to observe, have never witnessed. For instance, there may have been floods or volcanic eruptions in the past of greater magnitude than any that humans have observed. However, there was still a considerable bias in favor of gradualism, that is, in favor of explaining large-scale earth changes, like mass extinctions, in terms of gradual processes, like climate change or the slow decline of sea levels, instead of sudden catastrophes.

Catastrophism returned to geology with a vengeance in June 1980 with the publication in the influential journal *Science* of a paper by maverick physicist Luis Alvarez, his son, Walter, and their collaborators, Frank Asaro and Helen Michel. This paper blatantly violated Lyell's restrictions by explaining the mass extinction at the end of the Cretaceous Period—the famous "K/T" extinction that ended the dinosaurs—by hypothesizing that a massive extraterrestrial body, a comet or an asteroid, had collided with the earth, resulting in cataclysmic destruction. On this view, the dinosaurs, and upward of 50 percent of all living species on the earth at the time, were wiped out in a single stupendous blast and the extreme climate changes that followed. By 1980, some rather large impacts by extraterrestrial bodies had been observed, most notably the Tunguska Event, something that caused widespread damage when it fell in a Siberian forest. But burning up even a few thousand square miles of forest is one thing, causing worldwide catastrophe and mass extinction is something else entirely. Needless to say, humans had never experienced such a cataclysm or anything close to it. Worse, the Alvarez hypothesis postulated a very sudden mass extinction, one achieved in a few months at most, rather than a gradual decline taking millions of years.

Predictably, the paper by Alvarez and his collaborators instigated a scientific controversy of almost-unprecedented scope and nastiness. By 1994, over 2,500 articles had appeared in the scientific journals debating the impact hypothesis, as it came to be called. In some quarters, all dignity was lost as scientists snubbed each other in public and were reduced to name-calling. I personally attended a conference on the topic of mass extinction where one invited scientist refused to attend because he would not participate in the same forum as one of his opponents. So polarized were the contending parties that William Glen, a scholar who chronicled the debates, said that the term *incommensurability* was far too weak to capture the complete breakdown in communication that had occurred (Glen, 69). Actually, here we seem not to have had incommensurability so much as, like the South and the North prior to the Civil War, people who were just too angry to listen. Nevertheless, unquestionably one source of misunderstanding between the opposing sides was a differing set of assumptions about what kinds of answers science can give to questions like "What killed the dinosaurs?" For some, sudden catastrophes were acceptable; but for others, only gradual earthbound causes were permissible. Communication did break down, and, in part, for just the sort of reason Kuhn mentions. Still, if contending parties are not just too angry to talk, rational disagreement between contending scientists is possible even on topics like what constitutes a legitimate question or answer in science.

Getting back to the case of Newton, in effect, Newton was saying that scientists should not waste time addressing unanswerable questions about what causes gravity but should get on with the fruitful task of using his formulas to predict phenomena and subsume them under the laws he had discovered. To scientists of Newton's day, this reasoning seemed compelling. Newton convinced them that they had no good answer the question of what caused gravity, but they did have fantastically powerful tools, supplied by Newton, for understanding how it worked.

Scientists are an opportunistic lot, always willing to use whatever tools are available to get on with the job and leave aside, at least for the time being, questions that look too hard to answer. But this does not mean that those hard questions become literally meaningless or incomprehensible to them. Newton seemed to understand perfectly well the question about what causes gravitational attraction; he just thought that it neither had nor needed an answer. Further, Newtonians seem to have understood the old Aristotelian explanations of why things fall perfectly well; those answers just no longer satisfied them. To a Newtonian, to say that things fall because they have an intrinsic potency to move to the center of the cosmos is not an acceptable answer because (a) it presupposes a pre-Copernican geocentric cosmology, and (b) it does not really seem to *explain* at all. The playwright Molière satirized such "explanations" when he had one of his characters "explain" the fact that opium makes you sleepy by saying that opium has a "dormative [sleep causing] potency"—which is just a fancy way of stating the obvious fact that it makes you sleepy. Likewise, for Newtonians to say that earthy bodies have an intrinsic downward tendency just seemed to reiterate the everyday observation that they fall down. In short, Newton and his followers had good *reasons* for rejecting the old Aristotelian definitions and standards—reasons they could, and did, offer in debate with adherents of the old views.

In fact, by Newton's day, proponents of the new science, like Francis Bacon and Galileo, had largely won the battles against Aristotelian science. The chief rival of Newton's gravitational theory was the mechanistic theory of the French philosopher/scientist René Descartes. Descartes defended a theory that explained all physical effects in terms of "corpuscles," minute particles that mechanically interact with one another. Descartes postulated a vortex of fine celestial matter spinning around the earth, sun, and each other heavenly body. This celestial matter pervades the cosmos and is less dense than the matter that composes massy objects such as rocks. The spinning motion of the vortex produces centrifugal force, the outward-pulling force that you feel when you tie something to a string and then spin it around. The centrifugal force in the spinning vortex pulls the fine particles of celestial matter outward and away from the earth. Air is composed largely of the fine celestial matter. When you release a rock in the air, the air below the rock will strive, due to its centrifugal tendency, to displace the rock and move into the space it had occupied. Descartes also held that nature abhors a vacuum, so when any piece of matter vacates a space, it must be replaced by something else. It is the downward tendency of massy bodies to replace the upwardly tending celestial matter that we feel as the weight of those bodies.

Here, then, we have an explanation of gravity that does not postulate occult intrinsic propensities of motion but accounts for everything in terms of mechanical principles—centrifugal force and density. The reigning paradigms for all physical explanation in Newton's day required the postulation of such mechanical models, yet Newton violated that requirement in his theory of gravity. Prior to that, Newton had been as true an adherent of the "mechanical philosophy" as anyone. Did Newton undergo something like a conversion experience that suddenly propelled him into a new paradigm? That is, did he have a sudden and comprehensive shift of allegiance from the mechanical view to a new paradigm, so that his old view now seemed silly or incomprehensible?

No. On the contrary, Newton's suspicions about the "aether" (the light celestial matter that Descartes invoked to explain gravity—entirely different from Aristotle's aether, by the way) grew slowly, and he rejected it hesitantly and in a piecemeal manner. Newton came to doubt the aether because, no matter how fine or rarefied it was supposed to be, a matter that pervaded the cosmos should offer some resistance to the planets and moons moving through it, but Newton could discern no such effect. Heavenly bodies moved precisely in elliptical orbits, sweeping out equal areas in equal times, just as his theory predicted, and exhibited no effects of the sort you would expect if they had to move through a resisting medium. Still, because Newton was so deeply imbued with the mechanical view, the idea that all effects must be transmitted though physical contact, and because some experimental results seemed to support the existence of the aether, he tried hard to hold onto the idea. Only after devising a particularly subtle and ingenious experiment that failed to indicate the effects of any such aether, and after unsuccessful attempts to modify the aether hypothesis to save it from its empirical difficulties, did he finally conclude that most of "aetherial space" must really be void. He was not suddenly "converted" to the view that no aether was required; after much experiment and deliberation, he just could see no reasonable way to hold onto the idea. He rejected the aether hypothesis slowly, reluctantly, and apparently as a result of a perfectly straightforward (if brilliant) process of scientific reasoning.

The upshot is that the circumstances Kuhn cites in the above quote do not preclude meaningful and rational disagreement between advocates of competing paradigms, even on matters pertaining to the standards of science. Neither does it mean that individual scientists undergo sudden "conversions" to whole new sets of standards when they accept a new paradigm. Problems about standards, like what questions science should address or what sorts of explanations are required, seem to be topics that scientists can consider in a meaningful, rational way. Perhaps Kuhn means something different when he calls two theories "incommensurable" in the above passage. Perhaps he does not mean to assert that no rational disagreement over them is possible, only that there will exist no set methodology or foolproof technique for settling some issues. *If* this is all he meant to claim there, he is certainly right, but this weaker claim is fully compatible with the claim that disagreement between proponents of different paradigms can be settled rationally. In fact, in later writings, Kuhn admitted that meaningful, rational debate over such issues was possible, and he affirmed that he only meant that no formal decision procedure or algorithm could settle such disagreements.

INCOMMENSURABILITY OF VALUES

However, Kuhn has other passages in SSR that suggest a different notion of incommensurability:

> There are other reasons, too, for the incompleteness of logical contact that characterizes paradigm debates. For example, since no paradigm ever solves all of the problems it defines, and since no two paradigms

> leave the same problems unsolved, debates always involve the question: Which problems is it more significant to have solved? Like the issue of competing standards, that question of values can only be answered in terms of criteria that lie outside of normal science altogether, and it is that recourse to external criteria that lie outside of normal science altogether that most obviously makes paradigm debates revolutionary. (110)

Here Kuhn is saying that not only might competing paradigms disagree over standards—like what is a legitimate scientific question or what kinds of explanations physical phenomena must have—but also in basic *values*. Each theory, *even in terms of its own standards*, will have its own successes and its own failures. So, when two theories compete, which should we value more, the successes of one theory or the successes of the other? Which is the greater liability, the failures of one theory or of its rival? Should we regard the successes of a theory as outweighing its failures? Suppose that a new theory explains in a very plausible and natural way many phenomena that the old one does not, yet it requires a physical process that just seems impossible. Which should we prefer in this situation, the new theory that explains so much more so well, or the old one that explains less but does not seem to involve a physical impossibility? Should we go with the promising new theory and hope that the seemingly impossible process will be shown possible, or should we stick with the old one and hope to solve its outstanding puzzles someday? Kuhn says that there is no logical way of settling such disputes, so the proponents of rival paradigms still fail to make contact. Only a fundamental change in the basic values of a science will allow a new paradigm to replace an old one. Once again, let us evaluate Kuhn's claim in the context of an actual scientific example, this one from the science of geology.

Again, cases from the history of science show that, as Kuhn says, differing assumptions about values, the values we expect a good theory to embody, have sometimes led scientists to an impasse. Consider the theory of continental drift. Children looking at a world map or a globe often notice how the eastern part of South America fits into the western part of Africa like two pieces of a giant jigsaw puzzle. This fact also impressed a young German scientist named Alfred Wegener. Wegener took a closer look and found that not only did the shorelines of South America and Africa fit, but much more important, the continental shelves fit even better. Intrigued by the idea that the two continents might actually have once been connected, he began to search the geological literature for other pieces of evidence. He found other fits between shores of continents now separated by ocean. Such evidence, while suggestive, did not really convince him. More convincing were mountain chains and other topographic features that are geologically identical to corresponding features on another, now separate landmass. For instance, the Sierras of Argentina could easily be a continuation of the Cape Mountains of South Africa, though the southern Atlantic Ocean now separates them. Had these continents been together at one time, these mountains would have formed a continuous chain.

Paleontology also supplied evidence. The fossils of some species are found only in the sedimentary rock in corresponding locales on separate continents. However, had those continents once been together, the present fossil sites would have formed a single, continuous range for those organisms. If the continents were never connected, how could land-dwelling or freshwater organisms have traversed oceans to get to their habitats on other continents? Another puzzle had to do with ancient climates. In the Permian Period, about 275 million years ago, many areas that are now tropical or subtropical had been covered with glaciers, while parts of what is now northern Europe and Canada had been tropical or subtropical. What could account for this? Further, actual measurements of the location of Sabine Island off the coast of Greenland between 1823 and 1907 indicated that the island had drifted westward by several meters a year!

Wegener offered a theory that gave a simple explanation for all these facts, the theory of continental drift: The present continents were at one time fused into a single gigantic landmass that Wegener called Pangaea ("All-Earth"). Over vast geological time, Pangaea broke up, and different pieces eventually drifted far away from each other. If there had once been a single supercontinent, then it is perfectly reasonable that there should have been mountain ranges and habitats of organisms that were split apart as the continents separated, so they are now found on opposite sides of the ocean. If the continents do drift, it is understandable that parts that are now cold could have once been near the equator, and parts that are now on the equator could have once been in the arctic. Further, the measured drift of landmasses today is just the continuation of the drift that has occurred since Pangaea broke up.

Wegener's theory explained many phenomena that older theories could not. When Wegener proposed his drift theory, the two leading geological theories about the continents were permanentism and contractionism. Permanentism, as the name implies, held that the continents were original features of the earth and have had pretty much their present shape and location ever since. Contractionism, on the other hand, held that the present continents were fragments of once much greater continents. Over time, parts of these larger continents had broken off and subsided, forming ocean basins and isolating the current landmasses. These effects were caused by the actual contraction of the earth as it continued to cool from its original molten state. As the earth cools, its surface continues to shrink, and this causes the wrinkling, folding, and subsidence that created the earth's present surface features. Each of these theories had considerable backing and rested on much evidence. However, neither permanentism nor contractionism could easily explain the phenomena that Wegener accommodated with his drift hypothesis.

Despite its success in explaining things that the older theories could not, geologists did not rush to embrace Wegener's view. On the contrary, continental drift was not broadly accepted during Wegener's lifetime or for several decades after his death. In fact, many professional geologists treated the drift hypothesis with contempt; one eminent authority even derided it as "geopoetry." New theories are often scorned, sometimes for good reasons and sometimes for bad. The main motivation for the geological community's rejection of Wegener's drift hypothesis seems to have been that it seemed to require a sheer impossibility. Ocean floors are composed of dense, basaltic rock. The continents are composed chiefly of less-dense silicates. Wegener's theory therefore seemed to involve a geophysical impossibility—that drifting continents had somehow plowed through dense oceanic crust to reach their present positions. No known force was sufficient to push the continents around. Worse, any force that was strong enough to shove the lighter continents through dense seabed rock should literally have broken the continents to bits.

Here, then, we have the sort of case Kuhn was talking about, that is, a difference in basic values between Wegener and his more-conservative critics. Which should we value more, a new theory that in a very plausible and natural way may account for otherwise-inexplicable facts, or an old theory that, though it leaves these facts unexplained, does not entail an apparent physical impossibility? The empirical facts were not in dispute between Wegener and his critics so much as were their very different assumptions about theoretical values—explanatory success versus physical plausibility in this case. These differences meant that Wegener and his critics were simply at loggerheads, and continental drift, for all its promise, was for decades relegated to the margins of geological science.

However, the subsequent history of the continental-drift hypothesis shows that fundamental differences over theoretical values need not *permanently* block acceptance of a new theory, and that acceptance of that theory need not entail a "conversion" of the scientific community to a new set of values. In the 1960s, a remarkable thing happened in the science of geology. At the beginning of that decade, many geologists still looked upon the drift hypothesis as heresy. By the end of the decade, the consensus in the geological community (with some dissenters, of course) was that continental drift had occurred. What happened? Did the change involve a sudden shift in geologists' basic values?

Not at all. What really convinced most geologists was the development during the 1960s of the theory of plate tectonics and the startling confirmation of that hypothesis by powerful evidence. This new theory provided just what Wegener lacked, a credible mechanism for the drift of continents. According to plate tectonics, the continents rest upon massive crustal plates that themselves rest upon the earth's upper mantle. The rock of the earth's mantle is subjected to extreme heat and pressure. Under these very extreme conditions, the rock of the mantle becomes plastic and behaves in a semiliquid manner. In fact, the earth's interior heat can cause the mantle to flow with convection currents, like a very slow-motion version of water boiling in a saucepan. In places, like in the midocean ridge in the Atlantic, mantle material pushes up from deep inside the earth. As it does so, the ocean floor is pushed outward on both sides of the ridge. The ocean floors are part the great plates on which the continents rest, and the lighter continents float on the denser plates, so the spreading of the ocean floor can also move the continents. Startling and compelling geomagnetic evidence strongly confirmed this lateral movement of the seafloor over geological time.

The upshot is that the geological community did not have to undergo a massive shift in its basic values to accept continental drift. Rather, drift was subsumed under a well-confirmed wider theory that provided a highly credible mechanism for the migration of continents. At no point in this process did great chasms of logic or values arise between contending parties, chasms that scientists could only traverse with a leap of faith or a comprehensive conversion experience. In short, though the consequences of the emergence of plate tectonics were certainly revolutionary for the science of geology, decidedly *conservative* values and standards seemed to guide the change. Therefore, Kuhn's claim that the emergence of a new paradigm must involve a revolutionary change in scientific values does not seem to be borne out. Rather, there is far more continuity in scientific values across paradigm shifts than he admits.

INCOMMENSURABILITY OF MEANING

Still, there are other writings where Kuhn offers a more-radical and more-troubling notion of incommensurability. In the CCC essay mentioned earlier, Kuhn argues that some terms may be genuinely untranslatable from one theory to another because we can only learn the meaning of some terms in clusters with other terms. This is because the terms in such clusters are defined in relation to each other, so that we have to grasp their meanings all together and not separately. He means something like this: Suppose that you grew up in the Amazon rainforest and that you know absolutely nothing about carpentry. At some point, you see a hammer and you ask what it is. Someone tells you that it is called a "hammer," and its purpose is to drive nails. This is no help, however, because you have no idea what a nail is. Someone then shows you a nail and tells you that it is used to attach one board to another. However, you have only seen wood in its natural state, never as a finished piece of lumber, so you are still mystified. Only when someone shows you a couple of boards and demonstrates to you how the hammer and nail are used to attach one to the other do you begin to understand the purpose and nature of these strange things. So terms like *hammer* are only really understood when we learn their meanings together with terms like *nail* and *board*.

Likewise, Kuhn says that particular paradigms include sets of terms that are defined only in relation to each other in the context of that particular theory:

> In learning Newtonian mechanics, the terms "mass" and "force" must be learned together, and Newton's second law must play a role in their acquisition. One cannot, that is, learn "mass" and "force" independently and then learn empirically that force equals mass times acceleration. Nor can one first learn "mass" (or "force") and then use it to define "force" (or "mass") with the aid of the second law. Instead, all three must be learned together, parts of a whole new (but not wholly new) way of doing mechanics. (CCC, 44)

Newton's second law tells us how the motion of a body changes when a force is applied to it. This simple relationship is expressed by the formula $f = ma$, where f is force, m is mass, and a is acceleration. However, using the formula is one thing, really understanding the concepts "force" and "mass" in the context of Newton's mechanics is something else. Kuhn's point is that in the context of Newton's second law, you really cannot understand force unless you simultaneously have a grasp of mass (and *vice versa*) because the two ideas are conceptually dependent on each other. In fact, they are two sides of the same conceptual coin: force is understood in terms of how much acceleration it imparts to a given mass, and the mass of a body is understood in terms of how much force is needed to impart a given amount of acceleration to that body. Now, admittedly, we have *some* understanding of what mass and force are in everyday contexts. Anyone who picks up a dumbbell can feel that it has mass, and if he drops it on his toe, he has some idea of force. But Kuhn's point is that to understand mass and force the way that *Newton* intended, we have to grasp them as interdefined in the way the second law specifies.

Suppose we try to translate the Newtonian meanings of force and mass into the language of another theory—Aristotelian or Einsteinian mechanics, for instance—that does not have Newton's second law as part of its theoretical apparatus. The translation is bound to fail because without the second law to provide the conceptual context, *force* and *mass* cannot be interdefined in just the way Newton meant. If you take one such term out of the context that gives it meaning and try to translate it into the language of another theory, the effort will fail because the other theory lacks the interdefined terms that give it meaning. It is easy to show mathematically that when we are considering velocities much less than the speed of light, formulas from Einstein's theory can be reduced, very nearly, to $f = ma$. However, Kuhn's point is that even if the formula is the same and the terms are the same, the meanings of *force* and *mass* are so different that the formula expresses something entirely different in Einstein's theory.

Kuhn says that even natural languages are beset with problems of incommensurability. He uses the example of the word *doux*, which, for native speakers of French, has a variety of nuances that no single term can render into English. An expert translator will use different English words or phrases to capture the different senses of such words in other languages. No single locution in English can capture all of the nuances of many terms in other languages. To cite another example, in teaching Aristotle's *Nicomachean Ethics* to students in introductory classes, it is hard to explain just what Aristotle meant by his key term *eudaimonia*. Most translators render this as "happiness," but the English word *happiness* has many connotations that do not apply to Aristotle's term, and *eudaimonia* has various senses that *happiness* just does not capture. The conclusion that Kuhn draws is that speakers of other languages, such as French or classical Greek, employ concepts that structure the world in ways that simply do not map onto the concepts employed by English speakers.

What conclusion does Kuhn want to draw from these instances of seeming incommensurability between scientific and natural languages? Apparently it is this: Where incommensurability occurs, that is, when key concepts or clusters of concepts simply cannot be translated from one language to another, it is impossible to make a side-by-side comparison of those concepts and so determine which is more reasonable—the concepts employed in one language (or theory) or those employed in another language (or theory). For instance, if we tried to determine whether the Newtonian or Einsteinian concept of *mass* was better supported by the scientific evidence, our efforts would fail since the same word *mass* has different meanings in each theory, and those meanings cannot be translated from one theory to the other. The consequence is that even if a Newtonian and an Einsteinian appear to disagree—one says that mass is invariant through changes of motion and the other says that mass changes as bodies are accelerated—they are not really disagreeing but are asserting entirely different things.

Note that for Kuhn not only do the meanings of key theoretical terms change from theory to theory, observation terms change too. For instance, it may seem like a Newtonian and an Einsteinian can make contradictory predictions about the observed mass of bodies. Suppose that a Newtonian and an Einsteinian physicist each weigh a body and agree that it has 10 kg of mass. Now the body is accelerated to 99.999 percent of the speed of light while the two physicists remain behind as stationary observers. What does each predict that he will observe to be the mass of the rapidly moving body? The Newtonian physicist says that it will still be 10 kg; the Einsteinian physicist says that it will be a lot more. It looks like the two different theories therefore predict contradictory observational consequences, and simple measurements could confirm the one and disconfirm the other in a straightforward way. However, Kuhn regards this appearance as misleading. In his view, because *mass* means something so different in each theory, the observational statement "this body has a mass of 10 kg" would have meant two incomparably different things for Newton and for Einstein, even though they used the exact same words.

It seems to follow that supporters of different theories cannot always rationally disagree with each other because each means something radically different from the other even when they both employ the same terms like *mass* or *force*. It is this claim of Kuhn's, which we may call his thesis of "radical meaning variance," that is his most troubling notion of incommensurability. If two theories are incommensurable in this sense, then it does seem that proponents of the different theories are condemned on many occasions simply to talk past one another. In this case, choosing between the two theories seemingly cannot be a straightforwardly rational matter, at least not in the point-by-point manner that the H-D method postulated.

EVALUATING MEANING INCOMMENSURABILITY

But do not Kuhn's own words belie this very conclusion? In various writings Kuhn has explained very clearly and in detail the differences between Aristotelian, Newtonian, and Einsteinian ideas. In fact, elucidating the differences between the different theories that reigned at given times is a very large part of what historians of science do. Scientists are probably at least as smart as historians of science, so why cannot scientists also come to grasp all of the subtleties and nuances of different theories in much the same way that historians do? Further, in the CCC essay, Kuhn himself explains very clearly—in English—the various shades of meaning of the French word *doux*. Does this not show, though perhaps no single English word or phrase can translate the French word, that the meanings of that word are quite expressible in English? Why, then, cannot a physicist, using the same interpretive skills that historians of science regularly employ, grasp the meanings of two different theories and make a rational, responsible choice between the two? Kuhn's reply is that in learning the languages of two different theories, we are not translating them into the same language, but, in effect, becoming bilingual. A physicist might learn to speak both Newton-language and Einstein-language, and have an expert knowledge of both theories. But comprehending two different theories is not the same as being able to translate one into the other so that point-by-point comparisons are possible.

Now Kuhn certainly seems to be right that there are some theories that have key concepts that just do not precisely map onto the concepts of other theories, even if the two theories employ many of the same words. Let's consider another example that is probably familiar to more readers than Newtonian and Einsteinian mechanics. It is hard to think of two better candidates for incommensurable theories than Darwinism and Young Earth Creationism (YEC for short). These two different theories certainly embody two radically different and incompatible worldviews. Further, key terms in one theory just do not translate into the other. For instance, creationists often speak of *design*, by which they mean the plan imposed by a creator. However, a staunch Darwinian (and atheist) like Richard Dawkins can also speak of the "design" of organism. Dawkins clearly does not intend his term to imply in any sense the designing activity of a creator. For Dawkins, "good design" in an organism is fine-tuned adaptation to a particular environment and is the result of the operation of the "blind watchmaker"—natural selection. Clearly, though they both employ the word *design*, the creationist notion of design is not translatable into the language of Darwinism, since Darwinism lacks the concept of a creator. Also, just as *doux* cannot be translated by one particular English word or phrase, so the creationist term *basic kind* cannot be rendered into any one taxonomic category in Darwinian theory. When creationists use their taxonomic term *basic kind* (after Genesis 1:24, "Let the earth bring forth the living creature after his kind . . .") the nearest equivalent in Darwinian terminology is sometimes *species* but other times is *genus*, *family*, or *order*. So Darwinism and YEC certainly seem to be theories that Kuhn would regard as incommensurable since their key terms do not intertranslate.

Does this mean that no rational assessment of Darwinism *versus* YEC is possible, that the proponents of each theory are condemned simply to talk past one another? Now I personally have participated, on the Darwinian side, in a number of such debates. I will testify that the experience was frustrating in various ways, and that there were indeed times that my opponents and I seemed not to be communicating. Yet it certainly does seem to me that a fully rational critique of one theory by a defender of the other is possible and that powerful, indeed compelling, reasons can be given for accepting one theory and rejecting the other.

Consider a standard creationist gambit. Creationists frequently charge that evolutionary theory cannot be acceptable because it contradicts the second law of thermodynamics. The second law of thermodynamics states the principle of entropy—that in all closed systems there is a spontaneous and inevitable tendency for order to decay into disorder. They charge that evolution contradicts the principle of entropy because evolution implies that systems automatically develop from less-ordered to more-ordered states. After all, evolution says that a horse evolved from a distant one-celled ancestor, and a horse is obviously more complexly ordered than a one-celled organism. Now, the charge that evolution violates the laws of thermodynamics is dead wrong, but the point is this: *If* evolution really were inconsistent with thermodynamics, this *would* be a major, perhaps fatal, problem with evolution. The laws of thermodynamics are extremely well established and certainly are accepted by all evolutionists. The reason why there *cannot* be a perpetual-motion machine is that any such machine could not operate without flouting the laws of thermodynamics. If evolution were contrary to those laws, it would be just as impossible as perpetual-motion machines. The point is that although the creationist charge is in fact wrong, if it were right, evolution would be dead in the water, and evolutionists would have to admit this.

Consider now a serious problem with YEC. According to YEC, the fossil record was laid down suddenly, in the biblical flood of Noah. Why, then, are certain kinds of organisms found consistently only in the higher strata of the geological column and others consistently found only much lower? One answer that creationists give is that some animals were more mobile than others and so were able to escape to higher ground before the rising floodwaters engulfed them. Thus, horses are consistently found higher than trilobites because horses would have been able to run to high ground and so escape burial much longer. In response, evolutionists have gleefully pointed out that many kinds of rooted, completely immobile plants are found only in the lower strata, and other kinds only in the upper ones. Clearly, oak trees did not uproot themselves and head for the hills as the waters rose! So for this, and many, many other reasons, the creationist account is simply incompatible with the fossil record. The upshot seems to be that, though Darwinism and YEC are candidates for incommensurability if any two theories are, the advocate of one theory can, in principle, adduce powerful, perhaps fatal, criticisms of the other theory.

To make the above point stronger, let us look at some of the kinds of arguments that scientists actually use in their debates with colleagues. Our guide here is the book *The Discourses of Science* (1994) written by Marcello Pera. Pera agrees with Kuhn in rejecting the old idea of scientific method—that a single set of universal rules can prescribe foolproof procedures that automatically tell us which theories to prefer. Rationality is clearly a lot richer and messier than that. Pera argues that to understand scientific rationality we have to understand the rhetoric of scientific discourse. The term *rhetoric* now has unfortunate connotations. Political diatribes are dismissed as "mere rhetoric" by opposing politicians. *Rhetoric* in this sense means arguments based on emotional appeals and faulty logic that are used to whip up partisan sentiment and do not contribute to rational discussion. When Pera speaks of rhetoric, he means it in the sense that it was originally employed when rhetoric was recognized as one of the core disciplines of the liberal-arts curriculum. In this traditional sense, rhetoric is the art of constructing persuasive arguments; there is no connotation that those arguments must be fallacious or misleading.

Pera examines several works, including that masterpiece of scientific argument, Charles Darwin's *On the Origin of Species*. Darwin himself called the *Origin* "one long argument" and fortified his case with many different kinds of evidence and ways of arguing. For instance, he defends natural selection by an argument from analogy. Darwin notes that everyone is familiar with the fact that by selective breeding farmers can greatly alter the nature of crops and domesticated animals. The wild turkey, for instance, is smart, elusive, and tough. The domesticated turkey, the direct descendant of the wild turkey, is stupid, docile, and delicious. The corn on the cob now enjoyed at barbecues is descended from wild maize that had an ear smaller than the baby corn now served in Chinese dishes. Clearly, plants and animals have been greatly altered to suit the needs and desires of humans. By analogy, says Darwin, nature, acting over geological ages and on all the traits of organisms, can be expected to have produced much greater effects than agriculture working over a few thousand years to alter just a few traits. Darwin also frequently argued that the best explanation of many well-known biological phenomena—well known to opponents as well as supporters of evolution—is that organisms have inherited traits, often greatly altered in form or function, from distant ancestors. For instance, there are numerous examples in creatures, including humans, of what are called *vestigial organs*, structures that have no present purpose but make sense if they are seen as remnants of organs that were useful to ancestors.

For instance, people retain muscles and nerves that try to make fur stand erect when we are cold or frightened; we call this "getting goose bumps." The problem is that we lost the fur millions of years ago.

In the *Origin*, Darwin repeatedly puts his theory of natural selection head-to-head against creationism. Time and again, he argues that special creation can offer no explanation for numerous phenomena that are easily explained by natural selection. Why, for instance, would a creator, who would certainly not lack creativity, have employed the same framework of bones to create the hand of a man, the leg of a horse, the wing of a bat, and the flipper of a dolphin? Though these appendages look very different from the outside, a careful examination of the internal anatomy shows that they match up very closely, often bone for bone. These deep skeletal similarities are called *homologies*. Vestigial organs, homologies, and numerous other such facts have no explanation for creationists except the inscrutable will of the creator. Darwin argued that all these things and many more are easily explained if natural selection must work only on those features organisms have inherited from their ancestors. Note also that all these odd facts were discovered not by Darwin but by scientists who strongly believed in a creator.

Prima facie, these cases, and innumerable others that could be mentioned, certainly seem to show that proponents of rival theories have many legitimate argumentative strategies for attack and defense, even when the theories cannot be fully translated into a common language. Supporters of rival theories can argue, as did Darwin, that the opposing theory fails, *even on its own terms*, to account for numerous phenomena within its domain (what possible explanation could a creationist give for goose bumps?), phenomena that one's own theory easily accommodates. Scientists can argue that opponents must endorse claims incompatible with universally accepted background knowledge, like the laws of thermodynamics. These are just a few of the very many types of arguments that scientists can adduce against their rivals. Apparently, then, the fact that two theories are nonintertranslatable does not at all prevent them from being compared side by side to see which is the more reasonable in the light of scientific evidence and logical argument. It just does not seem necessary that one theory be fully translatable into the language of another in order for there to be a meaningful comparison of one with the other or to make a completely rational choice between them. Even point-by-point comparisons seem possible. For instance, evolution by natural selection apparently explains homologies and vestigial organs and creationism does not.

Kuhn could insist that appearances are deceiving here and that Darwinism and creationism are not really compared so easily *vis-à-vis* the evidence. Recall that for Kuhn radical theory change entails a change not only in *theoretical* terms but also in *observational* terms. The reason is that the vocabulary we use to describe observations is influenced by the theories we hold true (i.e., observation is "theory laden" as philosophers put it), so it might be hard for Darwinians and creationists to avoid equivocation even when discussing the supposedly neutral evidence and when using the same terms. For instance, the word *homology* means something different for creationists and for Darwinians. Leading comparative anatomist Richard Owen, staunch creationist (though not of the "young earth" variety) and bitter opponent of Darwinism, carefully identified and described the homologous structures of different groups of organisms. He showed in great detail how the underlying skeletal anatomy matched up bone for bone in very different creatures. Owen held that homologies reflected what he called "archetypes"—designs in the creator's mind that serve as generalized blueprints for organisms (Desmond, 42–44). For the Darwinian, homologies are not reflections of an ideal but are the result of modifications by natural selection of traits inherited from ancestors. More significant, Darwin and his followers came to define homologies in terms of their development from corresponding parts in the embryos of different organisms. Owen, however, always regarded homologies as primarily revealed in adult forms. Darwinians also rejected Owen's idea that parts within the *same* organism could hold homologous relationships with other parts. For Darwin and his supporters, homologies hold only corresponding anatomical features or different kinds of organisms.

In other words, to employ a bit of philosophical jargon, Owen and

Darwin disagreed over both the "intension" and the "extension" of the term *homology*. The "intension" of a term is the set conditions a thing must meet to be correctly called by that term. For instance, to be human, according to some philosophers, one must meet the joint condition of being an animal and being rational (a god is rational but an animal is not; a dog is an animal but is not rational). The "extension" of a term is the set of all objects referred to by the term. The set of all past, present, and future humans is therefore the extension of the term *human*. For Darwin and Owen, the intension of *homology* differed since they proposed different conditions for structures to qualify as homologous. Also, they differed over the extension of the term since Owen held that it could apply to different structures in the same organism and Darwin did not. So, when Darwin says that his theory explains homologies and creationism does not, does he not mean by *homology* something so different from what the creationist means that he and creationist opponents are simply equivocating rather than disagreeing?

But even if we admit that *homology* meant something entirely different for Darwin than for Owen, this objection does not wash. To confute Owen, Darwin need not claim that he means by *homology* the same thing or anything very similar to what Owen meant. Darwin could argue that his theory, *taken as a whole*—with all its theoretical *and* observational terms defined in *his* sense—is more satisfactory than Owen's theory as a whole. Darwin could argue that his theory postulates an intelligible and testable causal mechanism—natural selection—that offers a richer, more-detailed causal account, is broader in scope, and is more consistent with kinds of explanatory frameworks that have proven fruitful in other sciences. So, in effect, a sort of holistic "side-by-side" comparison of theories is possible, even if they are not translatable into a common language. Significantly, in some of his later writings Kuhn did concede that rival theories can always be compared in terms of shared theory-choice values such as simplicity, scope, accuracy, and fruitfulness.

But *did homology* mean something incommensurably different for Owen than it did for Darwin? Did *mass* mean something incomparably different for Newton and for Einstein? Kuhn assumes that when theories change drastically, the meanings of key theoretical and observational terms also change drastically, so drastically that, even when the new theory retains the same terms (like *mass* or *homology*), the terms of the new theory retain nothing or almost nothing of their old meanings. Also, for the doctrine of incommensurability to have any bite, the meaning of key observational terms must change completely or nearly so in all, or at least most, episodes of radical theory change. If genuine incommensurability does occur, but only very rarely, it will be hard for Kuhn and others to justify the importance they have attached to the idea.

That they are merely equivocating and talking past one another when they compare the observation consequences of rival theories would certainly be news to most scientists. On the contrary, the scientists who actually participate in incidents of radical theoretical transformation take it for granted that key observational consequences predicted by the new theory contradict the predictions of the old one—and if they contradict, they cannot be merely equivocal. It certainly did not seem to Darwin that he meant something *totally* different by *homology* than Owen did. Darwinians certainly thought that they were discussing many of the same anatomical parallels that Owen adduced when he talked about homologies. Are the scientists right in their intuitive sense that there is continuity of meaning across theory change, at least with respect to observational terms, or are they just deluded?

Why would anyone think that scientists are pervasively mistaken, so that they literally do not know what they are talking about when they engage in debates over opposing paradigms? Several commentators have noted that Kuhn and other defenders of incommensurability are assuming a particular theory about how theoretical and observational terms get their meanings. According to this theory, theoretical and observational terms acquire their *entire* meanings from how they are used in the context of a given theory. If this is so, then, of course, when you change theories, the meanings of all your theoretical and observational terms change too. If the theory change is radical, the meaning change is radical too. But this is a highly controversial theory of meaning that many philosophers reject. To my mind, a far more plausible theory, one supported by much empirical research in cognitive psychology, is that meanings are formed with reference to prototypes, standard exemplars of concepts. Thus, our concept of *bird* will depend largely on what kind of bird we regard as a typical example (a robin for most people in North America). In this case, I would expect that the prototypes of *massive body* would be pretty much the same for both Newton and Einstein so that their concepts of mass would not be so different after all (alas, we have no room to explore these fascinating ideas).

Suppose, though, that we concede that Newton and Einstein meant something very different by *mass*. What exactly do we mean by "mean" here? Recall the distinction made earlier between *intension* and *extension*. Again, the intension of a term is the set of conditions something must meet to be correctly called by a term; the extension is the set of objects to which the term refers. What Kuhn seems to be saying is that the intension of a term can change radically when we shift theories. Yet two terms with very different intensions can have the same extension. Let me illustrate: The phrases "the sixteenth president of the United States" and "Mary Todd Lincoln's husband" certainly have different intensions. An entirely different set of conditions must be met to qualify as Mary Todd Lincoln's husband than are required to be the sixteenth president. However, these phrases with very different intensions have the same extension—the particular individual Abraham Lincoln. Suppose I say, "the sixteenth president of the United States was over six and a half feet tall," and you say, "the husband of Mary Todd Lincoln was shorter than six and a half feet." We are still directly disagreeing with each other even though the phrases we have used to refer to Abraham Lincoln have very different intensions. Any evidence indicating that Lincoln was over six and a half feet tall would support my statement and undermine yours. Likewise, even if one theory means something different by *mass* or even *the sun* than another, as long as the sentences expressing their observational consequences have the same objects as extensions, the two theories can be directly compared. It certainly seems that Newtonians and Einsteinians would point to the same objects (like the sun) as referents of the term *object with mass*.

Suppose that defenders of the "radical meaning variance" notion of incommensurability argue that the observational consequences of radically different theories do not even refer to the same objects so that even the extensions of their terms have changed. In this case, where two theories do not even refer to the same objects, how can they be *rival* theories? For two views to clash, they have to be *about* the same thing. If someone says, "I think Muhammad Ali was the greatest boxer ever," and you say, "No, Grover Cleveland was not the only president to have a child out of wedlock," you are not disagreeing with the statement because you are not even talking about the same two people. Two theories with no referential overlap in the theoretical and observational entities they postulate cannot even provide rival accounts of what is in the world since, according to Kuhn, two such radically different theories constitute *different* "worlds"!

Arguments like these eventually led Kuhn to admit that incommensurability is never more than partial and that change radical enough to engender even partial incommensurability between theories is rare. But how partial is partial and how rare is rare? As philosopher and historian of science Stephen Toulmin notes, even the Copernican Revolution, Kuhn's paradigm of a paradigm shift, was worked out by rational argument:

> The so-called "Copernican Revolution" took over a century and a half to complete, and was argued out every step of the way. The world-view that emerged at the end of this debate had—it is true—little in common with earlier pre-Copernican conceptions. Yet, however radical the resulting change in physical and astronomical *ideas* and *theories*, it was the outcome of a continuing rational discussion and it implied no comparable break in the intellectual *methods* of physics and astronomy. If the men of the sixteenth and seventeenth centuries changed their minds about the structure of the planetary system, they were not forced, motivated, or cajoled into doing so; they were given reasons for doing so. In a word, they did not have to be converted to Copernican astronomy; the arguments were there to convince them. (Toulmin, 1972, 105)

The fact that Kuhn in his later years would probably have agreed with everything Toulmin says shows how little was left by then of the doctrine of incommensurability.

CONVERSION: A CONCLUDING CASE STUDY

Really to do justice to the discussion of incommensurability, we would have to get deeply into technical discussions of theories of meaning and reference. Here we simply do not have the space to do that. Instead, let us conclude this chapter by returning to an instance of real science and follow a particular scientist through an incident of radical theory change. Perhaps the best way to test claims about incommensurability is to follow such an individual scientist through an episode of paradigm shift. Optimally, we should look at a key player whose work was integral to the theoretical transformation. If Kuhn is right about incommensurability, a scientist instrumental in bringing in a new paradigm should experience some very significant changes in his or her psychological and cognitive makeup. As Kuhn says, such an individual would have to experience something like a "conversion." But talk of "conversion" is vague, and some of Kuhn's harsher critics have understandably dismissed it as "psychobabble."

What kinds of fundamental cognitive and emotive changes would we expect to see in one who has undergone a "conversion" from one paradigm to another? Drawing upon all the senses of *incommensurability* we have discussed in this chapter, we would expect such a scientist to display radical changes in his or her scientific standards and/or theory-choice values and/or the meanings he or she assigns to theoretical and observational terms. But what if we find that, post-"conversion," the scientist in question continues to go about science in much the same way, endorses pretty much the same standards as before, accepts the same basic values as before, seems to use scientific terms in pretty much the same way as before, and, in general, evinces no massive changes in worldview? What if we find no radical conceptual hiatus, no total displacement of a whole constellation of values, standards, or meanings? In short, what if we do not find that the post-"conversion" scientist lives in a different "world"? In this case, I think we will have to reject the Kuhnian claim that even radical theory change results in incommensurability in any sense.

Let's consider the case of David Raup, certainly one of the leading paleontologists of the twentieth century. Like other paleontologists, Raup had been deeply imbued with the idea that extinction must be due to gradual processes instead of a single, stupendous disaster. When the Alvarez hypothesis about the K/T extinctions came out in 1980, with its hypothesis of sudden eradication in a single cataclysm, paleontologists were generally, often vehemently opposed. During the 1980s, though, there were some significant defections to the catastrophist side. One of the most notable of these "converts" was David Raup. Raup even wrote about his switch from old-fashioned actualist views to a particularly radical form of catastrophism in a bestselling book—*The Nemesis Affair*.

In *The Nemesis Affair* Raup tells how he developed an exciting theory that explained mass extinctions by postulating periodic impacts of extraterrestrial objects. The book also tells a personal story. In 1980, Raup was one of the referees asked by the editors of *Science* to review the Alvarez article that introduced the impact theory. Raup records that he was first highly skeptical of the idea and rejected the paper with harsh and rather-dismissive remarks. As the controversy over the impact theory unfolded, however, Raup says that he soon had a complete change of view and became an ardent supporter of the impact hypothesis. In fact, he and colleague Jack Sepkoski went on to develop a particularly radical version of that theory—the Nemesis hypothesis. According to this hypothesis, the sun has a dim, distant companion star following a highly eccentric orbit that brings it into the solar system every twenty-six million years or so. When the companion, the Nemesis star, enters the Oort cloud, the hypothesized region of the extreme outer solar system that is the home of comets, its gravitational effects propel millions of comets into the inner solar system. So, every twenty-six million years or so there should be evidence of mass extinctions as comets plunge into the earth.

For a reader of *The Nemesis Affair* who is familiar with Kuhn's writings on paradigm shifts, it sounds like Raup is saying that he underwent a conversion experience. In a short period of time, he underwent a radical change of perspective—not just a change of opinion, but, he indicates, a visceral, comprehensive change in outlook. Like the apostle Paul, he "saw the light" and went from being a persecutor of the new movement to one of its most outspoken defenders. If Raup did undergo a Kuhnian conversion to the impact hypothesis, we would expect the post-"conversion" Raup to advocate very different scientific standards or values or to use terms in entirely new senses.

But a careful look at the writings Raup produced before, during, and immediately after the period of his "conversion," show no evidence of such radical disruptions. At no point does Raup begin to advocate whole new ways of doing science or evince wholesale rejection of methods and modes of reasoning he had previously employed. Though he reached radical new conclusions, the mathematical tools and scientific techniques he used to do so were not themselves radical or unlike any he had previously employed (in fact, as William Glen observes, defenders of the impact theory often appealed to *conservative* standards [89–90]). Nor is there any reason to think that the post-"conversion" Raup employed any terms in wholly new senses. Raup does come to understand the nature of extinction in a new way. In another book, *Extinction: Bad Genes or Bad Luck?*, Raup rejects the old idea that extinction must be understood in Darwinian terms, that is, as the elimination of the less fit from a given environment. Rather, he proposes that the really major instances of extinction, the mass extinctions, were due to causes so cataclysmic that the fit were wiped out with the unfit. In disasters of such magnitude, no Darwinian selection takes place. So, clearly, post-"conversion" Raup understood mass extinctions differently than he had before. Was his later understanding of mass extinctions incommensurably different from his earlier one? Not at all. In fact, Raup takes pains to emphasize that his understanding of extinction incorporates the Darwinian view instead of displacing it. The meaning of a key term in the new paradigm is not totally different but considerably overlaps the meaning of that term in the old one. Would Kuhn reply that scientists like Raup in situations of radical theory change simply fail to notice that the terms they employ have really taken on whole new meanings? If so, then we must conclude that scientists do not understand the meaning of their own theories and must wait for historians like Kuhn to enlighten them! Such a claim appears arrogant, to say the least, and imposes a burden of proof on Kuhn and other defenders of incommensurability that they have not met.

So, a careful examination of the case of a scientist who apparently has undergone a Kuhnian "conversion" if anyone has, fails to reveal evidence supporting any of the three theses of incommensurability examined in this chapter. Of course, Kuhnians could dig in their heels and insist that the change from gradualist to catastrophist extinction theories was not a true instance of paradigm shift, and therefore that Raup's case is not a genuine counterexample. The problem here is that one of the biggest difficulties with talk about paradigms all along has been its vagueness. Just when does theory change amount to paradigm shift? How big does it have to be? Raup and other earth scientists certainly did perceive the new extinction theories as bearing assumptions that contradicted 150 years of geological tradition. Lyell's work in many ways defined the modern science of geology, and the impact hypothesis seemed to be a spectacular resurgence of catastrophism. So, if this was not a paradigm shift, it needs to be explained why not.

In conclusion, a careful examination of each of the main senses that Kuhn says that theories may be incommensurable—with respect to standards, values, or meanings—fails to disclose any real breakdown in communication in any of these cases. True, standards, values, and meanings do sometimes change considerably from theory to theory, and the concepts of one theory often do not map onto those of a rival theory. However, on other occasions, scientists who adopt a new paradigm do not have to undergo a "conversion" in their basic standards, values, or concepts. Rather, continuing research shows that the new theory can satisfy conservative demands. Often there is much more continuity in standards, values, and meanings across theories than Kuhn, or at least the early Kuhn, seems willing to admit. Even when two theories are very different, we have seen that, as Kuhn claimed and contrary to the fears of his critics, in each such case scientists have an abundance of resources for the rational comparison of one theory to another. Further, even the sorts of "point-by-point" comparisons Kuhn says are impossible are often quite possible despite deep differences in meaning.

Perhaps Kuhn would agree that in cases of radical theory change, problems of incommensurability do not always arise but rather occur only in some cases. But how many would "some" be? Unless Kuhn shows otherwise, which he hasn't, we could just assume that instances of genuine incommensurability are few and far between in the history of science. Even a hard-bitten rationalist might concede that in a few cases in the history of science new theories were proposed that were so different that they could not be compared point-by-point with old ones. Even if such cases exist, we have seen, and Kuhn freely admits, that there are still many ways of rationally comparing such incommensurable theories. The lesson I draw from a study of radical theory change really is that science is rational across revolutions. That is, one paradigm can be judged as more reasonable than another *vis-à-vis* observational consequences and in the light of impartial reasoning. We do not have to see each paradigm as having its own "truth" or setting its own standards of rationality. So the question of relativism is not raised by paradigm shifts. Rather, just as scientists intuitively believe, empirical testing does give objective reasons for thinking that neutral evidence favors one theory over its rival.

One problem with instigating a revolution is that once you start one they are hard to stop, and they may go a lot further than you intended. This was the case with Kuhn. I think it is fair to say that when he published SSR in 1962 he had no idea that by the time of his death in 1996 scholars would be marching into intellectual battle under his banner. I mentioned in the preface the "science wars" that developed in the 1990s. These were bitter conflicts that arose as part of a wider "culture war" that pitted traditionalists against radicals in a battle for the soul of academe. Traditionalists defended old-fashioned ideas about the need for impartial, disinterested, and objective scholarship. The radicals, including social constructivists, postmodernists, and some (but not all) feminists, rejected these old ideals as a hypocritical sham and insisted that, since all knowledge is inherently biased and political, the ultimate aim of scholarship should be liberation of the oppressed. An icon of the radicals, one they never failed to mention when justifying their views on science, was Thomas Kuhn. They assumed that Kuhn had already exploded the myth of natural science as the exemplar of rational inquiry, leaving them only to mop up and fill in some details. We have seen in this chapter ample reason to think that this assumption is wrong, as Kuhn himself so often insisted. Nevertheless, Kuhn was the unwilling intellectual godfather of the radical science critics. In the next chapter we turn to an examination of some of the radicals who stormed the barricades of traditional philosophy of science wielding the banner of Kuhn.

FURTHER READINGS FOR CHAPTER TWO

Though outdated, Carl G. Hempel's *Philosophy of Natural Science* has never been surpassed (Englewood Cliffs, NJ: Prentice-Hall, 1966). Hempel was one of the foremost of the philosophers of science of the twentieth century, and his book remains a model of clear and insightful exposition. It is still the best source for beginners who want to see what philosophy of science was like prior to Kuhn. Chapters 3 and 4 give a very lucid statement of the hypothetico-deductive method.

I think the best general discussion of scientific rationality is W. H. Newton-Smith's *The Rationality of Science* (Boston: Routledge & Kegan Paul, 1981). Newton-Smith surveys the work of leading philosophers of science Karl Popper, Imre Lakatos, Thomas Kuhn, and Paul Feyerabend. He also tackles tough issues such as whether and in what sense observation should be regarded as "theory laden," or whether rival theories are ever genuinely incommensurable. The book is rather technical in places and expects the reader to keep track of many puzzling abbreviations for key terms (a bad habit of much philosophical writing). So parts of the book would be tough going for beginning students. However, Newton-Smith provides a cogent, and mostly quite readable, statement and defense of the rationality of science in the face of various challenges.

A terrific and I think greatly underappreciated book is Richard Bernstein's *Beyond Objectivism and Relativism: Science, Hermeneutics, and Praxis* (Philadelphia: University of Pennsylvania Press, 1983). While arguing for a number of important theses about human rationality, Bernstein devotes considerable attention to the "new" philosophers of science, such as Kuhn and Feyerabend. He argues, very persuasively in my view, that the true aim of these radical-sounding critiques is not to show that science is irrational, but to attack the idea, current through much of the twentieth century, that theory choice in science was a highly rule-governed activity that obeyed rigid norms and conformed closely to formal models of confirmation. Bernstein reads Kuhn as supporting the view that theory choice in science, while not irrational in any sense, is a much looser activity that involves a kind of reasoning that is more practical than theoretical. Bernstein draws on the distinction made by Aristotle between *theoria*, the faculty that perceives theoretical truth, and *phronesis*, the faculty of practical reasoning whereby we deliberate between various courses of action. On Bernstein's reading, Kuhn is saying that choosing between different theories in science is much more like deliberating between alternative courses of action than performing a formal inference in logic or mathematics. Bernstein's book is addressed to professional philosophers rather than students, but, after reading *It Started with Copernicus*, a student should be able to tackle Bernstein's sections on the philosophy of science.

Another outstanding work dealing with the rationality of science is Marcello Pera's *The Discourses of Science* (Chicago: University of Chicago Press, 1994). Many works on the rhetoric of science are debunking efforts that try to reduce scientific discourse to "mere rhetoric." Pera's book is very different. He shows, first, that formal models of scientific rationality fail to do justice to the richness of scientific debate. Further, there is no permanent, unchanging set of rules that define scientific method. As Pera observes, it would be bad if scientists did inflexibly commit to some such rigid set of methodological rules, because what could they do if *better* methods came along? He then shows, by careful case studies of the actual rhetorical practices of top scientists, that rationality and objectivity are achieved in the very process of articulating the best arguments and accumulating the best evidence in order to persuade colleagues.

As noted in the chapter, Kuhn in his later years attempted to clarify, extend, and modify his arguments in SSR. Some of his later reflections, in fact, sound like at least partial retractions of some of his more extreme-sounding claims. This is why Newton-Smith in his chapter on Kuhn says that over his career Kuhn moved from revolutionary to moderate social democrat. Several of the essays in Kuhn's *The Essential Tension* (Chicago: University of Chicago Press, 1977), particularly the essay "Objectivity, Value Judgments, and Theory Choice," present a far more balanced and nuanced view of theory choice than was given in SSR. The essay "Commensurability, Comparability, Communicability" (abbreviated here as CCC) is found in the collection of Kuhn's writings *The Road since Structure* (Chicago: University of Chicago Press, 2000). Kuhn is quite a clear writer, and many of his essays are accessible to beginning students. The main problem with understanding them is that Kuhn was a physicist, and the majority of his examples are drawn from the history of physics and may not be familiar to most philosophy students. Steven Toulmin's very insightful analysis and critique of Kuhn is found in his outstanding analysis of conceptual change *Human Understanding: The Collective Use and Evaluation of Human Concepts* (Princeton: Princeton University Press, 1972).

Isaac Newton's accomplishments dazzled his contemporaries, and even today it is hard to write about him without an effusion of superlatives. Merely to describe him as a genius seriously understates the case. His was quite possibly the greatest mathematical and scientific intellect the human race has yet produced. His intellectual incandescence and personal eccentricity, bordering sometimes on psychosis (some have speculated that his bizarre quirks were due to mercury poisoning), continue to make him an irresistible subject for biographers. A recent popular biography by a fine science writer is James Gleick's *Isaac Newton* (New York: Pantheon Books, 2003). The classic introductory account of Newton's accomplishment in the context of preceding physics and astronomy is chapter 7 of I. Bernard Cohen's *The Birth of New Physics* (New York: W. W. Norton, 1985). The details about Newton's rejection of the mechanical aether are found in "Newton's Rejection of the Mechanical Aether: Empirical Difficulties and Guiding Assumptions," by B. J. T. Dobbs in *Scrutinizing Science*, edited by Arthur Donovan, Larry Laudan, and Rachel Laudan (Baltimore: Johns Hopkins University Press, 1988), pp. 69–83.

Alfred Wegener stated his theories of continental drift in his book *Die Ehtstehung der Kontinente und Ozeane*, published in 1912. The fourth edition from 1929 was published in English translation as *The Origins of Continents and Oceans* (London: Methuen, 1966). A good short history of the development of the drift hypothesis and the final emergence of plate tectonics is given in A. Hallam, *A Revolution in the Earth Sciences: From Continental Drift to Plate Tectonics* (Oxford: Clarendon Press, 1973). A very interesting collection of essays telling the story of the development of plate tectonics, written by the scientists involved in developing and confirming the theory, is *Plate Tectonics: An Insider's History of the Modern Theory of the Earth*, edited by Naomi Oreskes (Cambridge, MA: Westview Press, 2001). One thing that makes the development of plate-tectonic theory so interesting is that we have here a truly revolutionary development in geology that is extremely well documented. It is therefore an excellent test case for claims that Kuhn made about Scientific Revolutions and theory change. For an analysis of the geological revolution in the light of Kuhn's and other theories of scientific change, see H. E. LeGrand, *Drifting Continents and Shifting Theories* (Cambridge: Cambridge University Press, 1988).

Many volumes have been written about the battles between creationists and evolutionists. This is one of those perennial battles arising from a clash of worldviews. In reading creationist literature, I have often had the creepy feeling that maybe Kuhn was right and that in some sense these writers and I do live in entirely different "worlds." My version of reality seemed hardly to match theirs at all. Upon reflection, though, I have to think that take about adherents of different "paradigms" living in different "worlds" really makes sense only as a metaphor for extreme differences of belief. But if creationists and evolutionists do live in the same physical world, enjoy the same endowment of human intellectual and cognitive capabilities, and equally have access to the tools and methods of modern science, how do we arrive at such divergent and apparently irreconcilable conclusions? Surely, Kuhn would say, if you reject my talk about different worlds, you at least have to answer this question. Naturally, any response offered by one of the contending parties in the debate will be tendentious; that is, it will reflect the party's conviction that reason is on his or her side. Recognizing this inevitability, I offer Nicholas Humphrey's *Leaps of Faith: Science, Miracles, and the Search for Supernatural Consolation* (New York: Copernicus, 1996). Humphrey explains very clearly why intelligent and otherwise quite rational people will accept supernatural explanations in the face of overwhelming contrary scientific evidence. Further references to literature relating to the creation/evolution controversy will be given in the "Further Readings" essay at the end of chapter 3.

One of the best of many books on Darwin and the Darwinian revolution is Michael Ruse's *The Darwinian Revolution: Science Red in Tooth and Claw* (Chicago: University of Chicago Press, 1979). The source for Richard Owen's views on archetypes and homologies is Adrian Desmond's *Archetypes and Ancestors: Palaeontology in Victorian London 1850–1875* (Chicago: University of Chicago Press, 1982). Desmond's book attempts to correct the neglect and disparagement of Richard Owen's works due to the fact that he was the loser in the debates over evolution. As Desmond notes, Owen was certainly a brilliant scientist, and his reputation should be restored. However, Desmond takes far too much of a "social constructivist" approach for my taste (more on social constructivism in the next two chapters). Richard Dawkins explains how natural selection "designs" organisms in *The Blind Watchmaker* (New York: W. W. Norton, 1987).

The loud and acrimonious debate over dinosaur extinction has resulted in the publication of several popular or semipopular works. Most of these take a strong stand and present a case either for or against impact theory. In my opinion, the best of the pro-impact books is James Lawrence Powell's *Night Comes to the Cretaceous* (New York: W. H. Freeman, 1998). The case against the impact theory is vigorously argued by Charles Officer and Jake Page in *The Great Dinosaur Extinction Controversy* (Reading, MA: Helix Books, 1996). Both books are fun to read, and you learn a lot whichever view you adopt. J. David Archibald's *Dinosaur Extinction and the End of an Era* (New York: Columbia University Press, 1996) presents a more balanced view that interprets the end-Cretaceous mass extinctions as due both to gradualistic factors, like marine regressions and habitat fragmentation, as well as catastrophic occurrences, like meteor impacts and massive volcanism. A serious book that traces the history of the K/T extinction debates, and presents analyses by historians, sociologists, philosophers and scientists, is *The Mass Extinction Debates: How Science Works in a Crisis* (Stanford: Stanford University Press, 1994). The editor, William Glen, also contributed two very stimulating chapters to the book. He examines the controversy in the light of Kuhn's analysis of scientific change. Even the subtitle of the book reflects Kuhn's examination of how science operates in the "crisis" phase when an old paradigm is dying and a new one is being born. Anyone wanting a succinct and balanced overview of the extinction debates should consult chapter 8 of my book *The Great Dinosaur Controversy: A Guide to the Debates* (Santa Barbara, CA: ABC-Clio, 2003), pp. 121–43.

Not all outstanding scientists are outstanding authors of science written for a nonspecialist audience. David Raup has enormous respect among paleontologists and also writes very readable and interesting popular science. His book *The Nemesis Affair* (New York: W. W. Norton, 1986), tells the story of how he and a colleague developed the Nemesis hypothesis and the controversy that resulted. The Nemesis hypothesis is no longer discussed; scientists were deadlocked on whether the statistical evidence for periodic mass extinctions was credible. Still, the book remains fascinating reading because it tells the inside story of a scientist apparently undergoing a Kuhnian-type "conversion" to a new paradigm. I analyze this "conversion" extensively in my book *Drawing out Leviathan: Dinosaurs and the Science Wars* (Bloomington, IN: Indiana University Press, 2001). Raup views mass extinction as more a matter of bad luck (cataclysmic events that wipe out whole classes of organisms) than bad genes (the traditional Darwinian view that the genetically fittest survive).

CHAPTER THREE

A WALK ON THE WILD SIDE

SOCIAL CONSTRUCTIVISM, POSTMODERNISM, FEMINISM, AND THAT OLD-TIME RELIGION

THE CONSTRUCTIVIST CHALLENGE

If there ever has been a "hero of science," it was Louis Pasteur. He is famous all over the world and is commemorated on every milk carton with the word *pasteurized*. In his native France, Pasteur was honored almost as a living saint during his lifetime. Now, well more than a century after his death, the Pasteur Institute in Paris remains a leading center for biomedical research. Though he had many brilliant accomplishments, he was most honored for developing vaccines, especially the vaccine against the dreadful disease rabies. Rabies had never been a great killer, like smallpox or cholera, but fear of the disease had always been far out of proportion to the number of its victims. Paul de Kruif's classic *The Microbe Hunters* contains an unabashedly heroic account of Pasteur's conquest of rabies. Here is de Kruif's description of Pasteur fearlessly risking his own life to combat the disease:

> And now Pasteur began—God knows why—to stick little hollow glass tubes into the gaping mouths of dogs writhing mad with rabies.

> While two servants pried apart and held open the jowls of a powerful bulldog, Pasteur stuck his beard within a couple of inches of those fangs whose snap meant the worst of deaths, and, sprinkled sometimes with a maybe fatal spray, he sucked up the froth into his tube—to get a specimen in which to hunt for the microbe of hydrophobia. (1926, 169)

After a long series of excruciatingly difficult experiments, Pasteur found a way to weaken the infectious agent of rabies, now known to be a virus. He concocted a vaccine that was administered in a series of fourteen injections, starting with the most weakened virus and proceeding to the most virulent. The vaccine worked perfectly on dogs, but would it work on humans? Pasteur was reluctant to try, but the story of how he did is one of the most famous in the history of medicine. In July 1885, Joseph Meister, a nine-year-old boy from the Alsace region, was mauled by a mad dog. His mother brought him to Paris and begged Pasteur to save her child. Moved by the plight of the terrified boy and his desperate mother, Pasteur agreed to try. De Kruif tells the story:

> And that night of July 6, 1885, they made the first injection of the weakened microbes of hydrophobia into a human being. Then, day after day, the boy Meister went without a hitch through his fourteen injections—which were only slight pricks of the hypodermic needle into his skin. . . . And the boy went home to Alsace and never had a sign of that dreadful disease. (179–80)

Pasteur's greatest triumph occurred when nineteen Russian peasants who had been bitten by a rabid wolf nearly three weeks before were brought in for his treatment. So long a time had passed since they had been attacked that few believed that Pasteur could save them. He took a terrible risk in trying; had he failed, his vaccine would have been blamed. Usually eight out of ten people bitten by rabid wolves got rabies, and once the disease strikes, death is inevitable. De Kruif records the result:

> And at last a great shout of pride went up for this man Pasteur, went up from the Parisians, and all of France and all the world raised a paean of thanks to him—for the vaccine marvelously saved all but three of the doomed peasants. . . . And the Tsar of All the Russias sent Pasteur the diamond cross of Ste. Anne, and a hundred thousand francs to start the building of that house of microbe hunters in the Rue Dutot in Paris—that laboratory now called the Institut Pasteur. (181)

De Kruif is unabashed in his hero worship, and the same sort of awed gratitude has been expressed by many of Pasteur's biographers. Then, in 1988, Bruno Latour—a Frenchman, no less—decided to take Pasteur down a few pegs. Latour's book *The Pasteurization of France* is anything but hero worship. In fact, it is a direct assault on the whole notion of scientist-as-hero. The Pasteur that emerges from Latour's work is not exactly a rascal, but he is certainly an opportunist and a grandstanding self-promoter whose successes were more theater than science.

Why does Latour want to expose Pasteur as a clay-footed giant? Latour is not motivated by envy or mean-spiritedness; he just does not see science as the noble, selfless pursuit of truth carried on by a few "great men" as de Kruif and other popular writers have depicted it. For Latour, science is war. Scientists may give lip service to the ideals of method and objectivity, but, just as the chaos of battle nullifies the generals' beautiful plans, so scientific battles make nonsense of such fine talk. For Latour, preaching standards and values in the middle of a scientific squabble would be like reciting the Ten Commandments in a barroom brawl.

Latour has a point. Anyone who thinks of scientists as serene truth seekers or emotionless Mr. Spock types has another thing coming. We have already mentioned in the previous chapter the enormous imbroglio that erupted over the impact theory of mass extinction. Scientists sometimes harbor personal animosities that border on mania and pursue vendettas with such tenacity that they harm science itself. The feud between Edward Drinker Cope and Othniel Charles Marsh, the two leading American paleontologists of the nineteenth century, is a case in point. They hated each other with a reckless intensity that it tarred their reputations and corrupted their science. According to reliable reports, Marsh ordered his workers to destroy fossil specimens rather than have them fall into Cope's hands. Cope launched a yellow-press newspaper attack on Marsh, leading to highly public mudslinging that dishonored both parties. In an effort to better his highly prolific rival, Marsh often rushed his findings into print, leading to errors that took the better part of a century to sort out.

The rancor between Marsh and Cope may have been exceptionally bitter, but in the history of science there has been no lack of conflict. Also, there is no question that some of the greatest scientists have been involved in some of the loudest disputes. But hardly any great achievement in any field has ever been accomplished without bitter, intransigent, and sometimes violent opposition. So, is science really less noble, or scientists any less worthy, because controversy always accompanies discovery? Latour puts his case this way:

> We would like science to be free of war and politics. At least, we would like to make decisions other than through compromise, drift, and uncertainty. We would like to feel that somewhere, in addition to the chaotic confusion of power relations, there are rational relations. . . . Surrounded by violence and disputation, we would like to see clearings—whether isolated or connected—from which would emerge incontrovertible, effective actions. To this end we have created, in a single movement, politics on one side and science or technoscience on the other. The Enlightenment is about extending these clearings until they cover the world. . . . Few people still believe in such an Enlightenment, for at least one reason. Within these clearings we have seen developing the whole arsenal of argumentation, violence, and politics. Instead of diminishing, this arsenal has been vastly enlarged. Wars of science, coming on top of wars of religion, are now the rage. (1988, 5)

The dream of the leading thinkers of the European Enlightenment of the eighteenth century, a dream inspired by the enormous achievements of modern science as epitomized by Newton, was that the rise of modern science had, at long last, brought truly objective knowledge and the one sure method for discovery. In the minds of Enlightenment thinkers, science had ushered in the age of reason that would displace the ages of dogma and superstition that had gone before. Humanity could finally outgrow the endless and divisive theological disputes, and the concomitant persecutions, inquisitions, and holy wars. Freed from the ever-finer hairsplitting of metaphysical speculation, the finest human minds could now turn to the production of useful knowledge. Some of the founders of modern science, like Francis Bacon and René Descartes, believed that the discovery of scientific method meant that such pointless controversy could end. Once we have the true method, our disagreements no longer will lead to bickering; rather, we will simply calculate. Scientific questions will be solved by appeal to universally accepted procedures and will be as calm, dispassionate, and certain as doing sums in arithmetic. It hasn't worked out like this. On the contrary, Latour argues, in science things are settled by rhetoric, negotiation, power politics, wheeling and dealing, grandstanding, *ad hominem* attacks, and intimidation the same as everywhere else. The most successful scientists were those who were best at forming powerful alliances, appropriating grants or scarce resources, browbeating opponents, or propagandizing their views.

There is considerable truth in Latour's gruesome depiction of scientific warfare. Scientists can play political hardball. Some, like Cope and Marsh, do stoop to character assassination to deal with scientific opponents. The old saying about whom you know being more important that what you know does often apply in science. But surely Latour's view is too cynical, most would still say. Surely, this is not *all* science is. True, scientists are merely human and they are subject to all the weaknesses and foibles—egotism, petty jealousies, spite, and narrow-mindedness—that prey on everyone else. Also, the hope that humanity has found a single, universal, scientific method is a pipe dream. Instead, there are many different methods for many different scientific disciplines; geology and particle physics just cannot be done the same way.

Also, within each discipline, methods are changing and developing all of the time. Yet it does seem that science has made some incontrovertible discoveries that have revealed much about the components and workings of nature, for example: the blood circulates; DNA is the genetic material and it has a double-helix structure; the solar system is part of a giant spiral galaxy we call the Milky Way; things are made of atoms, which are themselves composed of even smaller parts called electrons and quarks, and so on.

Latour is not impressed by such litanies of scientific achievement. In 1979, he coauthored, with Steve Woolgar, a book titled *Laboratory Life: The Construction of Scientific Facts*. Latour prepared to write this book by taking a menial job at the Salk Institute, a leading laboratory for biomedical research. This job gave him the opportunity to observe scientists at work in their native habitat, like an anthropologist who lives with a rainforest tribe to observe their customs and practices. In particular, he followed scientists through the laborious, tortuous process of trying to identify and isolate a highly elusive bodily substance called TRF for short. Latour charted the complex discussion and debate as scientists initially proposed the existence of TRF, encountered skepticism and opposition, responded to criticisms, engaged in a series of rebuttals and rejoinders, and finally succeeded in convincing their colleagues that TRF is real. Like a good anthropologist, Latour studied his subjects without accepting their worldview. Just as the anthropologist does not accept at face value the tribal shaman's claims about gods and magic spells, so Latour did not take for granted the truth of scientific claims.

In *Laboratory Life*, however, Latour and Woolgar do not simply give a detached anthropological description of the customs and beliefs of the scientific tribe. Rather, they offer an analysis and interpretation that radically undercuts the claims of science to discover objective facts about the natural world. Put bluntly, their aim seems to be to debunk science. For Latour and Woolgar, scientific "facts" are not discovered; they are constructed. According to this view, called "social constructivism," the so-called facts of science are mere artifacts of scientific culture, just as beliefs in gods, demons, and magical powers are artifacts of tribal cultures. According to *Laboratory Life*, scientists are in the business of generating fact statements, but nature—conceived as something that exists "out there" independently of our concepts—has virtually nothing to do with the generation of such fact statements. Such fact statements emerge when a given scientific community reaches consensus on the issue, and consensus is a *rhetorical* achievement (rhetorical in the sense of "mere rhetoric," where the goal is persuasion by any means necessary). In other words, all of the methods and techniques deployed in scientific debate are really just elaborate rhetorical devices, not, as scientists like to think, reliable means of testing theory against empirical reality. Consensus emerges, and so new "facts" are established, when some group of scientists employs such rhetorical devices skillfully enough to convince, or at least silence, all opponents.

For Latour and Woolgar, the means whereby scientists generate fact statements is really quite insidious. Every new "fact" begins as an innocent hypothesis. Everyone admits its tentative and speculative nature. But as the debate proceeds, proponents of the new "fact" use all the rhetorical means at their disposal to get skeptics to drop their opposition and accept the "fact." Once this has occurred, once consensus emerges in a scientific community, a curious process that Latour and Woolgar call "inversion" allegedly takes place. An "inversion" supposedly occurs when a scientific community forgets that its agreement on the new "fact" was achieved by rhetorical means and starts to think of the "fact" as "out there," that is, something that really has all along existed in the natural world. However, this is merely self-deception, Latour and Woolgar contend. The "out there"-ness is just a figment of the scientific imagination induced by a sort of collective amnesia whereby scientists conveniently forget the real process of rhetorical manipulation that got everyone to accept the new "fact." I say "conveniently" because it is greatly in scientists' interests to portray themselves as discoverers of objective reality rather than as, as Latour and Woolgar think they really are, just another set of tribal shamans pursuing their own myths and rituals. Biologist Matt Cartmill provides an unfriendly but accurate summary of the social-constructivist view:

> The philosophy of social constructivism claims that the "nature" that scientists pretend to study is a fiction cooked up by the scientists themselves—that, as Bruno Latour puts it, natural objects are the *consequences* of scientific work rather than its *cause*. In this case, the ultimate purpose of scientists' thoughts and experiments is not to understand or control an imagined "nature," but to provide objective-sounding justifications for exerting power over other people. As social constructivists see it, science is an imposing but hollow Trojan horse that conceals some rather nasty storm troopers in its belly. (1999, 49–50; emphasis in original)

In his more recent writings, Latour claims to have abandoned strict social constructivism. In *We Have Never Been Modern* (1992), he argues that scientific objects should be regarded as hybrid entities, neither as wholly natural, as scientists view them, nor as mere artifacts, as social constructivism holds. Rather, scientific objects, a virus, say, should be thought of as more or less natural or more or less constructed, depending on the context. Unfortunately, he never really clarifies just what it would mean to regard a virus as such a hybrid. Some commentators claim to have detected a "creeping realism" in Latour's later writings. They think that he begins to admit (sort of) that—just maybe—things like microbes really exist and have *some* bearing on the course of science. Whether or not this is so, with Latour, as with Kuhn, it was the earlier, more radical views that most impressed Latour's friends and critics.

Anyone imbued with a more traditional view of science might be tempted to dismiss social constructivism as a farrago of fuzzy thinking and exaggeration. The problem with such pat dismissal is that there are so many episodes in the history of science that do make us wonder how effective science really is at separating the real from the chimerical. How much of what we take to be facts about the natural world are really artifacts of our own making? One episode that raises this question is the famous wrong-headed dinosaur scandal.

The Carnegie Museum of Natural History in Pittsburgh houses one of the world's foremost exhibits of dinosaur fossils. For forty-five years, the Carnegie Museum displayed one of its prize specimens, the gigantic *Apatosaurus louisae*, with the wrong head. The head was not a little bit wrong, but way off, like a paleontologist of the distant future putting a giraffe's head on a horse's body. Worse, other paleontologists accepted the chimera as real. All of the top authorities accepted the wrong-headed creation as the real *Apatosaurus*. How did this happen?

Briefly, the problem had its roots in the feud between Marsh and Cope. Marsh published a reconstruction of *Apatosaurus* in 1883, which, to add to the confusion, he called by the familiar name *Brontosaurus*, mistakenly thinking it was a different kind of dinosaur than *Apatosaurus*. The problem was that *Apatosaurus/Brontosaurus* had been found without a head, and it just would not do to have a reconstruction without a head. So Marsh improvised and stuck on a cranium that he had found at a completely different site. It eventually turned out that the head he found belonged to *Camarasaurus*, a creature not closely related to *Apatosaurus*. When the Carnegie Museum mounted its prize *Apatosaurus* specimen in 1915, the museum's director W. J. Holland, an outstanding paleontologist in his own right, had deep reservations about the head Marsh had given the creature. Unfortunately, when the Carnegie Museum's specimen was found, it also lacked a head, so Holland had no definitive evidence against Marsh's reconstruction. Holland simply mounted the skeleton with no head.

Holland died in 1932, and in 1934 the new director of the museum decided that the *Apatosaurus* needed a head. Just who made the decision and on what grounds is not clear. Probably, since museums not only are research institutions but also are there for the edification and entertainment of the public, everyone felt that a headless *Apatosaurus* just made a terrible impression. So, following Marsh's precedent, a very robust *Camarasaurus* skull was attached to the *Apatosaurus* skeleton. The big skull looked great on the massive skeleton, and for over forty years everybody was satisfied, everybody except John S. McIntosh, perhaps the world's leading authority on sauropod dinosaurs like the *Apatosaurus*. McIntosh began to suspect that another skull, one already in the Carnegie Museum's possession, was the right one. That other skull, designated as CM 11162, was found with the *Apatosaurus* specimen, but not in position at the end of its neck. Skull CM 11162 looked too small to belong to so massive a creature as *Apatosaurus*. It looked like a somewhat-larger version of a *Diplodocus* skull, and *Diplodocus* was a much slimmer animal than the ponderous *Apatosaurus*. However, after a thorough review of the records of the discovery of the Carnegie Museum's *Apatosaurus*, and a careful examination of skull CM 11162, McIntosh decided that it had to be the right one. He published his conclusions with coauthor David Berman in 1978. Their argument was so convincing that the Carnegie Museum agreed to remove the *Camarasaurus* head. Finally, in 1979, after forty-five years of displaying an *Apatosaurus* with the wrong head, the Carnegie Museum held a ceremony to remove the old skull and attach a cast of skull CM 11162.

It is really quite shocking that one of the world's leading museums would display a prize specimen with the wrong head for so long. What if the Louvre had displayed a painting upside down for so long? Don't incidents like this make us wonder whether what we take to be scientific fact might not be artifact? Worse, there seemed to be no very good scientific reason for the Carnegie Museum's decision to mount the bogus head. It seems to have been a response to the demand to present a whole specimen for public viewing. Unlike the *Venus di Milo*, missing parts did not make *Apatosaurus* more appealing. Incidents like this lend credence to the claim that scientific decisions are often (strict constructivists would say *always*) made in response to social pressures and are not based on objective evidence.

Even if we admit, as we must, that science is often deeply influenced by social, political, and ideological pressures, must we accept the constructivist claim that, as Latour contends, the course of science is determined by power politics and rhetorical manipulation? Again, the issue of social and political influence on science is a very real and a very serious concern. It is indeed a serious matter when wealthy corporations aided by political ideologues manipulate or suppress legitimate scientific findings. But can the social-constructivist theory be the *whole* story about science?

A problem with assessing the claim that facts are social constructs is that the word *fact* itself is ambiguous. According to *The American Heritage College Dictionary*, one meaning of fact is "information presented as objectively real." In another sense, *fact* can mean "something having real, demonstrable existence." In other words, *fact* can refer to a *claim*, the assertion that something is really so, or it can refer to the *reality* that our factual claims are about. So, when one claims that scientific facts are constructed, this could mean either of two things: (a) our supposedly factual statements do not correspond to anything real (or, if they do, it is a sheer accident) but are mere artifacts of the scientific process, or (b) there is no objective, "out there" reality consisting of states of affairs that exist independently of our beliefs or concepts. Sometimes Latour speaks as if he means to assert (a) and sometimes as if he means (b). So far, I have assumed that he means (a), since it seems a more-plausible claim. That is, I have taken Latour to claim that, whether or not there is a real physical universe, such a putative natural world has no influence upon the "nature" conceived by scientists. The "nature" scientists study is therefore just an artifact, or, more bluntly, just a figment of the scientific imagination. Assuming that this is Latour and Woolgar's claim in *Laboratory Life*, how sound is it?

Latour and Woolgar present their conclusions as grounded on *empirical* (i.e., scientific) evidence. Latour based his conclusions in *Laboratory Life* on his research as an anthropologist of the laboratory. Anthropology is a science. If all scientific conclusions are social constructs, as *Laboratory Life* asserts, so are those of anthropologists like Latour. Practitioners of social-constructivist anthropology or sociology of science have only two choices when they confront this problem of self-reference, or "reflexivity" as it is called: They can bite the bullet and frankly admit that their "facts" are just as socially constructed as those of the natural sciences are alleged to be. In this case they have the burden of explaining why their so-called findings about the practice of science should be taken seriously by anyone skeptical of those alleged findings. The alternative is for social constructivists to argue, very implausibly, that social sciences like anthropology or sociology *do* draw upon reliable methods and objective evidence while the natural sciences, like organic chemistry or particle physics, do not. In other words, Latour's science was legitimate and Einstein's was not. Social constructivists have vigorously debated among themselves which horn of this dilemma to grasp, but neither option seems very appealing. The upshot is that if the social-constructivist thesis is taken in the above sense (a), it is hard to see how constructivism can debunk science without debunking itself.

Sometimes, however, Latour seems to be making the above assertion (b)—that there is nothing that just *is* so, but that reality is, in some sense, *created* by the beliefs or concepts we form. That is, he sometimes seems to be implying a metaphysical claim about the nature of reality. For instance, in his book *Science in Action* (1987), he considers the famous case of the French scientist René-Prosper Blondlot, who, in the early twentieth century, claimed to have discovered a new type of radiation he called N-rays. The way the story is usually told, Blondlot thought he could observe a previously undetected sort of radiation emitted by metal under strain. Other physicists were skeptical because they could not reproduce Blondlot's claimed observations. An American physicist, Robert W. Wood, visited Blondlot's lab to see the procedure whereby N-rays supposedly were detected. At one point in his visit, while Blondlot was engaged with his experimental apparatus, Wood quietly removed an essential piece of the equipment. Yet Blondlot continued to proclaim that he could observe the N-rays as they were being generated. For Wood, and soon the whole physics community, this was proof that Blondlot's N-rays did not actually exist and that the reported "observations" of them were delusions.

Latour indignantly insists that we should not interpret this incident as implying that Wood was right and Blondlot wrong:

> It would be easy enough for scientists to say that Blondlot failed because there was "nothing really behind his N-rays" to support his claim. This way of analyzing the past . . . crowns the winners, calling them the best and the brightest and . . . says that the losers like Blondlot lost simply *because* they were wrong. . . . Nature herself dis-

> criminates between the bad guys and the good guys. But is it possible to use this as a reason why in Paris, in London, in the United States, people slowly turned N-rays into an artefact? Of course not, since at that time today's physics obviously could not be used as the touchstone, or more exactly since today's state is, in part, the *consequence* of settling many controversies such as the N-rays. (1987, 100; emphasis in original)

That is, since the opinions of present physicists about the (non)reality of N-rays were shaped by the outcome of the N-ray controversy, those opinions cannot explain the outcome itself.

Well, why not? Of course, it was, in part, Wood's fine job of debunking that convinced everyone at the time that Blondlot's claims were false. But once we are convinced that there was nothing there for Blondlot to detect, doesn't this explain why he failed to see them (given that his equipment *would* have detected them had they been there)? Why did people in April 1912 fail to see the *Titanic* docking in New York? Because it never arrived. As for why Blondlot *thought* he saw N-rays, this is given a psychological explanation in terms of how wishful thinking and the inherent limitations of human perceptual abilities can make us "see" things that are not there.

For Latour, such a common-sense account of the outcome of the N-ray episode just will not do. Notice his language. He says that people "turned" N-rays into artifacts. This seems to imply that for Latour there was no fact of the matter, no way that things really were, before the controversy over N-rays was settled. It is not that Blondlot was deluded all along, and that Wood proved this to everyone else's satisfaction. Rather, Wood and others *turned* N-rays into artifacts. Does Latour think that N-rays could just as easily have been turned into real phenomena? It is hard to know just what Latour means here, but he seems to be saying that there simply was no fact of the matter about N-rays; they were neither real nor a mere artifact, until physicists *decided* the case. It is not just that nobody *knew* that Blondlot's purported observations were not real until Wood did his debunking;

such a claim would be boring even if true (and Latour is *never* boring). Rather, there was NO fact of the matter—nothing *there* to know—until the physics community agreed on its story!

Now such a claim may strike many as bizarre, but it is not obviously incoherent or self-defeating. It is reminiscent of the metaphysical idealism of British philosophy in the nineteenth century, which held that physical reality is a creation of the mind. I'm sure, though, that Latour would abjure any "idealist" label and certainly would disclaim any metaphysical agenda. Yet he seems to have fallen for an all-too-common fallacy of the sort that afflicted much idealist philosophy. This fallacious way of thinking begins with the innocent observation that we can only think with our ideas, but then it leaps to the conclusion that all that we can know are our ideas. Michele Marsonet explains this fallacy and points out its obvious flaw:

> We do not know reality *directly*, but only through representations such as ideas and mental images. If this is true, it follows [so the fallacious argument goes] that we only know our representations, while it is impossible to know an alleged reality in itself. However, it should be easy to realize that from the fact that we know *through* representations, it does not follow that we can only know representations and nothing else. . . . (1995, 59)

Of course, the philosophical debate about the relation of ideas to reality is extremely long and complex, and we cannot enter it here even in the most superficial way. So, let us assume that Latour does hold, as he certainly seems to, that all that we can know are our own ideas, and we cannot infer anything about any putative reality behind those ideas. He then has two options about how he thinks about any alleged mind-independent physical world: He can be agnostic about the existence of an objective physical universe, perhaps giving the stereotypical Gallic shrug when asked about it, or he can take the stronger position that there is no such world. We saw above that he sometimes speaks as though he takes the stronger line.

The problem with agnosticism about the existence of the physical world is that it is not clear how we are to explain the existence and the contents of our ideas unless we postulate physical objects as their causes. Surely René Descartes was right that our ideas have to be caused *somehow* and do not just pop into existence *ex nihilo*. Immanuel Kant postulated the existence of unknowable "*Dinge an sich*," "things in themselves," as the cause of our perceptions. Later philosophers complained, reasonably enough, that we cannot meaningfully say that something exists—and much less that it causes all our perceptions—unless we attribute *some* sort of nature or character to it. So it cannot be satisfactory to postulate physical objects as the unknowable "things in themselves" that cause our ideas. In *Laboratory Life*, physical objects do not even function as *Dinge an sich*. Instead, Latour and Woolgar invoke social and political factors as the sole and sufficient causes of scientific ideas. The problem with this option, as we have seen, is that it runs into the problem of reflexivity, that is, it undermines itself because if all truth claims are social constructs, so are those made by social constructivists.

Well, just what is wrong with denying the existence of the physical world? It is hard to say precisely how one could argue that there *is* an external, mind-independent physical world since any evidence you could mention would presuppose the existence of such a world. Could you convince someone like Latour of the reality of the external world by pointing to physical objects or waving them in his face? Should we attempt to prove the existence of an external world as British philosopher G. E. Moore famously did, by pointing to a hand and saying, "Here is a hand"? Steve Woolgar, Latour's collaborator on *Laboratory Life*, challenged his students to reveal a physical object to him without employing a representation of some sort. When students would point to a book or table, Woolgar would reply that pointing is itself a kind of representation. Maybe Woolgar was trying to make a point like the one artist René Magritte made when he painted a picture of a pipe and wrote under the depiction, "*Ceci n'est pas une pipe*" ("This is not a pipe"). Just as the picture of a pipe is not a pipe, so pointing to a pipe is a representation, not a pipe. Woolgar was (I think) trying to show that we live in a world of representations, signs, symbols, images, and ideas—not objects.

What the students should have done is to point out that Woolgar has set them a task that is by definition impossible. You cannot indicate an object without indicating it, and Woolgar will say that any such act of indicating is a representation. But while it is certainly true that you cannot indicate something without indicating it, it does not follow that indicate-able objects do not exist independently of our representations. Neither does it follow that we can never know objects but only our representations.

The approach to take with questions like the existence of an external world is to begin by noting that, as the philosopher John Searle puts it, the human mind comes with certain default settings. When you first boot up a personal computer, in order to function at all it must come with certain default settings that it will keep until you change them. Likewise, the human mind seems just naturally preset to take certain things for granted. One of those things we just take for granted is that there is an external physical world that exists "out there" independently of our consciousness. Now default settings can be changed on your computer and in your mind, but we humans have every right to demand a *very* good reason before we abandon a belief that is so spontaneous, natural, and (nearly) universal as belief in an external physical reality. In other words, to say the very least, a very heavy burden of proof is on those who would deny the existence of an external world. How could one make such an argument? Merely pointing out that we cannot think about things without using concepts or indicate objects without somehow representing them or make observations without appealing to some theory will not prove this at all. All of these claims may well be true (even truisms), but it just does not follow from any of them that there is no mind-independent reality or that we cannot know a great deal about it. At this point I shall simply cut to the chase and assert that, in my opinion, nowhere does Latour, Woolgar, or any other social constructivist offer arguments anywhere near strong enough to support so sweeping a claim as the nonexistence of a mind-independent physical reality. Nor do they show that we fail to have cognitive access—*through* our perceptions and concepts—to that reality. So Latour and Woolgar's version of social constructivism fails in its effort to debunk science.

POSTMODERNISM ATTACKS!

However, social constructivism is not the only recent program of radical science critique. There are also the postmodernists. *Postmodernism* is a term that defies precise definition. A variety of movements or styles in literature, art, and architecture may be called "postmodern." The "postmodernist" label has been attached to a wide variety of writers, including the philosopher Gilles Deleuze; his frequent collaborator the psychoanalyst Felix Guattari; sociologist Jean Baudrillard; psychoanalyst Jacques Lacan; and Luce Irigaray, whose writings deal with topics in many fields. So multifarious are these various manifestations of the postmodernist spirit that I can give only a very broad and impressionistic characterization of the attitudes and outlooks that tie them together. Anyway, postmodernist theorists themselves would probably reject any proffered canonical definition of *postmodernism*, since one thing postmodernists share is a distaste for canonical statements of anything. For them, anything that presents itself as canonical, authoritative, or definitive is something to be abused, ridiculed, or otherwise subverted. For the postmodernists, it is a dangerous delusion to think that we ever have the complete story or final answer about anything. We are all "on a darkling plain; swept with confused alarms of struggle and flight; where ignorant armies clash by night" (to quote "Dover Beach," by Matthew Arnold, one of the canonical "great poets" postmodernists love to hate).

To avoid confusion, let me make an important distinction: Practically all philosophers these days are fallibilists. That is, they recognize that even our best-supported theories and factual claims are fallible and may turn out to be wrong—as, indeed, they so often have in the past.

But fallibilism does not entail relativism. Even a thorough fallibilist can say that, so far as we can tell, some things just are so, some questions really have been settled, and some norms have at least *prima facie* validity. Not so for the postmodernists. For them all norms—whether ethical, aesthetic, or epistemological—have merely local authority and applicability and are radically contingent on such factors as gender and social class. Postmodernists are hostile to any claim that a standard is more than merely parochial, viewing all claims to objectivity as attempts by one group to impose its values on others. Hence, in all fields postmodernists celebrate a promiscuous eclecticism of standards, values, and norms.

Postmodernism often comes across as more of a style than a stance, more of a pose than a position. Sometimes they seem to take pride in outraging more traditional thinkers. Postmodernists scorn even ordinary speech, which they regard as polluted by oppressive standards of clarity and truth. For them, the requirement that words should have definite meanings is just another tool of oppression. Their prose style therefore is often verbose, paradoxical, allusive, convoluted, and, in general, intended to disrupt and frustrate our ordinary ways of thinking. The downside is that much of what they say sounds like gibberish to outsiders. Postmodernists typically also reject the idea that it is a legitimate function of language to *represent* a language-independent reality. Baudrillard argues that an image should no longer be regarded as a simulation of reality, but that our media-drenched world *is* now a world of rootless, free-floating signs.

From what I've said so far, you may have gotten the impression that postmodernists are a bunch of zanies who lack seriousness of purpose and whose views have no intellectual motivation. Such an impression would be mistaken; postmodernists are entirely serious in their aims, and their inspiration comes from thinkers who unquestionably were intellectual superstars. One such luminary was Friedrich Nietzsche (1844–1900). One aspect of Nietzsche's thought that the postmodernists have particularly emphasized is his view on the relation between knowledge and power. Nietzsche said that knowledge is an instrument of power, that is, that the motivation to acquire knowledge is to acquire more power. We want to "master" certain fields, thereby making that field of knowledge into a servant to promote our interests. For Nietzsche, as for the pre-Socratic philosopher Heraclitus, reality is an eternal flux that is always in a process of becoming. Knowing is not a matter of recognizing an objectively given reality; there is no such determinate, independent reality, only flux. Understanding therefore involves imposing our conceptual schemes, categories, and interpretations on the flux, thereby creating Being out of sheer Becoming. The interpretations we place on reality will reflect our vital interests and concerns, that is, the "reality" we create will be one that serves our purposes and enhances our power. Indeed, from an evolutionary perspective, only those "truths" that are useful survived. For Nietzsche, the idea that there is absolute truth, truth independent of all our interests and purposes, is a myth created by philosophers who hanker after a stable and permanent reality and are afraid to embrace endless flux. It follows that for Nietzsche, there is no one perspective, no "God's-eye view" that gives a comprehensive and authoritative view of the whole of reality.

Another, more recent, progenitor of postmodernism was philosopher Jean-François Lyotard (1924–1998). Lyotard argues that we should reject all "metanarratives." A "metanarrative" is any attempt to establish an absolute standard for any value or ideal—truth, rationality, goodness, or justice, for instance. Instead, we must recognize that there is an irreducible plurality of incommensurable narratives, each encompassing its own criteria for goodness, truth, and so on. Lyotard expressed these views most influentially in his work *The Postmodern Condition: A Report on Knowledge* (1979).

Another notable philosophical work published in 1979, Richard Rorty's *Philosophy and the Mirror of Nature*, articulated and promoted various postmodernist themes for an English-speaking audience. Rorty characterizes his philosophy as "pragmatist," that is, in the tradition of classical American philosophers such as William James and John Dewey, but many of his ideas are typically postmodernist. For instance, he strongly advocates that philosophers abandon the attempt to establish absolute foundations for knowledge and instead dedicate themselves to the facilitation of the "conversation of mankind." According to Rorty, all of the "voices" of humanity, from Hopi philosophy to Polynesian mythology to quantum physics, deserve to be heard and no one narrative should be preeminent. When it comes to grounding our beliefs, Rorty says that we can do no better than to say what our society lets us say, that is, when in Rome, follow the epistemological practices of the Romans. This does not mean that our beliefs are not to be subject to strict critical scrutiny, but Rorty thinks that the standards we employ when we thus examine our beliefs are contingent historical products of a particular time and place and lack universal authority.

It is hardly surprising, given such an intellectual background, that postmodernists do not like science very much. After all, there are some things that science says are just so, and others definitely not so—period. Heavy bodies in free fall in the vicinity of the earth's surface accelerate at a rate of about 9.8 meters per second squared. The sun has a mean distance from the earth of 149.5 million kilometers. The nuclei of human somatic cells contain forty-six chromosomes arranged in twenty-three pairs. Science does not say that such things are so from a given perspective, or according to some traditions, but that they are just so. Science claims to be the authority in answering certain questions. Where did birds come from? They evolved from theropod dinosaurs in the late Jurassic, say (many) paleontologists. Science says that it is simply false that God separately made birds, dinosaurs, and all the other organisms during a six-day creation. Further, science claims that its methods alone are the right methods for investigating the natural world and that valid methods are not, for instance, consulting horoscopes, gazing into crystal balls, or invoking the authority of ancient texts.

Because science so often claims to have *the* answer, and not simply to offer one among indefinitely many perspectives, it is bound to ruffle postmodernists' feathers. What right do paleontologists have to tell Christian fundamentalists and Orthodox Jews that the Genesis creation account is false? Is it not arrogant for anthropologists to tell

Native Americans, whose traditions teach that they are indigenous to North America, that their ancestors actually came across a land bridge from Asia? Following Nietzsche, postmodernists say that knowledge is power. They do not mean this the way that Francis Bacon did, as a recognition of the fact that knowledge gives us power over nature, but in the sense of one of their favorite writers Michel Foucault (1926–1984). When Foucault says that knowledge is power, he means it in the sense that the winners get to write the history books. For instance, the anthropological account of the origin of Native Americans is just the story that the winning white European culture gets to impose on the losing Native American culture. Postmodernists regard the reigning standards that define rational discourse—the standards that tell us what counts as a logical inference, objective evidence, or coherent speech—as potential tools of oppression. Small wonder that postmodernists want to challenge what they see as the intellectual hegemony of science. Though their jargon may be opaque, their intentions are clear. They aim to cut science down to size, to display it as just another form of discourse, and as no more "rational" or "objective" than any other. Postmodernist literature is vast and highly diverse. From these many writings I have selected two books to examine here: Donna Haraway's *Primate Visions* (1989) and W. J. T. Mitchell's *The Last Dinosaur Book* (1998). These two books offer postmodernist analyses of two fields of science that have much popular appeal, primatology and dinosaur paleontology.

Reading postmodernist literature, you can get the impression that they are obsessed with the electronic media. Because they reject the distinction between "high" and "low" culture, postmodernists repudiate the traditional academic disdain for popular entertainment. Papers by academic postmodernists will often have titles like "From Homer to Homer Simpson," where canonical texts like the *Iliad* and popular TV comedy get equal (and equally obscure) treatment. Not only do they collapse the distinction between popular and highbrow entertainment, they go even deeper, questioning the very distinction between a symbolic representation and the reality that it represents. Typically, postmodernists oppose what Haraway calls "binarisms"—paired concepts that, in their view, channel our thinking into narrow and misleading dichotomies. They say that rigid distinctions like subject/object, fact/fiction, same/other, and image/reality are embedded in our language and so lead us to box things into overly restrictive categories. Invidious value judgments go with such labeling, postmodernists argue, so that, for instance, people the "same" as us are good and those "other" than us are not.

Haraway attempts to subvert the distinction between science fact and science fiction. Likewise, Mitchell argues that it is impossible to maintain the distinction between the popular image of dinosaurs and what paleontologists think they really know. For Haraway, science is just another kind of narrative. True, primatologists have their story to tell, but it is just one of many and has no special authority over any other account. Similarly, Mitchell certainly feels that paleontologists should lend their "testimony" to our understanding of dinosaurs, but such "testimony" must be supplemented by the work of humanities scholars who are experts at the analysis of symbols and images. After all, echoing some of Latour's talk about hybrid objects, Mitchell says the dinosaur is not merely a natural object nor is it a pure fantasy, but is an irreducible composite. Mitchell holds that there is no way to make a workable demarcation between the extinct animal and the cultural icon.

For Haraway, the idea that science can be done in a neutral, disinterested, and impartial way is a pernicious myth. It is a myth because, she holds, all science is inevitably *political* science, since it always promotes the interests of some particular group. The myth of neutrality is pernicious because it obscures the fact that every scientific account, however purportedly "objective," or based on "logic," serves a hidden agenda. So far, that agenda has been the promotion of the interests of scientists, and their sponsors in government and industry, who are almost always white, male, and privileged. She strongly endorses the Latour/Woolgar view that all scientific "facts" are socially constructed and adds that it is vital to see for whose benefit they are constructed. She cites a well-known story about the history of primatology. According to this story, when primatologists began to analyze the social structure of primate groups, the (predominately male) scientists focused on the dominance of the so-called alpha male. The alpha male, like the silverback leader of a gorilla troop, is the dominant male. According to those early accounts, the story goes, the dominance of the alpha male was depicted as absolute. In particular, all of the females of the group were, in effect, the harem of the dominant male since he had the exclusive right to mate with them.

It is easy to see how such a representation of primate social groups could benefit males. Since primates are the closest animal relations to humans, the alleged dominance of the alpha male could be seen as the natural pattern for human society as well. That is, male dominance in human society could be justified as "natural" by pointing to the dominance of the alpha male in primate society. However, says Haraway, it fell to female primatologists to point out that the alpha male's dominance is far from absolute, and that female primates wield considerable power. For instance, among mandrills, when a new alpha male takes over by defeating the previously dominant male, he finds, no doubt to his intense chagrin, that the females are not instantly his to command. The females will defiantly refuse to mate with him until he meets their approval. Of course, Haraway cannot think that the female primatologists' conclusions were any more objective than those of their male counterparts. What matters about the stories science tells is not whether they are "true" (the scare quotes are needed because postmodernists do not buy the notion of just plain truth). What matters is whose interests those stories serve. As we shall see below, feminist theorist Sandra Harding picks up on this theme and runs with it.

Science has always assumed that the objects it studies are determinate entities that exist objectively and independently of the merely human activity called science. Science therefore gave itself the job of discovering such entities and understanding them as fully as possible. Even the weirdness of quantum mechanics has not really altered this fundamental goal of science. Quantum mechanics, at least as it is usually interpreted, tells us that there are some properties of particles that have no definite values until we interact with those particles in some way. But once the requisite interaction occurs and the particle assumes a definite value, then that value is as objective and determinate a fact as any other. For postmodernists, scientific objects lose all such status as determinate and independent entities. In postmodernist literature, a scientific object is little more than a nexus of multiplying interpretations, a blank screen onto which interested parties may project practically any image.

For paleontologists, a dinosaur was an *animal*, a creature that roamed the Mesozoic landscape and possessed distinct anatomical, physiological, and behavioral traits that we try to discover by the framing and testing of hypotheses. Mitchell treats dinosaurs as prefabricated metaphors, ready-made symbols that can stand for just about anything. For instance, dinosaurs can symbolize obsolescence, backwardness, and stupidity. On the other hand, dinosaurs can be cool and *chic*, like the sleek, fast, and deadly *Velociraptor* of *Jurassic Park*. According to Mitchell, dinosaurs can serve as the "clan sign" for just about any group. *T. rex* could symbolize unbridled ferocity while *Apatosaurus* might represent the gentle giant. A dinosaur can even be a plush purple TV figure who warbles saccharine ditties to preschoolers. For Mitchell, any attempt to strip away the layers of symbolism and get down to the *real* dinosaur would be like peeling an onion. You would never hit factual bedrock, only layer upon layer of symbol and metaphor. The upshot is that paleontologists cannot hope to understand dinosaurs, since they are under the illusion that they are studying an unambiguously *natural* object. Mitchell argues that to really understand dinosaurs, the researches of paleontologists must be supplemented by the work of humanities scholars, like himself, who are experts at the analysis and interpretation of symbols. Since dinosaurs are hybrid objects, irreducible composites of the natural and the symbolic, they must be studied by a hybrid discipline that combines the "testimony" of paleontologists with the interpretations of practitioners of cultural studies.

For Mitchell as for Haraway, scientific objects *always* have political overtones. For instance, he sees the paintings of battling dinosaurs done by Charles R. Knight at the beginning of the twentieth century as symbolic of the unrestrained capitalism of the Gilded Age:

> Knight's scenes of single combat between highly armored leviathans are the paleontological equivalent of that other war of giants, the struggles among the "robber barons" in late Nineteenth-Century America. This period, so often called the era of "Social Darwinism," economic "survival of the fittest," ruthless competition and the formation of giant corporate entities headed by gigantic individuals, is aptly summarized by the Darwinian icon of giant reptiles in a fight to the death. (Mitchell, 1998, 143)

Postmodernists also often claim to detect a sexual subtext in contexts where to others it seems hardly present. For instance, he comments on Henry Fairfield Osborn's dinosaur displays at the American Museum of Natural History in the early 1900s:

> Perhaps Osborn's most important contribution to the myth of the modern dinosaur was his linkage of it to questions of male potency. The connection between big bones and virility had already been established. . . . Big bones were also the trophies of the masculine ritual of the big game hunt, and the phallic overtones of "bones" need no belaboring by me. (150)

Even the greenish color given to dinosaurs in most depictions is full of symbolic import for Mitchell:

> So where does this leave greenness? Is it a symbol of the "colored" racial other, the savage, primitive denizen of the green world? Or is it an emblem of the white man's burden, the color of the military camouflage required for the Great White Hunter to blend in with the jungle and thus to dominate it? (147–49)

Can *everything* about dinosaurs really be bursting with political and/or sexual significance? Mitchell apparently thinks so, and he therefore recommends that Marx and Freud be invoked to analyze the political and sexual content of dinosaur images.

Any attempt at a straightforward point-by-point rebuttal of postmodern critiques of science will probably fail. This is not because those critiques are sound and therefore irrefutable. Rather, it is because almost anything a critic would take as a flaw of postmodernist analyses would be seen as a virtue by the postmodernists. Leading primatologist Matt Cartmill vents his frustration in attempting to criticize Haraway's *Primate Visions*:

> This is a book that contradicts itself a hundred times; but this is not a criticism of it because its author thinks contradictions are a sign of intellectual ferment and vitality. This is a book that systematically distorts and selects historical evidence; but that is not a criticism, because its author thinks that all interpretations are biased, and she regards it as her duty to pick and choose her facts to favor her own brand of politics. . . . This is a book that clatters around in a dark closet of irrelevancies for 450 pages until it bumps accidentally into an index and stops; but that's not a criticism, either, because its author finds it gratifying and refreshing to bang unrelated facts together as a rebuke to stuffy minds. . . . In short, this book is flawless, because all its deficiencies are deliberate products of art. (2003, 196)

Perhaps we have at last found a genuine example of incommensurable discourse: the debate between postmodernists and their critics!

Seriously, though, how do you meaningfully disagree with those who have apparently repudiated the very conditions of meaningful disagreement? How do you deploy objective evidence against those who regard objectivity as a myth? Perhaps the would-be critic would try to "deconstruct" postmodernist texts. "Deconstruction" is a kind of radically skeptical textual analysis frequently used by postmodernists. A deconstructive analysis turns a text against itself and attempts to show that it has no definite, distinct meaning, but lends itself to innumerable interpretations. Could we deconstruct Haraway's and Mitchell's texts? There seems to be no reason why not. That is, if we had the patience, we could no doubt go through their texts and pick out numerous passages that we could then interpret as meaning the exact opposite of what Haraway and Mitchell apparently intend. For instance, we could take Haraway's animadversions against objectivity and interpret them as ironical *defenses* of objectivity that work by showing the absurd consequences that follow when objectivity is repudiated. Likewise, we could take Mitchell's meditation on dinosaur's greenness as a demonstration of the silliness that inevitably results when basic distinctions are systematically conflated, like the distinction between an object and its image. In short, it looks like postmodernists are vulnerable to the same problems of reflexivity that plagued the social constructivists. If all texts can be deconstructed, so can the texts of postmodernists.

But such a quick, down-and-dirty dismissal of postmodernism completely misses the point since postmodernists emphatically reject the canons of rationality that underlie any such critique. They reject all demands that texts meet standards of consistency, coherence, or truthfulness. Postmodernists have no problem with reflexivity. They would be the first to admit that their own texts can be deconstructed! Perhaps, then, postmodernist texts should not be regarded as rational arguments; their goal is not to arrive at truth, or even to achieve coherence; such notions are for them passé. Postmodernist writing is above all a *performance*. That is, perhaps it is best to regard postmodernist science critique as a genre of confrontational performance art; its goal is not to persuade but to provoke. Some critics of postmodernism have therefore concluded that instead of wasting rational argument on such provocateurs they should play tricks back on them.

This is precisely what physicist Alan Sokal did when he wrote a spoof of postmodernist science critique, intentionally filled it with arcane postmodernist jargon and absurd arguments, and passed it off as a serious article to *Social Text*, a periodical that prominently features postmodernist writers. Sokal gave his piece a suitably portentous title: "Transgressing the Boundaries: Towards a Transformative Hermeneutics of Quantum Gravity." The text was a farrago of ludicrous claims about the political implications of recent developments in physics spiced with particularly opaque passages from leading postmodernist writers. In 1996, *Social Text* published Sokal's parody as a serious article, and the joke was on them. Sokal revealed the hoax in the periodical *Lingua Franca* and contended that his successful sting had exposed the ignorance and laziness of the postmodernist science critics. He charged that such critics had shown that they would endorse anything, no matter how incompetent, that supported their view. Needless to say, many of the postmodernists were embarrassed and outraged and responded to Sokal with considerable asperity. Stanley Fish, a noted literary scholar and onetime editor of *Social Text*, castigated Sokal and accused him of creating a spiteful Trojan horse to embarrass colleagues. Such behavior, Fish charged, only undermined the basic trust necessary for scholarship as a cooperative enterprise. Even some philosophers of science who are sympathetic to Sokal's view feared that his hoax would lead only to polarization when bridge-building between various disciplines is sorely needed.

There are places in postmodernist writings where they do seem to be making straightforward claims backed by evidence and argument. For instance, what are we to make of Mitchell's proposal that paleontology be replaced by a hybrid discipline that combines the expertise of paleontologists and the skills of "cultural scientists," as he thinks specialists in his field should be designated? Our reaction to this proposal will depend on how we take Mitchell's claim that dinosaurs are inevitably hybrid objects and that it is impossible to scrape off the accretion of symbolism and get down to rock-solid, literal truth about dinosaurs. Now, admittedly, there are some things about dinosaurs we do not and very probably never will know. For instance, the colors of dinosaurs will probably remain conjectural. We just have no way of knowing whether dinosaurs were the greenish color that got Mitchell's interpretive juices flowing or whether, maybe, they really were purple. We hardly know everything about living creatures, so how could we ever know everything about extinct ones?

Yet we seem to know that some things about dinosaurs just *are* so, and Mitchell never offers any good reason to doubt that we do. Just because an object has potent symbolic import for us is no reason to think that we cannot know many things that are literally true of that object. For instance, a cross naturally has deep symbolic significance for devout Christians, but Christians can still understand the cross as the instrument of torture and death that it actually was. Mitchell tells us, "Nature *is* culture, science is art. We don't ever 'see nature' in the raw, but always cooked in categories and clothed in the garments of language and representation" (58; emphasis in original). Of course, since it is true by definition, we must admit that we cannot think about nature without using language, categories, and representations (we cannot think about something without thinking about it). But we can admit this and still think that we do, on occasion, *get things right*.

How would the workaday scientist react to postmodernist writers? He or she would probably would think that writers like Haraway and Mitchell, whose academic careers have involved them exclusively in a world of symbols, tropes, and texts, have simply lost contact with the intractable, obstinate, downright recalcitrant world of physical fact that scientists confront daily. Even scientists who never leave their air-conditioned labs must struggle daily to square their conjectures with the hard constraints imposed by an unyielding cosmos. When it comes to telling stories about dinosaurs, Mitchell says, "There is no limit to the stories that can be made up. . . ." (48). For the paleontologist, coming up with even *one* story can be devilishly difficult. The reason for this difference is that nothing constrains Mitchell's storytelling except the limits of his own imagination. Paleontologists' stories are severely constrained both by background knowledge and by physical fact. For Mitchell, the stories we tell about dinosaurs should be full of "fantasy, unbridled speculation, and utopian imagination" (284). Science also thrives on speculation and imagination, but in science fancy must sometimes be allowed to crash into the hard rock of empirical reality. Because of these fundamental differences between Mitchell's cultural-studies approach to dinosaurs and the paleontologist's, a hybrid discipline that yokes these two disciplines is not feasible.

IS "OBJECTIVITY" WHAT A MAN CALLS HIS SUBJECTIVITY?

A major intellectual movement of the last three decades has been the rise of feminist scholarship. Science has drawn the attention of many feminist writers. These writers certainly found much about science that rightly concerned them. When you look at the index of any history of science, you find that scientists have been a very diverse lot. Over the past five thousand years, significant scientific discoveries have been made by Egyptians, Babylonians, Greeks, Chinese, Indians, Arabs, Jews, Mayans, Italians, Germans, English, Scots, Russians, Hungarians, French, Danes, Americans . . . and on and on. Great scientific work has been done by pagans, Christians, Jews, Muslims, Hindus, Buddhists, Confucians, atheists, and agnostics. Scientists of my personal acquaintance run the gamut from conservative Republicans to Marxists. Scientific journals are published in dozens of languages. In a given year, a scientist might attend professional conferences in London, Rio de Janeiro, or Riyadh. A "Who's Who" of scientists would have representatives of almost every race or ethnicity. Truly, science seems to be a characteristically *human* enterprise and not the domain of just one group.

Or is it? One jarring fact about the names we find in an index of the history of science is that the overwhelming majority of those scientists were *male*. Of course, there have been distinguished women scientists from ancient times to the present day. The list would include such names as Hypatia, Caroline Herschel, Marie Curie (winner of Nobel Prizes in chemistry *and* physics), Irene Joliot-Curie, Lise Meitner, Cecelia Payne-Gaposchkin, Barbara McClintock, Rosalind Franklin, Vera Rubin, and Lynn Margulis. But these were very much the exception to the rule. One reason for the lack of female names in the lists of notable past scientists is unquestionably that women in the past simply were not given due credit for their scientific work. For instance, Caroline Herschel was certainly a notable astronomer, but her work is usually mentioned as a footnote to the achievements of her more-famous brother William Herschel. Still, there can be no doubt that the apparent paucity of women scientists largely reflects the historical reality of women's exclusion from science. The inescapable conclusion is that over the centuries a vast reservoir of scientific talent was hardly tapped at all. Things are better today but still far from ideal. Some fields, like medicine, have basically achieved parity in numbers (if not in status and power); other areas, such as many engineering fields, are still over 80 percent male.

Why has science historically been, and to a large extent still is today, so predominately a male enterprise? For feminists, the answer is obvious: Science is a boy's club. It is run not only by but also for men. Women are still subtly, and sometimes not so subtly, discouraged from entering science. Even when women succeed in becoming scientists, they sometimes are marginalized and relegated to lower-status jobs or just "left out of the loop" by male colleagues. A woman of my personal acquaintance, a nuclear engineer, was told by her boss that women are "weak links" who must be driven from the profession. Further, male scientists have certainly entertained some odd, even bizarre, notions that it is hard to imagine female colleagues—had there been any—taking seriously. For instance, many physicians of the nineteenth century addressed the problem of hysteria, allegedly a nearly universal ailment of women that was supposed to cause them to experience uncontrollable emotions (*hysteria* is the Greek word for "womb"). The diagnosis of a troublesome woman's behavior as hysterical was certainly a convenience for men. If a woman made a public scene (like demonstrating for the right to vote), the problem was medical; she was hysterical.

Feminists are rightly concerned about such issues. Insofar as they aim to redress past wrongs and ensure that women have equal opportunities to enter and advance in scientific fields, their efforts can only be welcomed. But for more radical feminists, such efforts merely scratch the surface. The deeper problem, as they see it, lies not just with the obstacles that have been put in the way of girls and women who might enter science but also with the very ideals and standards of science themselves. The focus of much radical feminist science critique is the idea, which scientists have championed since the Scientific Revolution of the seventeenth century, that science should be value-free. Here we must make an important distinction between epistemic values and nonepistemic values. As we noted in an earlier chapter, *epistemic* means "relating to knowledge," so, the epistemic values of science are those conducive to the aim that science produce genuine knowledge. For instance, scientists place great value on the rigorous empirical evaluation of hypotheses since they hold that only such stringent testing can eliminate false hypotheses and lead us toward the true ones. Nonepistemic values are those which, although most important for human life, do not set norms for good inference or the correct evaluation of evidence. For instance, moral values can tell us that murder is wrong, but they do not tell us the right way to conduct a homicide investigation. A homicide investigation is an empirical inquiry and is guided by epistemic values. Nonepistemic values also include political values, spiritual values, aesthetic values, and so forth.

Traditionally, scientists and philosophers of science have held that science should, insofar as possible, make its practice independent of nonepistemic values. The reason why seemed obvious. People are passionate about their moral, political, and religious values. The "hot button" issues that enliven political campaigns and editorial pages are hot because they involve such basic values. Budget deficits might doom our children to a future life of hardship, but that distant prospect gets people less excited than whether kids should say "under God" while reciting the Pledge of Allegiance. Scientists are people too and are just as likely as anyone else to have strong feelings about political, religious, and moral issues. If, therefore, the results of science were not made independent of our strong feelings about moral, political, and religious values, scientists feared that scientific objectivity would be badly compromised. Objectivity is the goal of telling it like it is, even when our scientific conclusions run against deeply entrenched convictions.

Unquestionably, science does often conflict with such entrenched convictions. Darwinism is probably the most obvious example. There is the famous anecdote about the aristocratic Victorian lady's reaction upon first hearing of Darwin's theory. She cried out to her husband in horror: "Oh, my dear! Descended from apes! Let us hope that it is not true, and if it is that it does not become generally known!" Philosopher Daniel Dennett has rightly characterized Darwinism as "universal acid," an idea so corrosive that it threatens to dissolve any dogma or ideology it contacts. The only problem with Dennett's claim is that it is not broad enough. Any scientific theory, not just Darwinism, can and often does undermine entrenched beliefs. Inevitably, then, science will often run into entrenched ideological opposition and the inevitable pressure to modify or reject scientific conclusions when it does. To prevent such ideological obstruction, and to permit science to tell it like it is even when the truth hurts, scientists have embraced norms and methods intended to identify the polluting taint of ideology and insulate science from its influence. For instance, though scientists like everyone else often have strong political convictions, they are supposed to follow norms and practice methods that prevent their convictions from distorting their science. Thus, *science* is supposed to be dispassionate, disinterested, and politically neutral even though the people who practice it, being merely human, cannot be.

Many feminist writers reject the idea that science should, or can, be freed from the influence of nonepistemic values. They argue that science is inevitably and pervasively influenced by such values and that these shape both the practice and the conclusions of science. It is therefore pointless, and in fact disingenuous, for science to pretend to be value-free (from now on, by "value," I'll mean nonepistemic value). Rather, it should explicitly embrace the *right* values. Science should begin to serve the interests of the down-and-out and stop catering to the up-and-in.

Sandra Harding is one feminist writer who argues this way, and she gives this argument perhaps its most uncompromising expression. She says that science has so far claimed to follow the ideal of neutral, disinterested, and dispassionate inquiry. She calls this ideal "weak objectivity" and says that, though scientists and philosophers of science pay lip service to this ideal, in reality it never has and never will be actually practiced. She cites Kuhn and the social study of science to back her claim and concludes:

> Modern science has again and again been reconstructed by a set of interests and values—distinctively Western, bourgeois, and patriarchal. . . . Political and social interests are not "add-ons" to an other-

> wise transcendental science that is inherently indifferent to human society; scientific beliefs, practices, institutions, histories, and problematics are constituted in and through contemporary political and social projects and always have been. (Harding, 2003, 119)

In short, Harding endorses the slogan that heads this section: "Objectivity [as traditionally construed] is what a man calls his subjectivity." That is, the values that guide science have been those that men personally endorsed because they served male interests. Men tried to disguise the self-serving and subjective nature of those values by calling them "objective," "disinterested," and "impartial." Further, the influence of social and political factors on science cannot be blocked by adopting stricter methods and tighter controls. The idea of a value-free science is therefore not only a myth but a dangerous myth since it employs the language of impartiality to obscure the hidden agendas that science has always served.

Harding recommends that science instead pursue the ideal of "strong objectivity," which it can achieve only when it starts "thinking from women's lives." That is, "women's experience," specifically as interpreted by feminist analysis, must inform all of the standards, values, and methods of science. Feminist analysis turns women's experience of sexist oppression into a source of insight by raising victims' consciousness. Men, as the beneficiaries rather than the victims of sexist oppression, will not have such insights. It is like the situation where the slave knows all of the master's moods, whims, and quirks perfectly, but the master is largely oblivious of his slaves' lives. To get the benefit of women's experience (correctly interpreted), and realize the ideal of "strong objectivity," science must explicitly adopt the feminist standpoint.

Let me emphasize here that Harding is *not* making the rather-bland recommendation that science should be open to people from many different backgrounds so that scientific communities will contain persons with a variety of perspectives arising from differences of "life experience." The idea that there should be a diverse "web of knowers" whose different perspectives will lead to a fairer evaluation of knowledge claims seems to be a reasonable suggestion. Harding, though, wants science to adopt a *specific* perspective and set of values—hers.

She realizes that critics will point to notorious cases when people adopted doctrinaire assumptions and tried to do "politically correct" science. Perhaps the most famous incident involved Soviet pseudoscientist Trofim Lysenko, whose attempt to introduce a Marxist/Leninist brand of genetics destroyed the legitimate practice of that science in the Soviet Union. By currying favor with Stalin, who imposed a totalitarian stranglehold on every aspect of Soviet life, Lysenko was able to get his rivals banished to the Gulag and ensure that only his own views were taught. The attempt to apply Lysenko's crackpot theories to agriculture resulted in disaster, and, perhaps, ultimately contributed to the fall of the Soviet Union. Maybe equally notorious were the attempts in Nazi Germany to pursue an "Aryan Science," which repudiated "Jewish Science," such as the theories of Einstein.

Harding's reply is to reiterate her allegation that that a value-free, apolitical, impartial science is a myth, and again to state that *all* science must be done from *some* political perspective and work to promote *some* set of values. Therefore, we must choose whether science will serve a liberating set of values or the values of an oppressive doctrine. Immediately, though, there is a problem. Just who gets to decide which doctrines are "liberating" and which are not? Why cannot evangelical Christians, for instance, insist that *their* doctrine is the liberating one and that science should be based on *their* perspective and values? Harding says that the feminist standpoint is preferable because it leads to a science that is less "partial and distorting" than other doctrines. However, this implies that we can have some *impartial* means of telling which doctrines are more "partial and distorting." That is, we have to have some dependable methods for recognizing whole, undistorted truth and distinguishing it from the half-truths and distortions of ideologues. Those methods, whatever they are, of distinguishing truth from distortion cannot themselves presume the feminist standpoint. Assuming the feminist standpoint to prove the legitimacy of the feminist standpoint is obviously arguing in a circle. If Harding admits this, she must admit that we have reliable means of recognizing whole, undistorted truths, and that those means do not depend upon adopting the feminist standpoint or any other ideology. But this seems tantamount to admitting that we *can* have a disinterested, impartial, value-neutral science.

A deeper issue is what exactly it would mean for science to adopt the feminist standpoint. Most important scientific discoveries were credited to male scientists. It is a good bet that most of these men of science shared the common prejudices of their day. Some were even blatantly misogynistic. However, with many of these discoveries it is very hard to see how the incorporation of "women's experience" into the information available to scientists would have aided the discovery process or made the results less "partial and distorting." Objects fall at the same rate for women as they do for men. A cup of hot coffee cools at the same rate for women as it does for men. The speed of light is the same for both men and women. It is hard to think that Newton's laws and principle of universal gravitation, the laws of thermodynamics, or the principles of special relativity would have been any different had they been formulated by feminists. Perhaps Harding would concede this, and she would now say that it is only in the more "human" and social sciences such as anthropology, sociology, primatology, psychology, and so forth that adopting the feminist standpoint would make a significant difference.

Now it is certainly true that bias of various sorts has at times distorted science. Harding is quite right that the banner of scientific objectivity has often been unfurled to hide the ugliness of bigotry. Stephen Jay Gould's 1981 book *The Mismeasure of Man* is both amusing and horrifying when it recounts how nineteenth century anthropologists pursued craniometry, the measurement of the size of human skulls. They simply assumed that a bigger skull would house a bigger brain and therefore indicate higher intelligence. They measured the cranial capacity of the skulls of many races—Europeans, Sub-Saharan Africans, Eskimos, Semites, Native Americans, Australian Aborigines, and East Asians. Which group did these European researchers find to have the greatest cranial capacity and therefore the highest intellectual ability? You get one guess. German researchers even found that German skulls were more capacious than other European skulls. Other prejudices have distorted other sciences. Until the 1970s, homosexuality was listed in psychiatry texts as a mental disorder. The silly things that male scientists have said about women could (and did) fill volumes. So, would these sciences have been made more objective had they adopted the feminist standpoint?

The answer depends, first, on whether Harding has shown that all science is inevitably and inextricably bound to political agendas and the promotion of nonepistemic values. If she has not, then perhaps the way to remove bias from science is not to bring in a new ideology but to pursue the old-fashioned goal of a more impartial science. Second, we have to ask whether the feminist standpoint would itself introduce its own form of bias so that there will be no net gain in achieving a less "partial and distorted" science. Quite frankly, there are tenets of feminist doctrine that could bias science by placing an *a priori* ban on possible scientific results. For instance, perhaps the leading school of feminist thought today is called "gender feminism." Gender feminists argue that, while sex is a biological fact, gender is a social construct. That is, it is a natural fact that women bear children and men do not, but the various social roles that have traditionally been assigned to women and to men are, in their view, *entirely* products of culture. As gender feminists see it, the stereotypes about men and women's behavioral dispositions, such as that men, as a matter of biological fact, tend to be more physically aggressive and women more nurturing, are all false. In their view, in a restructured society, one engineered around feminist values, these supposedly "natural" behavioral differences would wash out.

But what if gender feminists are wrong on these points? What if there really are natural differences in the behavioral dispositions of the sexes? Of course, some women are more physically aggressive than some men, just as some women are taller than some men. But what if, just as men are, on average, naturally taller than women, they also, on average, are naturally more physically aggressive? Steven Pinker in his recent book

The Blank Slate (2002) argues very cogently that there *are* innate, biological dispositional differences between the sexes. Pinker could be wrong; debates over these points are far from settled. But the point is that he *could* be right. There is no way we can know ahead of time; we have to carry out the research and see. If Pinker is right, that is, if the natural facts are as he says, then gender feminists might be casting themselves in the role played by Pope Urban VIII when he proscribed Galileo's findings because they contradicted sacrosanct beliefs. Even if gender feminism turns out to be consistent with all empirical findings to date, we should continue to presume it true only so long as it continues to face all challenges. John Stuart Mill (himself a strong advocate of feminism) spoke what should have been the final words on this matter:

> There is the greatest difference between presuming an opinion to be true, because, with every opportunity for contesting it, it has not been refuted, and assuming its truth for the purpose of not permitting its refutation. Complete liberty of contradicting and disproving our opinion is the very condition which justifies us in assuming its truth for purposes of action; and on no other terms can a being with human faculties have any rational assurance of being right. (1952, 276)

It is hard to avoid the impression that Harding would be most displeased if science had "complete liberty of contradicting and disproving" her version of feminism.

IS SCIENCE GODLESS?

The science critics we have so far considered—social constructivists, postmodernists, and gender feminists—are representatives of the "academic left." Postmodernism, for instance, seems to have sprung from the failure of the French radical movement of the late 1960s. Yet it should come as no surprise that conservatives have also recently challenged the claimed objectivity of science, at least as it is now practiced. Science has traditionally clashed more often with conservative thinkers than with left-wing ones. The most notorious conflicts occurred when science and religion clashed. Science and religion are not always opposed; sometimes they even cooperate in a symbiotic relationship. Yet conflicts are inevitable. Stephen Jay Gould, the author of *The Mismeasure of Man* mentioned above, argues that science and religion do not clash because they are what he calls "non-overlapping magisteria." That is, science deals with matters of physical fact and theory while religion is the realm of value and spirit. But this is merely wishful thinking. There is no self-evident principle that relegates science and religion to different spheres. There is no *a priori* reason why religion cannot have something to say about the physical universe or why science cannot say something about value.

In fact, there are any number of ways that science and religion can clash. For instance, many religions teach that humans have an immortal soul, the seat of mind and consciousness, that will survive (and, according to some traditions, predates) its incarnation in a human body. Yet the flourishing field of neuroscience takes it for granted that mind and consciousness, those phenomena previously thought to be the province of the soul, are due entirely to the physical functions of neurons—brain cells. Of course, neuroscience has not—and may never—explain precisely how the firing of neurons creates consciousness. Nevertheless, the marvelously entertaining books of neurologist Oliver Sacks show in fascinating and sometimes-disturbing detail just how intimately our deepest thoughts and feelings, indeed our whole conception of ourselves and our world, are related to brain function. Apparently, if you change your brain you change your *self*. Religion might also clash with the social and human sciences. It is a central tenet of Christian belief that humans are sinners and that sin is a matter of the conscious choice of morally responsible agents. Yet some psychological theories deny that humans have such freedom and interpret human behavior as caused by, for instance, conditioning (behaviorism) or subconscious motivations (psychoanalysis).

Of course, the most famous conflict between science and religion, one that occasionally erupts even in the present day, is the clash between conservative Christianity and evolutionary theory. As historian James Moore showed, it is false that Christian theologians consistently opposed Darwinian evolution from the beginning. As Moore indicates, many Christians were quickly reconciled to Darwinism and even embraced it enthusiastically. Today most Christian denominations officially accept evolution. Pope John Paul II stated that evolution is "more than just a theory," and that it is clearly the correct account of the origin of biological species. Still, many conservative Christians have simply never been able to accept evolution or square it with beliefs that are for them essential elements of Christian doctrine. Perhaps the main problem is that evolution contradicts a straightforward reading of the creation accounts in the book of Genesis. For today's fundamentalists, like the Young Earth Creationists, this is undoubtedly the basis for much of their animus against evolution.

Not all opponents of evolution are fundamentalists. Phillip E. Johnson, a professor in the School of Law at the University of California at Berkeley, is perhaps the best-known current critic of Darwinism. Johnson is not a Young Earth Creationist. He is perfectly willing to admit that the earth is billions of years old, just as geologists and evolutionary biologists claim. Further, he is not committed to a view of scripture as inerrant, so it is not the conflict of evolution with a literal reading of Genesis that bothers him. He does think that the evidence for evolution is shoddy, and he has written extensively trying to show that this is so. What really bothers him, though, is not evolution itself but what he sees as an even deeper problem with the reigning assumptions and practice of science. Why, he asks, if the evidence for evolution is so weak, is it so nearly universally accepted among scientists? His answer is that science has been adulterated by a philosophical dogma, the doctrine of metaphysical naturalism. Johnson explains metaphysical naturalism as follows:

> Naturalism assumes the entire realm of nature to be a closed system of material causes and effects, which cannot be influenced by anything from "outside." Naturalism does not explicitly deny the mere

> existence of God, but it does deny that a supernatural being could in any way influence natural events, such as evolution, or communicate with material creatures like ourselves. (1991, 114–15)

In other words, metaphysical naturalism assumes that all natural things have only natural causes and therefore rejects out of hand any hypotheses postulating supernatural causes, like a divine creator. It is this *philosophical* bias against the supernatural, says Johnson, not anything necessary for good science, that leads scientists to accept evolution and reject creationism:

> Creationists are disqualified from making a positive case, because science by definition is based upon naturalism. The rules of science also disqualify any purely negative argumentation designed to dilute the persuasiveness of evolution. Creationism is thus ruled out of court—and out of classrooms—before any consideration of evidence. (2001, 67)

So, Johnson's argument is that science has compromised its objectivity by ruling out supernatural hypotheses, like creationism, without a hearing while accepting naturalistic theories, like evolution, that have little going for them. How good is Johnson's case? This is not the place to enter into the evidence for evolution. Many good books have done that already. Also, this is not the place to rehash the whole creation/evolution debate. Many fine books, a few of which are listed below in the "Further Readings" section, have done that job admirably. Here we shall address just three questions: (1) *Is* philosophy ever relevant to the evaluation of scientific hypotheses? (2) *Does* science assume metaphysical naturalism? (3) *Are* supernatural hypotheses like creationism dismissed by philosophical fiat and without a thorough empirical evaluation?

The three questions asked at the end of the previous paragraph raise some even more basic questions: Why should there be any *philosophical* debate about scientific hypotheses? Why not just run every proposed hypothesis through a good empirical test? After all, isn't the whole point of scientific method supposed to be that we can test hypotheses rather than engage in longwinded philosophical debate? The simple fact of the matter is that far too many hypotheses can be thought up than can possibly be tested. The empirical evaluation of hypotheses is an exacting, time-consuming process that requires meticulous planning and frequently involves the use of very expensive equipment that is often available only on a very competitive basis. Astronomers sometimes have to wait months to get one night on one of the big telescopes, and if it turns out cloudy on their night to observe—too bad. Besides, scientists are very, very busy people. So, before scientists can even begin to consider a hypothesis for testing, it has to show considerable promise. Scientists get ideas all the time, the vast majority of them bad. Like the White Queen in *Through the Looking-Glass*, scientists can often think of six impossible things before breakfast. How do we distinguish the hypotheses with promise, the ones that we actually will consider testing, from those that are throwaways?

Partly, scientists judge on the basis of track record. If a new hypothesis is just a variant of a kind that has been tried and has failed repeatedly, scientists are likely to give it short shrift. Now this may be unfair in many cases; some worthy hypotheses may be judged guilty by association. After all, *no* sort of hypothesis has a good track record until one of that sort actually does succeed. But nobody ever said science had to be completely fair, and there just does not seem to be any other way to proceed.

Philosophical considerations also inform our judgments about whether a hypothesis shows promise of test-worthiness. For instance, until well into the twentieth century, many professional biologists advocated vitalism. Vitalists held that life and living processes could not be completely explained in terms of the laws of chemistry and physics, so they postulated an additional "vital force" or animating "principle" that was supposed to permeate every tissue of living things. Vitalism was not a silly or obscurantist doctrine. Many of the leading figures in the history of the life sciences, including Pasteur, advocated some form of vitalism. However, vitalism was eventually abandoned in part because it lost repeatedly when placed head-to-head against mechanistic hypotheses. Another reason it was abandoned was that biologists came to see "vital force" as an explanatory dead-end. Instead of explaining organic phenomena, invoking "vital force" just seemed to deepen the mystery.

The question of what constitutes a legitimate scientific explanation is a philosophical question, one pursued at considerable length by philosophers of science. Because this is such an important question in the philosophy of science, let's digress a bit to review briefly what some philosophers have said about it. It is widely agreed that one goal of science is to explain the observed features of the physical universe. Science therefore asks questions like these: Why are certain zones of the earth's crust particularly susceptible to earthquakes while others hardly ever have even a tremor? Why are galaxies arranged in clusters and superclusters rather than just spread randomly through space? Why are island faunas so unique, often displaying a range of adaptations not found in related faunas of the nearest mainlands? Scientific explanations relieve our puzzlement about such questions by showing why these particular phenomena were to be expected.

The classic modern model of scientific explanation was articulated by philosophers Carl G. Hempel and Paul Oppenheim in 1948. Hempel later devoted much effort to refining and extending this model. The basic idea is that what sets scientific explanation apart from other ways of achieving elucidation or enlightenment is that scientific explanations have a distinct *form*. A scientific explanation has the form of an *argument* where a conclusion is drawn from a set of premises. The datum to be explained, what—philosophers call the *explanandum*—is scientifically explained when it is correctly inferred from particular kinds of premises. For Hempel and Oppenheim, at least one premise needs to state a *natural law*. Another premise states *initial conditions*, that is, an appropriate set of concrete physical circumstances. Propositions that state natural laws are called *nomological propositions* (from the Greek word *nomos* meaning "law"). A *Deductive Nomological* (DN) model of explanation is therefore one in which an explanandum is explained by deducing it from premises that state a natural law and a set of initial conditions. Because natural laws play a vital role in this model of explanation, it is called a "covering law" model.

According to Hempel and Oppenheim, many explanations in science have a DN form. Here is a simple example:

Natural law: When water freezes, it expands with enormous force.
Initial conditions: The water in the pipes froze solid overnight.
Conclusion: The pipes burst.

Given that freezing water exerts an enormous expansive force—a force too great for household plumbing to contain—and given that the water in our pipes did freeze last night, we can *deduce* that our pipes burst. So, if we want to know why we have burst pipes, and a terrible plumber's bill to pay, we gain scientific understanding (but not much solace) when we know that freezing water expands and that the water in our pipes froze.

Many explanations in science fit the DN model and the allied *Inductive Nomological* (IN) model, which is much the same as the DN model, except that we infer the *probable* occurrence of the explanandum from a law and initial conditions. However, as various philosophers have pointed out, not all legitimate scientific explanations conform to the DN or IN models. Here is an example of one that does not:

Any unvaccinated person exposed to a live influenza virus has a 20 percent to 40 percent chance of developing a case of the flu within seventy-two hours. Sam has had no flu shot this year, and he sat for two hours in a movie theater next to Sarah, who was just coming down with an active case of the flu. Two days later, Sam developed a case of the flu. Therefore, Sam got the flu because he was exposed to live influenza virus he got from Sarah.

Surely this is a legitimate explanation of why Sam got the flu, but it fits neither the DN nor the IN model. You cannot *deduce* that Sam will get the flu from the fact that he is unvaccinated and has been exposed to flu virus. It is not *certain* that Sam will get the flu; there is only a 20 percent to 40 percent chance. Also, this explanation does not fit the IN model because it is not even *probable* that Sam will get the flu. If there is a 20 percent to 40 percent chance that Sam will get the flu, there is a 60 percent to 80 percent chance that he will not. So it is *prob-*

able that Sam will *not* get the flu despite his exposure to the virus. Still, if Sam *does* get the flu, exposure to the virus is the explanation.

To deal with cases like this, philosophers developed the *Causal Statistical* (CS) model of scientific explanation. According to the CS model, we understand an event when we spell out the physical factors statistically relevant to the event and also specify the underlying causal processes that brought about the event. For instance, we above explained why Sam came down with influenza by noting that he is unvaccinated and that he came in contact with another person with an active case. To say that exposure to influenza is statistically relevant to Sam's getting the flu does not mean that such exposure makes it *likely* (i.e., more than 50 percent probable) that Sam will get the disease. It only means that such exposure makes it *more likely* that Sam will get the flu than if he had not been exposed. A 20 percent chance of the flu is greater than a 0 percent chance. We expand our explanation by specifying just how the influenza virus does its dirty work on the body. Viruses display a malign ingenuity in the way that they invade body cells and hijack the cell's genetic machinery to make more copies of themselves. Such knowledge about how viruses operate greatly expands our understanding of the facts about infection.

Other philosophers reject all such models of scientific explanation and recommend a pragmatic approach. According to these philosophers, all we can really say about a good explanation is that it answers our "why" questions about a given topic of concern by telling why *this* particular outcome was to be expected rather than one of the other members of that event's "contrast class." The "contrast class" consists of all of the other events that conceivably could have occurred in that situation but did not. For instance, a satisfactory scientific explanation of the dinosaurs' demise would tell us why the dinosaurs went extinct, but crocodilians sailed right through the K/T mass extinction and are with us today.

Let's return from this (all-too-brief) review of what some philosophers have said about scientific explanation to the main question: Are philosophical considerations ever relevant to theory choice? In particular, might a philosophical consideration, like our ideas about what constitutes a good scientific explanation, reasonably guide us in deciding which hypotheses are promising candidates for further testing and which are nonstarters? It is important to note that concern about the nature of scientific explanation is not merely an armchair amusement for philosophers. As we saw in chapter 2, Kuhn notes that scientists themselves often engage in vigorous debates over standards, like what should constitute a legitimate explanation for some range of natural phenomena. It seems, therefore, that philosophical considerations should sometimes guide us in choosing which hypotheses look promising enough to go to the trouble of testing. We cannot test every hypothesis that anyone proposes. If a hypothesis looks like it does not even offer us a good explanation, it is not unreasonable or unfair to pass it over, at least for the time being, in favor of something more promising.

Well, what about supernatural hypotheses? Does philosophical bias prevent them from receiving due consideration? One complaint often made against hypotheses that invoke God's acts is that they do not explain things but only hide our ignorance behind a theological fig leaf. For skeptics, saying "God did it" does not enhance our understanding of some strange phenomenon—a sudden, unexplained remission of metastatic cancer, for instance—but only drapes it in deeper mystery. Is this accusation fair? Do hypotheses that invoke God, or perhaps a more nebulous creator, offer legitimate explanations, or are they only markers for our ignorance, placeholders for explanations we hope someday to get?

Defenders of supernatural hypotheses could strengthen their case if they could show that their hypotheses offer explanations that conform to one or more of the recognized models of scientific explanation. Let's consider whether there could be supernatural explanations that conform, for instance, to the DN or CS models. There do not seem to be any "laws of supernature" to serve as covering laws to explain particular events, so the DN model seems to be out. For instance, we just do not know the general circumstances in which God is likely to perform a miracle. We cannot articulate any general laws of the form "every time God's people are in dire-enough need, God performs a public miracle to deliver them." Putative beings like gods and ghosts are not constrained by natural law; their actions are unpredictable, so it is hard to know what effects of those actions are to be expected. Whether supernatural hypotheses specify statistically relevant factors for the occurrence of events is a matter of debate among philosophers of religion. Some theistic philosophers argue that the existence of the universe, or of a particular kind of universe, is more likely if there is a God than if there is not. However, nobody can specify the particular causal processes whereby God is supposed to bring about his effects; after all, God's ways are proverbially mysterious. Nobody can be much more specific than to say that when God created something—birds, let's say—he just said, "Let there be birds," and *POOF!* there were birds! Since supernatural explanations say nothing about causal mechanisms, they do not conform to the CS model of explanation.

Supernatural explanations do not even meet our pragmatic explanatory needs very well. For instance, if we try to explain the sorts of anatomical homologies mentioned in the last chapter by saying that God created organisms according to a plan, this leaves all of our questions unanswered. Why this plan rather than one of the indefinitely many others that an all-powerful, all-knowing creator could have enacted? Why just this instance rather than one of an indefinitely large contrast class?

Defenders of supernatural hypotheses will counter, correctly I think, that the models of scientific explanation so far developed do not necessarily exhaust all the legitimate possibilities. While philosophers of science may have identified *some* kinds of good scientific explanation, it is highly questionable whether they have identified *all* possible types. Therefore, the fact that supernatural explanations do not conform to any "model of scientific explanation" so far proposed does not mean that they cannot be legitimate scientific explanations. Fair enough, but surely the burden of proof is on the defenders of supernatural hypotheses. *Prima facie*, such hypotheses do not seem to offer much elucidation. Again, if we say, for instance, that God created birds, what does that tell us? How did He do it? For what reason? Why birds?

Why didn't He stick with the highly successful flying reptiles? Surely, any kind of acceptable scientific explanation should *show* why the explanandum—the existence of birds in this case—was to be expected. As philosophers Karel Lambert and Gordon G. Brittan Jr. note (1987, 22), invocations of God's will, like appeal to signs of the zodiac, just do not provide such information.

The upshot is that there *is* a philosophical motivation behind the scientific practice of giving short shrift to supernatural hypotheses, just as Johnson says. But until defenders of supernatural hypotheses can show that such hypotheses promise legitimate scientific explanations—and do not just disguise our ignorance—such practice is neither biased nor unfair. Please note that this does *not* mean that supernatural hypotheses cannot be true or that we cannot have *very good reasons* for thinking they are true (more on this below). Maybe we will just have to admit that some things do not have a scientific explanation and are due to the mysterious acts of a creator. Maybe on some topics, like the origin of life, say, scientists may someday come to the point where they should just throw up their hands and say that they will never explain some things and concede that there are ultimate mysteries, facts attributable only to the unfathomable and inscrutable actions of a creator. But it is far from clear that that day is today.

Let's move to the second question raised by Johnson's critique: *Does* science assume metaphysical naturalism? This charge has been made many times, and the standard reply is that the naturalism science assumes is methodological, not metaphysical. The difference is this: Metaphysical naturalism is a doctrine about the nature of reality. It can take the strong line that only natural things are real or the weaker line that supernatural things might exist but they cannot causally interact with the natural world. Methodological naturalism does not offer opinions about the nature of ultimate reality; it merely requires that, as a matter of good scientific practice, we consider only naturalistic hypotheses. T. H. Huxley, Victorian scientist and man of letters, was very emphatic that metaphysical questions about the nature of ultimate reality were none of the business of science. Huxley said that you might as well inquire into the politics of extraterrestrials as to ask whether ultimate reality is material or spiritual. Yet he strongly advocated naturalism as a methodological requirement because he held that naturalistic explanations are comprehensible while supernatural explanations only hide mysteries behind a veil of theological obscurity. In a similar vein, contemporary philosopher Rob Pennock argues that science should be godless in the same sense that plumbing is godless. Good plumbing practice obviously does not involve grandiose metaphysical assumptions but proceeds on the assumption that the cause of a problem is in the pipes. Pennock argues that the requirement that scientific hypotheses be testable entails that they involve only natural objects that follow predictable laws. As noted earlier, putative supernatural entities, like gods and ghosts, are not bound by natural law and so are notoriously difficult to test.

Johnson does not buy these arguments and insists that methodological naturalism is only a dishonest front for metaphysical naturalism. I think we should concede that *in principle* good science could confirm supernatural hypotheses, however difficult they might be to test *in practice*. Nineteenth-century English scientist Francis Galton proposed a test for the efficacy of prayer. He noted that members of the royal family certainly were the beneficiaries of more prayers for their health than any other British family. He concluded that, if prayer works, the royal family should be healthier than other comparable families (he found that they were not healthier, by the way). Now a legitimate test of the efficacy of prayer is probably impossible to achieve in practice. How could Galton rule out that many disgruntled people may have been praying that God strike down the royal family? Still, this seems to be a practical difficulty and not an indication that an experimental test of prayer is in principle impossible.

Interestingly, the Bible tells of an incident that would be about as good an experimental test of God's power as anyone could devise. 1 Kings chapter 18 tells the story of Elijah and the priests of Baal. Elijah challenged the priests of Baal to a contest to see which god was real, Baal or the Lord, the God of Abraham, Isaac, and Israel. The priests of Baal built an altar and placed a sacrifice upon it. All day they cried for Baal to send fire to burn their sacrifice, but nothing happened. At the day's end, Elijah erected an altar, placed a sacrifice on it, and had everything thoroughly soaked with water. He then called upon the Lord, and according to 1 Kings 18:38, "Then the fire of the Lord fell and consumed the burnt offering and the wood and the stones, and the dust, and licked up the water that was in the trench." Now this would certainly seem to be about as good an example of a crucial experiment as any scientist has ever devised. If it occurred today, the churches and synagogues would fill with former doubters. Of course, such things apparently do not happen today, but the point is, again, that there seems to be nothing *in principle* impossible about an experimental test of God's power.

So, is it simply a matter of ideological prejudice that supernatural hypotheses are rejected by science? No, for two reasons (besides the doubts raised earlier about supernatural "explanations"): First, though it is not a methodological *requirement* of science, naturalism has unquestionably proven a valuable *heuristic*. A heuristic is a presumption that serves as a guide for inquiry. An example of a heuristic principle that guides science is the principle of simplicity, the postulation that physical reality is ultimately simple, and that science should therefore seek simple theories. The idea that things will ultimately turn out to be simple is, of course, a speculation. Absolutely nothing guarantees that at bottom physical reality is simple. Yet no one can deny that the presumption that deep simplicity underlies the surface complexity of nature has been an extremely valuable heuristic guiding science.

Similarly, naturalism has been a very successful heuristic principle. Unquestionably, much of the progress of science is due to the fact that it doggedly sought natural hypotheses and excluded those postulating gods, souls, angels, demons, ghosts, fate, magic, astrological influences, hexes, spells, good luck charms, and the like. So long as a heuristic continues to deliver the goods, scientists are fully justified in sticking with it. Is there any indication that a naturalistic heuristic has served its purpose and now leads science in the wrong direction? For instance, does naturalism induce scientists to accept evolution despite a dearth of evidence? For decades, antievolutionists have charged that evolution is a "theory in crisis" and that Darwin is once again "on trial." They have insisted repeatedly that the evidence for evolution is so shoddy that the whole edifice of evolutionary science is about to come crashing down and that the only thing propping it up is naturalistic bias.

Let's pause for a second and consider just how strong a claim this is. A recent, thorough electronic search of the professional, peer-reviewed scientific literature over the previous twelve-year period turned up over one hundred thousand articles with *evolution* as a key word; and, by the way, practically none referring to concepts of supernatural design. So, if evolutionary theory has been "in crisis" and "on trial" for decades, this news has yet to reach the writers of the professional scientific literature. Evolutionary biology looks extremely spry for a field supposedly on its deathbed! Johnson and other antievolutionists have to attribute evolutionary biology's appearance of health and vigor to a massive intellectual fraud perpetrated on science by a cadre of ideologues. But no ideology, not even when backed by the power to burn dissenters at the stake, has ever held science down for long. Not even the enormous power and intellectual influence of the seventeenth-century church could hide the bankruptcy of the old Ptolemaic system for long. So it is just hard to believe that nothing but ideological obscurantism keeps scientists from recognizing the alleged weakness of evolutionary theory.

A second and more important reason for denying that negative attitudes toward supernatural hypotheses are due to bias is that Johnson's charge is simply false. The answer to the third of the questions we are addressing in this section is: No, creationism has not been dismissed by philosophical fiat. Dozens of books and hundreds of articles, many available on the Internet in the magnificent archives of the talk.origins site, have subjected creationist claims to careful, extensive, point-by-point empirical critique. In the *On the Origin of Species*, Darwin himself showed time and time again that natural selection better explains the natural facts than special creation does. Therefore, the creationist hypotheses have not been rejected by philosophical fiat and without a fair and thorough hearing.

CONCLUSION

In this chapter we have extensively examined critiques of scientific rationality and objectivity from both the left and the right. Our conclusion has to be that, though science is far from perfect, as any human enterprise must be, there is still something left of the Enlightenment ideal derided by Latour. There is a physical world "out there," and we can know some things about it. That is, we can say of the natural world, without qualification or apology, that some things really just *are* so, and are not artifacts of our percepts, concepts, or categories. Further, our observations of the physical world can be used to rigorously evaluate our theories, so that our theoretical beliefs are shaped and constrained by nature, and not merely by politics, rhetorical manipulation, or ideology. Disinterested knowledge really is possible, and is, in fact, achieved far more often than cynics suppose.

Yet even their harshest critics must admit that the social constructivist, postmodernist, and feminist science critics have performed a valuable service. These critics have certainly succeeded in disposing of what might be called the "passive spectator" stereotype of scientific knowledge. According to this stereotype, modern science began when the pioneers like Copernicus, Galileo, and Darwin stopped bowing to ancient authority and opened their eyes to the world around them. Once people started looking at *nature* rather than at old books, the story goes, scientific knowledge flowed into open scientific minds like water pouring into an empty bucket. Now, of course, this is a comic-book version of the history of science, and no serious scholar has ever thought that it really happened this way or that we gain scientific knowledge merely by the passive reception of information. Still, this has been a very influential stereotype. A powerful image can influence our thinking more than all the careful arguments of scholars. One of the indelible images of our intellectual culture is the picture of Galileo boldly scanning the heavens with his telescope, eager to discover whatever his eyes revealed to him, while his ecclesiastical oppressors, besotted with scripture and Aristotle, refused even to look through the instrument. The founders of Britain's Royal Society, the preeminent British scientific body, were so impressed with this image of the scientist as the ideally objective and open-minded observer that they adopted as the society's motto "*Nullius in Verba*." This motto is hard to translate precisely, but it means that you should take nothing on authority. Instead, you should look and see for yourself.

Scientific discovery requires active engagement, however, not just passive seeing. Galileo didn't just look through the telescope and report what he saw; he interpreted, theorized, speculated, measured, analyzed, and argued. Darwin did not just go to the Galapagos Islands, see some odd finches and tortoises, and then awaken to the truth of evolution in a flash of blindingly obvious insight (scientific discoveries are always "obvious" only in hindsight). Darwin's private notebooks, written as he struggled to define his ideas on evolution, reveal a complex process of questioning, argument, and counterargument, with tentative conclusions drawn and then repeatedly rejected or refined. Scientists do not just *absorb* a picture of the world; they *create* a picture and then do their best to see how accurate it is. Unavoidably, when we create our theories of the natural world, we must employ the only cognitive tools we have—the concepts, language, perspectives, interpretive and observational skills, and presumed background knowledge that we possess. Inevitably, multifarious biases lurking in our language and concepts sometimes—all too often—slip unnoticed into our theories. Our only way of dealing with this problem is to continually refine and revise our ideas through ongoing interaction with the natural world and the effort to devise stricter methods, more rigorous tests, and more accurate measures. The work of Kuhn examined in the last chapter, and that of the radical science critics considered in this one, unquestionably succeeded in debunking the simplistic stereotype of the scientist as ideally objective, open-minded, and passive observer.

What we need, then, is a balanced view of science, one that rejects both the excessively cynical and the unrealistically idealized stereotypes of science. David Young, in his excellent book *The Discovery of Evolution*, strikes just the right note of balance in our interpretation of science; his words can serve as a coda for this chapter:

> The picture of the scientist as an objective spectator has died a natural death, thanks to the work of historians and philosophers of science. It is now clear that even simple observations are not imbibed passively from the external world but are made by a human mind already laden with ideas. The shaping of these ideas is a human activity carried out in a particular social context, with all the frailties and limitations that that implies. This has led some people to the other extreme, in which scientific knowledge is viewed as no more than the expression of a particular social group. On this view there are no such things as discoveries in science, only changes in fashion about how we choose to view the world. However, such a view cannot account for the fact that scientific understanding does not merely change but is progressive. . . . A sensible view of scientific theory must lie somewhere between these two extremes and embody elements of both. Certainly, scientific discovery does not involve a one-way flow of information from nature to a passive, open mind. It involves a creative interaction of mind and nature, in which scientists seek to construct an adequate picture from what they see of the world. (1992, 219–20)

Young's view is neither novel nor especially profound. It lacks all of the edgy excitement of the radical science critiques. It only has one big advantage over those accounts: It is true.

FURTHER READINGS FOR CHAPTER THREE

Paul de Kruif's *Microbe Hunters* (San Diego: Harcourt Brace, 1926), is still in print eighty years after it was written. I remember being fascinated as a child reading an old, dog-eared paperback copy. I am sure that it has inspired many readers to enter medicine or biomedical research. As I say in this chapter, de Kruif regards Pasteur and the other microbe hunters with unabashed hero worship and he treats the pursuit of science as the noblest and most selfless of activities. We now, of course, realize that these views are naïve. Ambrose Bierce once defined a saint as "a dead sinner, revised and edited," and no doubt well-meaning admirers like de Kruif have likewise redacted the stories of the "saints" of science. Historians of science perform a valuable service when they present the story, "warts and all." However, "warts and all" does not mean "nothing but warts." In my view, undercutting one myth, the myth of the saintly scientist grappling with the demons of disease, should not lead to the creation of more-pernicious stereotypes, such as the image of the scientist as cynical self-promoter, rampant ideologue, or stooge of vested interests.

Upon reading a draft of this chapter, one referee said that I had introduced Bruno Latour as the "villain" who insulted the memory of France's national hero of science, Pasteur. No. I present Latour as a radical revisionist, a characterization I have no reason to think he would repudiate. In fact, Latour boasts of the deflationary intention of his work. In a letter to the editor of the (now, sadly, defunct) magazine *The Sciences* (vol. 35, no. 2, [1995]: 7), Latour compares his work in science studies to the work of Darwin in biology: "Those of us who pursue science studies are the Darwins of science, showing how the exquisite beauty of facts, theories, instruments and machines can be accounted for without ever resorting to teleological principles or arguments by design." I see no other way to read this passage than as a statement of Latour's intent to replace traditional representations of science as motivated by reasons with a reductionistic sociological analysis. And that is how I have presented him in this chapter.

There is a lot to be said for the slogan "it is more important that an opinion be interesting than true." Some errors are uninteresting because they are due to silly mistakes in reasoning; others are interesting because they involve deep confusions in our concepts or language. Latour's errors are *never* dull. When he is wrong, you learn a lot about science and how it operates in society even as you try to pinpoint his errors and sort out what is really behind his conclusions. Latour made his big splash with *Laboratory Life: The Construction of Scientific Facts* (Princeton: Princeton University Press, 1979; reprinted with new postscript and index in 1986). This book, co-written with Steve Woolgar, was one of the founding documents of the whole "science studies" movement. It is rather technical in places, and the prose is often muddy. Still, the aim of going into a laboratory as an anthropologist to observe scientists in their native habitat as one would the Yanomamo or the Inuit was a brilliant idea and makes for fascinating reading.

Latour examines Pasteur and his influence in *The Pasteurization of France* (Cambridge: Harvard University Press, 1988), translated by Alan Sheridan and John Law. Latour's most ambitious and comprehensive work is *Science in Action: How to Follow Scientists and Engineers through Society* (Cambridge: Harvard University Press, 1987). Latour's thorough took an interesting turn in 1993 with the publication of *We Have Never Been Modern* (Cambridge: Harvard University Press), translated by Catherin Porter. Here Latour claims to abandon social constructivism and aims to explore a middle course between constructivism, the view that scientific facts are cultural artifacts, and the view of scientists that such objects are objective truths about nature. Latour defines what he calls "quasi-objects" that are neither wholly natural nor wholly constructed. Unfortunately, just what he means by a "quasi-object" is not made entirely clear.

Steve Woolgar develops his views in confrontational style in *Science: The Very Idea* (London: Tavistock Publications, 1988). One very well-known scientist and writer who, at least sometimes, seemed to endorse social constructivism was Stephen Jay Gould. His book *The Mismeasure of Man* (New York: W. W. Norton, 1981) is often cited by social constructivists, postmodernists, and feminists as proof of how bias and bigotry shape science. Matt Cartmill's perceptive but acerbic characterization of Latour's social constructivism is found in his review of *Mystery of Mysteries: Is Evolution a Social Construct?* by Michael Ruse. This review appeared in *Reports of the National Center for Science Education* 19, no. 5 (1999): 49–50. The story of the wrong-headed dinosaur episode is found in chapter 1 of my book *Drawing Out Leviathan* (Bloomington: Indiana University Press, 2001). Michele Marsonet's very interesting discussion of the way that philosophy can slide into "linguistic idealism" is found in his *Science, Reality, and Language* (Albany: State University of New York Press, 1995). John Searle's discussion of the "default settings" of the human mind are found in his book *Mind, Language, and Society* (New York: Basic Books, 1998).

As I say, the roots of postmodernism can be traced at least back to Nietzsche. Nietzsche is an exciting but challenging thinker. His writings often have a declamatory or oracular character, which puts some readers off. Also, he is very easy to misread. Sometimes he sounds like an anti-Semite, a misogynist, or a protofascist, though his defenders insist that he was none of these things. Because of some of the difficulties with reading Nietzsche, it might be good to start by reading a reliable introduction to his thought. A good, succinct, and readable account is *On Nietzsche* by Eric Steinhart (Belmont, CA: Wadsworth, 2000).

An essential document for understanding postmodernism is Lyotard's manifesto, *The Postmodern Condition: A Report on Knowledge* (Manchester: Manchester University Press, 1979), translated by Geoff Bennington and Brian Massumi. Also, Richard Rorty's *Philosophy and the Mirror of Nature* (Princeton: Princeton University Press, 1979) introduced many characteristic postmodernist theses to English-speaking philosophers. Rorty's work attracted quite a bit of notoriety because he had previously been regarded as one of the really tough-minded philosophers of the analytical tradition, and it seemed to many of his contemporaries that he was simply abandoning philosophy. A fun and very accessible introduction to postmodernism is Glen Ward's *Postmodernism* (N.p.: NTC/Contemporary Publishing, 1997). Postmodernism is largely a development of recent literary theory, and a perceptive critique of postmodernism by an expert on such theory is *The Illusions of Postmodernism* by Terry Eagleton (Oxford: Blackwell Publishers, 1996).

The two works I selected to represent postmodernist commentary on science were Donna Haraway's *Primate Visions: Gender, Race, and Nature in the World of Modern Science* (New York: Routledge, 1989) and

W. J. T. Mitchell's *The Last Dinosaur Book: The Life and Times of a Cultural Icon* (Chicago: University of Chicago Press, 1998). What makes these books particularly interesting is that they deal with primatology and dinosaur paleontology, which are branches of science that are easier for most people to relate to than, say, particle physics. When scientists encounter Haraway's and Mitchell's books, they often are nonplussed or outraged. Matt Cartmill's trenchant review of *Primate Visions*, published in the *International Journal of Primatology* 12, no. 1 (1991), must surely express the exasperation many primatologists would feel toward Haraway's book.

The quote from Cartmill's review of *Primate Visions* in this chapter is from the reprint of that review in my anthology *The Science Wars: Debating Scientific Knowledge and Technology*, ed. by Keith M. Parsons (Amherst, NY: Prometheus Books, 2003), pp. 195–207.

I coauthored an essay with geologist Peter Copeland, "Toward a Postmodernist Paleontology?" in *Academic Questions* 17, no. 2 (spring 2004) that examines and criticizes Mitchell's claims in detail. Alan Sokal's *faux*-postmodernist essay, "Transgressing the Boundaries: Towards a Transformative Hermeneutics of Quantum Gravity," is mostly conveniently found in *The Sokal Hoax: The Sham That Shook the Academy* (Lincoln: University of Nebraska Press, 2000), edited by the editors of the magazine *Lingua Franca*. This volume is a lot of fun, with fierce polemics and much outraged harrumphing on both sides.

A good place to start with the feminist philosophy of science is the entry "Feminist Accounts of Science" by Kathleen Okruhlik in *A Companion to the Philosophy of Science*, edited by W. H. Newton-Smith (Oxford: Blackwell Publishers, 2000). Okruhlik provides an authoritative overview of the diverse views of feminist philosophers of science. As I state in both prefaces, Sandra Harding appears in this chapter not as a "typical" representative of feminist philosophy of science (though I do not consider her too atypical, either) but as a controversial figure whose opinions are bound to excite discussion. If Harding has piqued your interest in feminist philosophy of science, you should probably next read *Elizabeth Fox Keller's Reflections on Gender and Science*

(New Haven: Yale University Press, 1985) and Helen Longino's *Science as Social Knowledge: Values and Objectivity in Scientific Inquiry* (Princeton: Princeton University Press, 1990). The Harding quote in this section comes from her *Whose Science? Whose Knowledge? Thinking from Women's Lives* (Ithaca, NY: Cornell University Press, 1991). For convenience, I quotes Harding from the selection "Feminist Standpoint Epistemology and Strong Objectivity" from the book *The Science Wars* (bibliographical details below).

For critiques of the feminist philosophy of science, see Cassandra Pinnick, "Feminist Epistemology: Implications for the Philosophy of Science," in the journal *Philosophy of Science* 61 (1994): 646–57; Janet Radcliffe Richard's "Why Feminist Epistemology Isn't," and Noretta Koertge's "Feminist Epistemology: Stalking an Un-Dead Horse." Richard's and Koertge's essays are found on pages 385–412 and 413–19, respectively, in Paul Gross, Norman Levitt, and Martin Lewis, eds., *The Flight from Science and Reason* (New York: New York Academy of Sciences, 1996). A book-length critique of feminist epistemology and philosophy of science is Ellen R. Klein's *Feminism under Fire* (Amherst, NY: Prometheus Books, 1996). Steven Pinker's provocative discussion of gender differences is found in his *The Blank Slate: The Modern Denial of Human Nature* (New York: Viking, 2002). The quote from John Stuart Mill is from his classic essay, *On Liberty*, in *Great Books of the Western World*, Robert Maynard Hutchins, editor-in-chief (Chicago: Encyclopedia Britannica, 1952).

The debates over social-constructivist, postmodernist, and feminist accounts of science reached the boiling point in the mid-1990s as scientists began to fire back at what they perceived as attacks on the aims, methods, and values of science by critics of the "academic left." The "science wars" really erupted with the 1994 publication of Paul Gross and Normal Levitt's splendidly pugnacious *Higher Superstition: The Academic Left and Its Quarrels with Science* (Baltimore: Johns Hopkins University Press). The academic left responded in equally bellicose fashion in the collection of essays *Science Wars*, edited by Andrew Ross (Durham, NC: Duke University Press, 1996). Alan Sokal, of the Sokal hoax, and his collaborator, Jean Bricmont, states their case in *Fashionable Nonsense: Postmodern Intellectuals' Abuse of Science* (New York: Picador USA, 1998). By 2000, the rhetorical temperature of the science wars had cooled a bit and books appeared that were less polemical in tone. A clear and insightful survey of the main issues debated in the science wars in the context of the recent history of the philosophy of science is James Robert Brown's *Who Rules in Science: An Opinionated Guide to the Wars* (Cambridge: Harvard University Press, 2001). I offer an introduction to some of the main writings and debates in the anthology *The Science Wars: Debating Scientific Knowledge and Technology*, edited by Keith M. Parsons (Amherst, NY: Prometheus Books, 2003).

Stephen Jay Gould's views on the relation between science and religion are found in his book, *Rock of Ages: Science and Religion in the Fullness of Life* (New York: Ballantine Publishing, 1999). A solid, thorough overview of the relations between science and religion is John Hedley Brooke's *Science and Religion: Some Historical Perspectives* (Cambridge: Cambridge University Press, 1991). A detailed yet quite readable examination of the theological response to Darwinism in Britain and America is James R. Moore's *The Post-Darwinian Controversies: A Study of the Protestant Struggle to Come to Terms with Darwin in Great Britain and America, 1870–1900* (Cambridge: Cambridge University Press, 1979).

The recent controversy over "intelligent design," actually a continuation of the controversy over creationism of the 1980s, was kicked off by the publication of Phillip E. Johnson's *Darwin on Trial* (Washington, DC: Regnery Gateway, 1991). A history and overview of the intelligent-design movement written by a sympathizer is Thomas Woodward's *Doubts about Darwin: A History of Intelligent Design* (Grand Rapids, MI: Baker Books, 2003). An anthology of writings by advocates of intelligent-design theory with responses by critics is *Intelligent Design Creationism and Its Critics: Philosophical, Theological, and Scientific Perspectives* (Cambridge: MIT Press, 2001), edited by Rob Pennock, which includes the essay by Johnson quoted in this chapter, "Evolution as Dogma: The Establishment of Naturalism."

The volume of literature on evolution is simply stupendous in its quantity (and highly variable in its quality). Here I shall simply recommend one book that seems to me the best presentation of evolutionary theory for the nonspecialist, Colin Patterson's *Evolution*, second edition (make sure you get the second edition, it is much better than the first) (Ithaca, NY: Comstock Publishing, 1999). Patterson's treatment is crystal clear, and while it presents the evidence for evolution cogently, it is very undogmatic in tone. Really to understand Darwinism, you need to see it presented in the context of its historical development. A superbly written, insightful, and beautifully illustrated history of evolution is David Young's *The Discovery of Evolution* (Cambridge: Cambridge University Press, 1992). Perhaps the best nontechnical statement of the intelligent-design position is still the Johnson book mentioned above. I think that for most readers the best critique of Johnson's view and the claims of intelligent-design creationism is Robert T. Pennock's *Tower of Babel: The Evidence against the New Creationism* (Cambridge: MIT Press, 1999).

One very good introduction to the topic of scientific explanation is the chapter "Scientific Explanation" by Wesley C. Salmon in Salmon et al., *Introduction to the Philosophy of Science* (Englewood Cliffs, NJ: Prentice-Hall, 1992). Salmon, one of the top philosophers of science of the twentieth century, made many seminal contributions to our understanding of scientific explanation. He was also a gifted expositor who could make difficult ideas very clear for beginners. Another very clear and helpful overview of the topic of explanation is in *An Introduction to the Philosophy of Science*, third edition, by Karel Lambert and Gordon G. Brittan Jr. (Atascadero, CA: Ridgeview Publishing, 1987).

T. H. Huxley's comments on methodological naturalism are found in his essay "On the Physical Basis of Life," in *Selected Works of Thomas H. Huxley* (New York: D. Appleton, n.d.), pp. 130–65. Huxley's prodigious learning, wit, and trenchant style are as enjoyable now as they must have been discomfiting for his nineteenth-century opponents. "Darwin's bulldog" still has considerable bite. For full details on the search of scientific literature that turned up one hundred thousand articles with *evolution* as a key work, see J. R. Staver, "Evolution and Intelligent Design," *Science Teacher* 70, no. 8 (2003): 32–35.

CHAPTER FOUR

ASCENDING THE SLIPPERY SLOPE

SCIENTIFIC PROGRESS AND TRUTH

How do the natural sciences differ from other fields of human intellectual and creative endeavor? Most people would say that science progresses in ways that, for example, the visual arts do not. This is not to say that the visual arts do not progress. The discovery of perspective by Renaissance painters gave artists a new technique that they used to create some of the most memorable masterpieces, like Raphael's *The School of Athens*. However, it would be rash indeed to claim that the art being produced now is better, in any absolute sense, than the art of the Renaissance masters. Yet we generally have no qualms at all about saying that we have far more scientific knowledge now than we did centuries ago. That is, our scientific beliefs are not just regarded as different, but as *better*. For instance, when the Black Death swept Europe in the middle of the fourteenth century, killing approximately 40 percent of the population, the best "scientific" explanation that the most learned medical authorities could offer was that the catastrophe was caused by a malignant conjunction of the planets Jupiter, Saturn, and Mars. For most people, the cause of the calamity was beyond any scientific understanding and was simply due to the wrath of God. Now we know that the Black Death was an outbreak of the bubonic plague, a highly infectious disease caused by the bacterium *Yersinia pestis*. The plague is spread by rats that carry fleas infected with the plague germ, and wherever people lived in the fourteenth century, there were also rats. When it came to treating the plague, medieval people were completely helpless. In the present day, the few people who do get bubonic plague can be quickly and effectively treated with antibiotics. So we clearly seem to know more about the plague and how to treat it than they did 650 years ago.

How then is it that science is supposed to progress in ways that most other human endeavors do not? Quite simply, science is supposed to progress toward *truth*. Of course, many others, writers for instance, also strive for truth. But it is decidedly ***not*** the case that writers of the present day are any better at telling us the truth about ourselves than Shakespeare was 400 years ago or Euripides was 2,400 years ago. Yet it is a commonplace that schoolchildren of today can tell us things about the universe that Aristotle never dreamed of. In other words, the best writers of today cannot expect to surpass the greatest of the past, but any scientifically literate person of today can give us more accurate information about the nature of the physical universe than the greatest Greek or medieval scientist. At least, this is how the story goes.

THE EVILS OF WHIG HISTORY

The story of science has usually been depicted as a triumphant march from darkness into light. According to this kind of story, we could graph the history of science by letting the horizontal axis of our graph stand for the last ten thousand years of human history and the vertical axis could stand for the level of scientific knowledge. The line plotting the rise of scientific knowledge would be fairly level and not too far above zero during most of human history. It would bump up a little with the start of agriculture and the invention of the wheel. It would start to rise about 2,500 years ago—slowly at first but steadily rising. There might be a slight decline during the Dark Ages of Europe when some knowledge was lost, but then there would be an explosive rise starting with the Scientific Revolution of the seventeenth century. The learning curve would get steeper and steeper in subsequent centuries until, in our day, it is just a few degrees from vertical.

The story of science as an ever-steeper learning curve is an attractive picture. As Latour said in the last chapter, we would like to think that amid the general chaos and confusion of human life there exists an oasis of steady progress. The story of science as a steady march of progress is called "Whig history." Just how it got this name needs some explaining. The Whigs, of course, were the British political party that opposed the Tories. The historian Herbert Butterfield coined the term *Whig history* to describe the kind of history written by Whig partisans. These partisans would always depict the course of English history as a struggle between the progressive and enlightened Whigs and the backward and benighted Tories. Naturally, on their account, the historical figures who supported Whig programs were the good guys and the villains were those who stood in the way. Butterfield said that the history of science is told in this Whiggish vein. Those who formulated or anticipated the theories we now accept were lauded as the true scientists while those who took an opposing line are ignored or are castigated as obscurantists or pseudoscientists.

The problem with the Whig approach to the history of science is that it is inaccurate and unfair. To graph the history of science as an ever-steeper learning curve, we have to be very selective about the data points we chart. We have to make sure that we chart only those theories and discoveries that we continue to accept as valid. What about all the blind alleys, theories that were quite successful at the time but eventually led nowhere? What about the scientists who went down those blind alleys? Many of them were among the most accomplished scientists of their day. Does the fact that they turned out to be wrong make them "bad guys"? In judging any historical figure it is manifestly unfair to expect him or her to know everything we know today. As the proverb puts it, hindsight is always 20/20. All that anyone can be expected to do is to make the best judgments he or she can, given the information and investigative tools available at a particular time and place. Those learned doctors who attributed the Black Death to the malignant conjunction of planets were not being stupid. Lacking all of the advanced equipment and techniques that biomedical science now enjoys, and lacking the knowledge we have gained in the last 650 years, there was no way that they could have guessed the truth.

As we said in chapter 1, the past is another country with different ways and customs. But the ways they did things in the past seemed as sensible to them as our ways seem to us today. They made just as serious an effort to base their beliefs on the facts and accurate reasoning as we do. Just as it would be ignorant and foolish to laugh at the odd-seeming customs of a rainforest tribe, so it is equally presumptuous to scorn past theories merely because they are outmoded or past scientists because they advocated those theories.

To see how unfair Whig history can be, recall Richard Owen, the scientist whose archetype theory opposed Darwinism. Owen was not a lovable man. Prickly, egotistical, and arrogant, he could be both petty and spiteful in his dealings with other scientists. However, in the eyes of posterity, his anti-Darwinism was his greatest sin. Owen engaged in long and bitter controversy with T. H. Huxley, "Darwin's bulldog." Because Darwin won, Whig historians painted Owen as the arch obscurantist, a poor scientist who opposed Darwinism on ideological rather than scientific grounds. Owen's criticism of Darwin was particularly rancorous. Some of Darwin's letters reveal that he was deeply hurt by the vicious attacks of one whom he had previously considered a friend. But Owen's reputation as a scientist remained under a cloud for many decades until more recent historians set the record straight. In recent years, historians of science such as Adrian Desmond have rehabilitated Owen's reputation as one of the outstanding scientists of the Victorian period. Of course, Owen did have an ideological ax to grind, but so did Huxley. Whatever his personal shortcomings, Owen was perhaps the greatest anatomist of his day, and his criticisms of Darwin were not frivolous but were serious objections that had to be addressed by Darwin's defenders.

Professional historians of science now scrupulously avoid writing history in the Whig style, though it is still often encountered in popular accounts of science. Historians now make every effort to place past scientists and theories squarely in the context of their original historical setting. They also strive to tell the story of science "warts and all," not just recounting the successes of the theories that ultimately triumphed but also giving evenhanded accounts of those that did not. In general, the stand professional historians of science have taken against the Whig style is laudable, but any good thing can be taken too far. In fact, I think that the fear of writing Whig history has grown from a healthy caution to an unhealthy phobia in the minds of some historians. Historians now are so afraid of falling prey to what writer C. S. Lewis called "the parochialism of the present" that they shy away from making any judgments about particular episodes as promoting or inhibiting the growth of science. In consequence, the history of science has become particularly susceptible to the allure of social constructivism.

SOCIAL-CONSTRUCTIVIST HISTORY

Social-constructivist history of science certainly is not history done in the Whig style. On the contrary, social constructivists, as we saw in the last chapter, interpret *all* scientific episodes, the "successful" and the "unsuccessful" alike, as ultimately caused by ambient social and political circumstances, not the considerations of evidence and logic traditionally thought to drive science. Or, more accurately, they see even the canons of evidence and logic employed in the sciences not as embodying universal norms of rationality but merely as "rules of the game" radically contingent on historical factors. If the very standards of rationality and evidence employed in science are themselves socially constructed through and through, then science cannot be a march toward truth, but only a random walk in response to the shoves and jerks of historical forces. For the social constructivists, therefore, no episodes in the history of science can be seen as "progressive," in the sense of approaching closer to truth or providing objectively better methods for the discovery of truth.

Perhaps the best-known example of social-constructivist history of science is the 1985 book *Leviathan and the Air-Pump* by Steven Shapin and Simon Schaffer. Shapin and Schaffer's book focused on an important controversy from the mid-seventeenth century, the nasty dispute between scientist Robert Boyle and irascible philosopher Thomas Hobbes. The dispute centered on the efforts of Boyle and his allies to introduce experimental methods into science. These changes required the radical reorganization of the way that science was practiced and the imposition of tight controls over who was recognized as a legitimate practitioner. Today experiment is such an integral part of scientific procedure that it is hard to imagine that science was ever practiced without experimental methods. Yet such methods had to be invented and incorporated into science. Although he was a skilled dissector, Aristotle performed no experiments in the modern sense. For Aristotle, and his followers right up into the modern era, scientific method was a matter of performing "inductions" from particular instances to first principles, and then making "deductions" from those principles back to particular kinds of things (at least this was Aristotle's stated method; his own scientific practice did not always conform to his official line). Francis Bacon famously proposed the use of experimental methods in the place of Aristotelian methods, but the closest he ever came to actually performing an experiment was an attempt to preserve a chicken by stuffing it with snow (Bacon became a martyr to science when he caught a fatal case of bronchitis after his chicken-stuffing experiment). Galileo performed brilliant experiments testing the acceleration of falling bodies by rolling balls down an inclined plane. Galileo's most famous experiment was one that never took place, the famous story about him dropping cannon balls from the Leaning Tower of Pisa. However, it was Robert Boyle's experiments with the newly invented air pump—which for the first time allowed scientists to create a fairly good vacuum—that placed the experimental method front and center in physical science. Thus it was only in the 1660s, the time of the Restoration (the restoration of the monarchy after the English Civil War) that experiment became essential to science.

Surely, it seems, the adoption of experimental methods was an instance of scientific progress if anything was. Again, it is hard for us to imagine things any other way. Shapin and Schaffer say that when Boyle first began his promotion of experimental methods in the 1660s, things were not so clear. In particular, noted philosopher Thomas Hobbes objected to the experimental method on various grounds. Among other complaints, Hobbes charged that the experimental method restricted the practice of science to a specialized elite, a privileged few who were granted access to expensive equipment that only they knew how to build and operate. With Boyle we see the beginning of the professionalization of science, that is, the recognition of a properly credentialed professional elite as the only legitimate practitioners of science. Today, of course, the certification process is quite strictly controlled. If you want to have a career in science, the doctorate is your union card. With the exception of astronomy and paleontology, where amateurs still often make valuable contributions, science is done by the professionals. In Hobbes's view, plain, everyday observation and careful deductive reasoning were all that should be needed for good science. For Hobbes, armchair science was still very much a live option. As for Boyle's research into pneumatics, the study of the properties of air, Hobbes objected that Boyle's pump was leaky and unreliable. It had to be, Hobbes argued, since he thought he could demonstrate that a vacuum was, in principle, impossible.

Boyle and his allies, on the other hand, were deeply suspicious of the grand speculative systems of philosophers like Hobbes, which they saw as castles on the clouds. For Boyle, as for scientists today, the problem with philosophical systems is, to paraphrase a quip of Mark Twain's, that they deliver such a wholesale return of conjecture from a trifling investment of fact (Twain originally said this about *science*, but it fits philosophy much better). For Boyle, science must be based on high-quality facts, and only rigorously controlled experiment could provide facts with the precision and certainty necessary for good science. Uncontrolled, everyday observations are just not good enough because they are so subject to uncertainty, vagueness, and error. Further, only those with the specialized and rigorous training in scientific theory and practice have the competence to conduct experiments.

Of course, Boyle and the experimentalists won this debate. Whig historians of science therefore depicted Boyle as the hero and Hobbes as a crackpot. Shapin and Schaffer object that this is far too facile an evaluation of the debate, and they contend that at the time Boyle was not so clearly right and Hobbes was not an obvious crank. In fact, they argue that Boyle's case was just not strong enough to have carried the day on rational grounds. Why, then, did Boyle win? Shapin and Schaffer answer that Boyle won because he played the political game much better than Hobbes, and because Hobbes was swimming against the tide of history. They argue that the emergence of a new way of organizing science was an integral part of the emergence of the new social order of Restoration society.

Hobbes was already at a disadvantage when he entered the dispute. He had the reputation, not entirely undeserved, of a cantankerous and crotchety old man who was always eager to keep a dispute boiling even when he was demonstrably wrong. As a philosopher, Hobbes endorsed an uncompromising materialism, and his classic political treatise *Leviathan* supported absolute authoritarianism in government. Worse, Hobbes was suspected of atheism; he certainly made a number of statements that made the orthodox nervous. Boyle, on the other hand, was fervently and famously pious, and that mattered a lot in those days. Boyle also had many connections with the rich, powerful, and influential. Whom you know may not count for everything in science, but it certainly counts a lot. In the end, the well-connected Boyle just had too many resources and too much influence for Hobbes to counter.

Shapin and Schaffer generalize from their case study and assert that the methods of science are always a product of local social and political influences. The very methods of science, the methods that scientist think will lead them to truth, are actually historically contingent conventions adopted on the basis of political expedience. The consequence, Shapin and Schaffer state very plainly, is that *we ourselves* determine the content of our science, not some putative external nature:

"As we come to recognize the conventional and artifactual status of our forms of knowing, we put ourselves in a position to realize that it is ourselves and not reality that is responsible for what we know" (1985, 344). In other words, the methods of science may change, but they do not *improve* in any sense other than perhaps having more social or political utility for the people who endorse them. Clearly, if the methods of science are really just arbitrary conventions and stipulated "rules of the game," there is no clear sense in which science progresses—certainly not toward truth. So it looks like if we go too far trying not to tell the story of science as a triumphant march of steady progress, we wind up not being able to see how it can progress at all.

Shapin and Schaffer attempt to give a *causal* account for the fact that scientists sometimes adopt a new set of methods. For their account to be adequate they have to rule out the opposing hypothesis, that scientists adopt new methods for the simple reason that they see that they *are* better ways of doing science. No, Shapin and Schaffer contend, scientists adopt new methods because they are politically and socially expedient. So, for their explanation to work, Shapin and Schaffer must argue either (a) that no scientific method really is any better than any other, or (b) that scientists cannot distinguish better from worse methods.

How could anyone argue (a), that no method is better than any other? Scientists are constantly trying to work out new methods that will permit them to better test their theories against data. A new method can be better than an old one by providing a greater quantity of data, a greater variety of data, data more directly relevant to the hypothesis being tested, clearer data with more signal and less noise, or more precise data allowing for a stricter, more rigorous test. Other methodological innovations might provide more incisive mathematical or analytical tools for making better sense of data. How can anyone say that no method can be better than another in any of these regards? The only way to argue this in a comprehensive way would be to adopt a universal skepticism about method. The most famous case for such skepticism is the 1975 book *Against Method* by maverick philosopher of science Paul Feyerabend. Feyerabend appeals to the history of science to argue that no methodological prescription has ever been consistently followed in science. Every methodological rule, he claims, has been violated at one time or another by even the most famous and successful scientists. Further, he asserts that the biggest scientific discoveries *could not* have been made had not scientists broken the rules and thought "outside the box." Feyerabend concludes his attack on method with the ironic comment that the only rule that science had consistently followed is "anything goes."

Now Feyerabend is certainly right that no simple set of exceptionless rules has guided every scientific judgment. By analogy, no set of simple, unbreakable rules covers every possible situation where we have to make a *moral* decision. Even good rules like "do not lie" and "do not steal" admit of well-known exceptions where it would be permissible or even morally obligatory to lie or to steal. Also, methodological rules change over time (If they didn't, how could they ever improve?). However, Feyerabend's methodological anarchism goes way too far. As philosopher of science W. H. Newton-Smith notes, to show that a scientific method is worthless, it is not enough to cite anecdotes about its failure on one occasion or another. You would have to show that it has failed more often than not, and this task Feyerabend did not even begin to accomplish. Further, to make any judgments about a particular rule, we must, at least for the time being, assume that other rules are reliable:

> For how do we know that a particular rule has led us to make unfortunate choices? We have no omniscient God to whisper the answers in our ears. Trapped as we are within the scientific enterprise without such a divine road to knowledge, we have no recourse but to make such judgments on the basis of other principles of comparison. Thus any historically based attack on a particular methodological rule of the sort being envisaged will presuppose the viability of other such rules. The best one can do through an historical investigation is to take up a single plank of the ship of methodology while the rest remain, for the moment at least, firmly in place. (Newton-Smith, 1981, 134)

Newton-Smith says that rejecting a methodological rule is like taking a plank out of a ship at sea. We can replace a bad plank, but only if we leave the other planks in place for the time being. If we try to take all the planks out at once, we have no ship left and sink to the bottom. Likewise, we can reject a particular methodological prescription only if we rely on others that, for the time being, we do not question. Trying to reject all methods at once would leave us helpless to make any judgments. Therefore, the idea that we could reasonably be skeptical about *all* methods at once is just incoherent.

So Shapin and Schaffer cannot reasonably maintain a global skepticism about all methods. Therefore, they would have to claim (b), that, though improved methods might be proposed, scientists are incapable of recognizing them as superior to other methods. Outside of postulating a universal curse laid on all human cognitive powers—in which case their *own* reasoning would be in doubt—it is hard to see how Shapin and Schaffer could argue this. Suppose that double-blind trials really are better ways of testing the effectiveness of new drugs than gazing into crystal balls is. What could prevent sane, intelligent humans from learning this fact?

Perhaps, then, Shapin and Schaffer would concede that some methods are better than others, and that scientists can sometimes recognize that a particular method is better than another. However, they would argue that in real life scientists seldom adopt new methods because they see that they are better than the old ones, but do so instead because they give in to political and social pressures. In other words, perhaps Shapin and Shaffer's position is not so much skeptical as cynical. However, such cynical conclusions would have to be argued on a case-by-case basis. Shapin and Shaffer certainly cannot justify universal cynicism about science by generalizing from a single case study—though this is precisely what they try to do in *Leviathan and the Air-Pump*.

Another problem with the cynical thesis is that there are so many apparent counterexamples from the history of science, that is, instances where scientists without political clout or social standing proposed new methods that soon were widely accepted. For instance, in 1908,

Henrietta Swan Leavitt, a low-level employee at the Harvard Observatory, noticed something remarkable about a particular kind of star. She was examining stars known as Cepheid variables in the Small Magellanic Cloud, an irregular companion galaxy to the Milky Way. These stars were called variables because they go through a regular cycle of getting brighter and then dimmer and then brighter again. She noticed something peculiar, namely, that the brighter a Cepheid variable was, the longer was the period of its cycle from brightest to dimmest. Since all the stars in the Small Magellanic Cloud are about the same distance from us, the stars there that *look* brighter really *are* brighter. So, Cepheid variables can serve as "standard candles," that is, when you see a Cepheid variable in a distant galaxy, you can measure the period of its brightness variation, and then you know how intrinsically bright the star is. Knowing how intrinsically bright a star is, and comparing that to how bright it looks, can tell us how far away the star is, and so how far away its galaxy is. So Leavitt had discovered facts that gave astronomers an invaluable tool for measuring celestial distances. Astronomers had nothing socially or politically to gain from adopting the method based on Leavitt's discovery of the period/luminosity relationship; she was a woman when women had little status in science and she certainly had no important political connections. Clearly, the reason that astronomers soon adopted methods based on Leavitt's discovery was that those methods were *good methods*.

So have historians of science, perhaps due to excessive fear of writing Whig history, failed to address the issue of progress in science? Noted philosopher of science Larry Laudan thinks so. He says that if a historian admits that some theories have had more empirical success than others (like, e.g., making more accurate predictions), and this is hardly deniable, then the issue of progress should be addressed:

> If he once grants that certain scientific theories or approaches have proved more successful empirically than their rivals, then the historian who disavows any interest in scientific progress is confessing there are some facts about the past which he has no interest in explaining.

> Given that no history can be complete, that alone would not be very distressing. But the progressiveness of science appears to everyone *except the professional historian of science* to be the single most salient fact about the diachronic development of science. That, above all else, cries out for historical analysis and explanation. (Laudan, 1990, 57; emphasis in original)

A historian of science who ignores the issue of progress is like a historian of the American South who ignores the issue of race. One of, if not *the*, most important issue of the field is simply dismissed.

But how can we designate certain episodes in the history of science as more progressive than others without privileging some set of criteria about what constitutes progress, and wouldn't such judgments inevitably be Whiggish? By *whose* standards are we to judge that progress has or has not taken place? By *our* standards, of course. As Newton-Smith noted above, we hold, and are bound to hold, that some methods for obtaining knowledge are better than others. The only alternative is an attitude of complete indifference about all knowledge claims. Such an attitude is not in evidence among historians of science, not even—perhaps we should say "especially not"—among those inclined toward social constructivism.

So, if past scientists reached a conclusion we still think is true, or approximately so, and if they based their conclusion on reasons that still seem good to us today, we should regard this as an instance of progress in science. Is such a judgment Whiggish? The problem with Whig history is that it is unhistorical; it judges people by standards inappropriate for that time and place. But if we judge the theory-choice decisions of past scientists in terms of the reasons available *to them*, the fact that those reasons are still good for us today does not make such judgments unhistorical. It can never be unhistorical to tell the truth about history, and, as Laudan asserts, perhaps *the* salient truth about science is that it has progressed. If so, to approach the history of science as the social constructivists do, with the grim determination to explain away any appearance of progress, is as blatant an instance of ideologically blinkered ahistoricism as anything ever perpetrated by the Whigs. Laudan states this point eloquently, referring back to Herbert Butterfield's original definition of *Whig history*:

> One can sympathize with Butterfield's concern that a tale of victory, told only by the victors, makes for bad history. But in denying that historians are ever justified in recognizing that certain parts of science are better than others, and in asserting that it is no part of the historian's task to explain the conditions which made them more successful, Butterfield (and those historians of science who follow him) would appear to be abandoning the programme of telling the full story of the past to which they are otherwise so deeply committed. (1990, 56–57)

DOES SCIENCE CONVERGE TOWARD TRUTH?

Let us suppose then that science does progress. How does it progress? Laudan, who is so eloquent in defending the progress of science, is equally eloquent in arguing that there is little evidence that it progresses toward *truth*. Science could progress in various ways without getting closer to truth. It could progress in the purely pragmatic sense that our science today gives us greater control over nature and more advanced technology than past science. Science could progress by producing theories that did a better job of saving the appearances, that is, by generating more and better predictions. As we saw with Ptolemy's astronomy, a false theory can often save the appearances very successfully. When Ptolemy said that Mars would be found at a particular part of the sky at a particular time, he was always very close to being exactly right. Science might also progress in the Kuhnian sense that a new theory can accommodate the anomalies that stumped the old theory. Traditionally, though, science was thought to progress toward truth. On this view, Newton gave us a truer view of the universe than Aristotle, and Einstein's was truer than Newton's. Do we have the final truth yet? No, because our two fundamental theories of physics, relativity and quantum mechanics, contradict each other, so something is wrong somewhere. Still, scientific optimists are convinced that successive theories *converge* toward truth, and hope that someday, perhaps, we will have a complete "theory of everything." Let us therefore call "convergent realism" the view that over the history of natural science, successive theories have converged toward truth, that is, that later theories are better approximations of truth than earlier ones.

Laudan argues vigorously that the history of science confutes the idea of convergent realism. Why have people usually thought that science progresses toward truth, that is, that as new theories replace old ones over time, the new theories get closer and closer to the whole truth? The reason usually given is that in any field of science the more recent theories are almost always more empirically successful than the earlier ones. A theory is more "empirically successful" than another if it makes more accurate predictions or passes more stringent tests or gives better explanations of puzzling phenomena or allows more effective intervention in the natural world (as, for example, in better preventing disease). (By the way, by "prediction," philosophers of science just mean an observational consequence of a theory. A prediction in this sense does not have to forecast the future; it can "retrodict" something that has already happened, and even something already known.) Thus, Newton's theory made many more, and more accurate, predictions than the old Aristotelian scheme. It also permits us to do a lot more; calculations drawing on Newtonian laws are still used to send spacecraft to the outer planets. Einstein's theories incorporated and improved on the empirical successes of Newton's theories. To many minds, the spectacular increase in the empirical success of science over time is just incomprehensible unless science is converging toward truth. As a number of philosophers have put it, the ever-growing empirical success of science would simply have to be a stupendous miracle unless science were also growing toward truth. The tremendous technological advances of the past four centuries depend on the accuracy of theoretical predictions. How then is it that we can send rockets to distant planets, cure diseases once thought incurable, and probe the cosmos as it was billions of years ago, unless we have some inkling about what is really going on? If we don't really know more than people did five hundred years ago, how can we *do* so much more?

The above argument certainly has great intuitive appeal—how could anyone resist it? The most direct way would be to show that empirical success and truth have not been connected in the history of science. That is, if we could show that many past theories enjoyed considerable empirical success in their day but are now rightly regarded as false, this would decouple empirical success from truth. In this case, the increasing empirical success of successive theories will not show that science is progressing toward truth. On the contrary, the fact that so many theories have been shown wrong, though they enjoyed great success and nearly universal acceptance in their day, should lead us to conclude that the majority of our present theories are likely false. This argument has been called the "pessimistic meta-induction." Reasoning from particular instances to a general conclusion is a form of induction. The induction here generalizes from the history of scientific theories to the pessimistic conclusion that our current theories are probably false (it is called a "meta" induction because it does not generalize from particular facts but from theories that are already general statements—but this isn't really important). Laudan asserts that the fact that so many successful theories of the past have turned out false is good inductive evidence that our present ones are false also, and this is why the induction is "pessimistic."

Laudan refers to case after case of past theories that enjoyed considerable empirical success in their day, even making accurate predictions that were novel and surprising (which counts a lot in a theory's favor), yet these theories are now universally rejected as false. A good example would be the theory of phlogiston. In the late eighteenth century, chemists wanted to know exactly what happens when something burns. When something burns, it seems to lose substance; matter flows out, leaving only ashes. Eminent English chemist Joseph Priestly hypothesized that the "fire stuff," the volatile matter that flows out of things as they burn, is something called "phlogiston." The phlogiston hypothesis had a lot going for it. It accurately predicted that a burning candle placed in an airtight container would soon go out. The explanation was straightforward and made perfect sense: As the candle burns, the air inside the container becomes saturated with phlogiston and the phlogiston still in the candle has no place to go, so it stops flowing out. Phlogiston theory could also explain the process of smelting whereby crude ores are refined into metals. These and other empirical successes made phlogiston theory a prominent and widely accepted chemical theory in its day. Nevertheless, phlogiston theory was ultimately rejected in favor of Antoine Lavoisier's hypothesis—that combustion is not a matter of something flowing out but of something combining with the burning substance, namely oxygen. We still accept Lavoisier's account of combustion today, so phlogiston theory is regarded as false.

Laudan says that such examples can be adduced *ad nauseam*, that is, we can list indefinitely many instances of past theories that enjoyed considerable empirical success in their day but are now universally rejected. We cannot even regard theories like phlogiston as partially true, says Laudan. Often a new theory does not completely supplant the old one but retains some portion of the old theory or perhaps even absorbs it as an approximation or a limiting case. As we saw in chapter 2, many claim that Newtonian theory remains as a limiting case of Einstein's theory of special relativity. At relative speeds far below the speed of light, Einstein's laws of motion reduce, very nearly, to Newton's. Often, also, a theory can be considered partially true because its central terms refer to objects still recognized as real. For instance, when Gregor Mendel first published his theory of inheritance in 1866, he postulated the existence of certain "factors" or "elements" as units of heredity. Though the science of genetics has progressed enormously since Mendel, transformed fundamentally by molecular biology, Mendel's "factors" or "elements" are still thought to refer to what we call "genes" today. Yet with theories like phlogiston, *nothing* is retained in later theories; they are rejected lock, stock, and barrel. "Phlogiston" does not refer to anything still considered real any more than does "Zeus." Therefore, nobody can say that phlogiston theory was partially true and therefore a step up the ladder of convergence toward the true theory. The history of science is a graveyard of theories that were mighty in their day but are now nothing but dust occasionally shaken from old books.

Apparently, then, a theory can enjoy considerable empirical success even if it is not even partially true. Actually, we can see this rather easily without having to make reference to arcana from the history books like phlogiston theory. The earth is not flat, but if you want to build a house, or even a large building, the hypothesis that the ground underneath the building is flat works perfectly well. The reason why, of course, is that the earth is so big that its curvature, even over the area of a large building, is negligible and the space can just be thought of as flat. The sun, planets, and stars do not revolve around the earth, but for purposes of navigation, a geocentric theory works as well as a heliocentric one. So, false theories can have true consequences, and this shouldn't really surprise anyone. Any student taking an introductory logic course soon realizes that false premises can entail true conclusions. From the false premises "All Democrats are Republicans" and "Ronald Reagan was a Democrat," it follows as a valid deductive consequence that "Ronald Reagan was a Republican," which is true.

Defenders of convergent realism are, of course, aware that false theories can have true consequences, and therefore that false theories can be expected to have some empirical successes. Their argument is that the truer a theory is, the *more likely* it is to have true consequences, and so, the *best explanation* of the increasing empirical success of science over time is that successive theories are better and better approximations of truth. Further, convergent realists argue that new theories typically do retain and improve on elements of the theories they replace. They conclude that the best explanation for the empirical success of earlier theories is that the parts of those theories retained in later theories express at least partial truth about physical reality. Einstein did not retain all of Newton's ideas. For instance, Newton claimed that space and time are absolute, which Einstein explicitly and emphatically rejected. Convergent realists think that the parts of Newtonian theory that Einstein did retain as a limiting case are the parts that give us a very good approximation of the truth about the motion of bodies at relative velocities far lower than the speed of light. They also think that it is those parts that were retained, not the ones Einstein discarded, that explain the great empirical success of Newton's theories.

Laudan replies that the connection between partial truth and empirical success has not been established. True statements cannot entail a falsehood, but Laudan thinks that no one has shown that, in general, a theory that is 50 percent true (whatever that means) is more likely to have true consequences than one that is 0 percent true. Further, he denies that as a rule new theories retain significant elements of the theories they replace. So, Laudan challenges convergent realists to prove the connection between partial truth and empirical success and to show that the history of science supports their factual claims about successive theories.

ASSESSING LAUDAN'S CRITIQUE OF CONVERGENT REALISM

Whether or not partially true theories are more likely to enjoy empirical success than wholly false ones, it is clear that the history of science has *not* been a tale of steady ascent toward truth. Such a tale is what we earlier called Whig history, and we have already seen that such an interpretation of the history of science is distorted and inaccurate. Science is not a straight-line ascent to the peak of Mount Truth. Even those who think that science has acquired some deep truths about the cosmos must view the history of science as a difficult scramble up a slippery and treacherous slope, with many halts, detours, slips, and stumbles along the way. So, if convergent realism requires a steady march of scientific progress toward truth, then the history of science refutes convergent realism, just as Laudan claims. But is it possible to take a view more consistent with the actual history of science and still see science as progressing toward truth? How can we generalize about the history of science without running into the sorts of rather-obvious counterexamples Laudan adduces—theories very successful in their day but now recognized as wholly false?

As Laudan interprets the case for convergent realism, it claims that the best, perhaps the only, explanation for the empirical success of science is that current theories (at least in the "mature" sciences like chemistry or physics) are typically approximately true, and more recent theories are truer than their predecessors. We can interpret Laudan's critique as making a strong claim and a weaker claim. The strong claim is the pessimistic metainduction: that even our best-supported current theories should not be regarded as approximately true given the historical record of highly successful theories that were wholly false. If, typically, even the most successful past theories turned out completely false, then most of our currently accepted theories are probably not even approximately true, despite their enormous empirical success. Laudan's weaker claim is that convergent realists cannot argue that the increasing empirical success of theories over time supports their assertion that science converges toward truth. He contends that the argument fails because convergent realists have not established the necessary correlation between degrees of truth and degrees of empirical success. Laudan complains that we have no philosophical reasons to expect such a correlation, and he contends that the facts of the history of science are against it.

Starting with the stronger claim, we first have to be clear on what convergent realists are claiming. They are happy to concede that, strictly speaking, many of our best current theories are false. Convergent realists claim that these theories, though false, are nevertheless approximately true. For instance, convergent realists can regard the currently accepted Standard Model of particle physics as false, strictly speaking, but still as approximately true. Critics such as Laudan often complain that no one has yet articulated a philosophically satisfactory notion of approximate truth. Actually, since philosophers have yet to agree on the meaning of *truth* it is hardly surprising that there is no consensus about the meaning of "approximate truth." Still, we have very deep and undeniable intuitions that some claims are truer than others, or, in the words of Isaac Asimov, that wrong is relative. It is wrong that the earth is a sphere; actually, it is an oblate spheroid that bulges slightly in the middle and is slightly flattened at the poles (from space it still looks like a perfect sphere, though). But is it just as wrong to say that the earth is a sphere as to say it is a cube? Clearly, we have a very strong intuitive sense that the statement, "the world is a sphere," while false strictly speaking, is more "truth-like" than the statement "the world is a cube."

Philosopher Stathis Psillos defends a notion of approximate truth that builds on these intuitions about truth-likeness (1999). Psillos notes that in science, theories are seldom even presented as *exactly* true (more on this in the next chapter). Rather, scientific theories generally offer representations of the world that are idealizations, simplifications, or approximations rather than assertions purported to be precisely true. Even statements of the laws of nature contain clauses that bracket off and exclude distorting factors, like the effects of air resistance on free fall. Yet a theoretical description can still be highly truth-like, even if not strictly true, when it *fits* reality to a high degree of approximation. Psillos spells out these intuitive ideas more precisely as follows:

> A description D *approximately fits* a state S (i.e., D is approximately true of S) if there is another state S' such that S and S' are linked by specific conditions of approximation, and D fits S' (D is true of S'). (1999, 277)

In other words, a truth-like theoretical description will be one that is *strictly* true of an imaginary state, but that imaginary state is one that is very similar to the relevant state of the actual world.

Consider the imaginary world in which Newton's laws of motion hold *precisely* in all circumstances; let's call this imaginary world "Newton-World." Newton-World is not the real world. We now know that Newton's laws fail in various circumstances, and physicists have given us relativity theory and quantum mechanics to deal with the realms where Newton's laws fail. Yet Newton's laws remain indispensable tools for applied physics. For instance, space scientists employ Newtonian calculations to plot the courses of spacecraft through the solar system. When you send a probe to Neptune, and it arrives within seconds of the predicted time, you are doing *something* right. So it looks like Newton-World is, in many important respects and under many different conditions, very much like the real world, and Newton's theory, though false, still has a high degree of truth-likeness.

Now Psillos admits that such an intuitive formulation of the nature of truth-likeness will not satisfy philosophers who want a formal and rigorous statement of truth-likeness. Yet he does not regard the lack of such a rigorous formal statement as an insuperable problem for realists. I agree with him. If we had to wait until philosophers formulated rigorous accounts of all of our key concepts, we would not get much done. So, I shall assume that the notions of truth-likeness and approximate truth are clear enough for our discussion to proceed.

Getting back to Laudan's pessimistic metainduction, all such inductions are based on samples. Such arguments conclude that what is true of a sample is, to a high degree of approximation, true of the whole from which the sample is drawn. For instance, Laudan offers instances of successful theories that turned out wholly false and bases conclusions about the whole history of science on his sample. To reach a true conclusion, inductive arguments of this sort must draw on samples that are truly representative of the whole. Inductive arguments can go disastrously wrong if proper steps are not taken to ensure that the sample is representative of the whole. Defenders of the pessimistic metainduction therefore need to show that empirically successful theories of the past *typically* turned out to be *totally* false. Only if they show this can they effectively argue that today's successful theories are likely to be wrong. Laudan has not shown this. Even if we can adduce examples of successful totally false theories *ad nauseam*, as Laudan claims, this does not prove that this list is representative of all past theories. We can also list *ad nauseam* instances of past theories that, while woefully wrong on some things, got other things nearly right or at least referred to entities science still regards as real.

However, even if we for the moment concede, purely for the sake of argument, that the majority of successful past theories were totally false, the pessimistic metainduction is not necessarily warranted. The reason is that the quality and quantity of the empirical tests theories must pass has greatly increased over the history of science. One way that science unquestionably progresses is that new methods and techniques are frequently found that allow for ever more stringent tests of theories. Judged by the kinds of empirical tests and analytic tools we have today, many past theories were not very well tested, and so were not really very successful compared to current theories. On the other hand, the Santa Claus hypothesis is highly empirically successful for five-year-olds, but, of course, five-year-olds are very limited in their ability to test hypotheses. The point is that a theory's degree of empirical success is highly dependent on the severity of the tests it must pass and the stringency of the standards it must meet.

Even in a "soft" science like paleontology, the number and the rigor of empirical challenges that theories must meet are far greater than in the past. For instance, the idea that dinosaurs might have been warm-blooded, rather than cold-blooded like typical reptiles, was first proposed by T. H. Huxley in the mid-nineteenth century. At the time, Huxley's proposal was not much more than a speculation since data bearing on dinosaur physiology were sparse and very indirectly relevant. In the 1970s, Robert Bakker revived the idea of warm-blooded dinosaurs, and vigorously prosecuted his case. By then the quantity and quality of evidence that could be brought to bear, and the sophistication of methods and techniques for judging Bakker's claim, vastly exceeded anything available to Huxley. Bakker employed microscopic studies of bone histology to argue that dinosaurs had bone architecture closer to that of (warm-blooded) mammals and birds than to cold-blooded reptiles. He appealed to predator/prey ratios in fossil assemblages to argue that the proportion of predators to prey among dinosaurs was like the ratio of (warm-blooded) lions to prey in the Serengeti, not the ratio of (cold-blooded) komodo dragons to their prey. More recently, tests on the ratios of oxygen isotopes in dinosaur bone have been used to test the hypothesis of warm-blooded dinosaurs. Needless to say, Huxley could not even have dreamed of having such evidence. Even whole new subdisciplines—like taphonomy, the study of the process of fossilization, and paleoichnology, the study of fossil tracks and traces—have been developed to provide more tools for the evaluation of theories in paleontology.

The upshot is that the success of false past theories is due largely to the fact that they were not tested nearly as severely as we test theories today. Hence, the generalization the pessimistic metainduction rests on, the claim that many, perhaps most, successful past theories turned out to be wholly false, is not strong enough to show that our currently accepted theories are not approximately true. The "success" of many of those past theories is just not comparable to the success of today's accepted theories. Laudan, for all his emphasis on scientific progress, seems not to appreciate the degree to which, in any given field of science, the sophistication and rigor of hypothesis-testing practices can greatly increase in a relatively short time. Does this mean that we have crossed some sort of threshold and that our tests are now so rigorous that we can be *sure* that we will keep any totally false theory from slipping by? Of course not, but convergent realists can say with considerable confidence that a totally false theory would have a much tougher time hiding its false consequences today than two hundred years ago.

What convergent realists would therefore expect to see when they look at the history of science is that wholly false theories would proliferate when empirical tests are weak, but would enjoy less and less success as tests get progressively stronger. Instead, they would expect that when more recent theories are overturned, after reigning for a time in their fields, they would not so often be totally rejected, but would more often be partially preserved in the new theories. Does the history of science conform to these expectations? This is far too big a question to be settled here, but I think we can say that *prima facie* it does, or at least that the burden of proof is on Laudan to show that it does not.

What about Laudan's weaker claim that the history of science fails to support the convergent realists' argument that the best explanation of the increasing empirical success of science is that it converges toward truth? If *convergence* is interpreted in the Whig way, Laudan's argument is airtight. The history of science has not been the steady accumulation of ever-closer approximations of theories now regarded as nearly true. Rather, in each field of science there were long periods dominated by empirically successful theories (successful by the standards of their day) but now wholly rejected. But could a realist, one who thinks that at least some of our best current theories are approximately true, still find support from the history of science? What would a more realistic realist, one not mired in Whig fantasies about steady progress, expect to see in the history of science?

The history of science is not one of steady cumulative progress, but neither is it a succession of mutually exclusive paradigms where each new theory totally wipes the slate clean and starts all over again. *If* we regarded *all* past theories as *totally* false, then the pessimistic metainduction probably should make us doubt our present theories, however empirically successful they are. But the history of science is *not* always like the famous Peter Arno cartoon from the *New Yorker*: A test flight has just ended in a horrendous crash. The aircraft designer turns his back on the ensuing chaos, claps his hands together, and blithely chirps, "Well, back to the old drawing board." Science does not have to go back to the drawing board with every superseded theory. Rather, when we look at the history of any field of science, a few theories will stand out as major breakthroughs. Once these breakthroughs occur, they are retained, in one form or another, through all subsequent theory changes, even through major conceptual revolutions. For instance, the mathematician and physicist James Clerk Maxwell (1831–1879) formulated a small set of simple equations that explained all the diverse phenomena of electricity and magnetism. He concluded that electricity and magnetism were different aspects of the same force, electromagnetism, and that light is actually a form of electromagnetic radiation. Maxwell's *Treatise on Electricity and Magnetism* was published in 1873, well before the two major revolutions in twentieth-century physics, relativity and quantum mechanics.

The revolutions of twentieth-century physics overthrew some of Maxwell's ideas. For instance, he thought that since light was a wave, it had to be carried by some medium, the "luminiferous ether," an idea rejected by subsequent theory. However, light is still regarded as electromagnetic radiation, and Maxwell's equations, in modified form, are still regarded as valid for a given range of electrical and magnetic phenomena. Likewise, Newton's famous law of universal gravitation is retained in current physics as correctly applying to things not moving too fast and to gravitational forces that are not too strong. So, many of Maxwell's ideas, like Newton's, have survived the enormous conceptual upheavals of the relativity and quantum revolutions, revolutions that overthrew so many of the ideas of "classical" physics. Within limited contexts, Maxwell's and Newton's theories are just as valid as they ever were. Other breakthrough theories have shown similar staying power in other fields of science. Lavoisier's theory that combustion is an oxidation reaction is still accepted despite the revolutions in chemistry and physics since the eighteenth century. Darwin's theory of natural selection and Mendel's genetics have been absorbed, considerably modified and augmented, into current biological theory. The retention of key elements of past theories into the present does not always happen; it may not even be the typical case. But the fact that it has happened as often as it has—that so much has been retained so often, even across Scientific Revolutions and through paradigm shifts—is a remarkable fact about the history of science. What accounts for it?

Realists have a simple explanation for the remarkable persistence of certain theoretical ideas: Newton, Darwin, Lavoisier, Mendel, and their peers were on the right track. They did not have the whole truth, but they had some of it. Human thought has enormous centrifugal force. Ideas tend to fragment and to disperse into mutually exclusive and usually mutually hostile schools, dogmas, ideologies, and isms. We see this often in the history of philosophy, where each major philosopher saw his system as the one true one, displacing all earlier theories. Religious doctrines are notoriously brittle, easily fragmented by schism and heresy. This is why religious bodes have so often employed heavy-handed tactics like inquisitions and crusades in attempts to impose unity. In science, to a greater degree than in any other human intellectual endeavor, we find a centripetal force that is strong enough to overcome the centrifugal pull of clashing ideas. Time and time again, scientists are pulled back to certain ideas. Indeed, they are often hauled back kicking and screaming.

Consider Friedrich Ostwald, a prominent German chemist of the nineteenth century. Ostwald held that scientists should believe only in what they could directly measure, and therefore should regard undetectable theoretical entities such as atoms as useful fictions. In 1905, the same year (and in the same issue of the same journal) that he published his first paper on relativity, Einstein also published a paper explaining the puzzling phenomenon of Brownian motion. Tiny particles of dense materials do not sink to the bottom when placed in water but remain suspended and jiggle about in a random manner; this random jiggling is called *Brownian motion*. Einstein assumed that the Brownian motion of the particles was caused by the collision of the particles with innumerable, invisible, and randomly moving water molecules. His equations, based on that assumption, gave precise predictions about the expected distributions of tiny particles when suspended in water. A few years later, French physicist Jean Perrin performed careful observations of the distributions of tiny particles suspended in a drop of water and found that these precisely matched Einstein's predictions. Not only that, but Perrin was able to calculate how big water molecules would have to be to produce the observed effect. When confronted by such startling evidence, Ostwald threw in the towel and admitted the reality of atoms.

Although it might surprise us now that atoms are so fundamental to our view of the world, Ostwald was hardly the only antiatomist in the history of science. A very fat book could be (and probably has been) written on the history of antiatomism. The idea that matter is ultimately composed of tiny, discrete bits was abhorrent to many thinkers who opposed atomism with might and main. Religious thinkers have often associated atomism with atheism, so atomism often had to face both scientific and religious opposition. The two founding figures of Western philosophy, Plato and Aristotle, were ardent antiatomists. But the idea just would not go away. It survived every attempt to discredit it, even reemerging, as Laudan notes, after long periods when opposing theories enjoyed greater empirical success. Needless to say, in the 2,400 years since the idea of atoms was first proposed by the Greek philosophers Leucippus and Democritus, the concept of the atom has been radically altered, even losing its original defining feature of being impossible to split ("a/tom" means "uncuttable" in Greek). Yet the idea has persisted through all of the revolutions in science over the last 2,400 years.

For realists, the only reasonable explanation of the peculiar persistence of the atomic idea, its triumph through revolutions and in spite of the efforts of some of humanity's best minds to discredit it, is that there is something right about it. There must really be atoms, and we must really know something about them. It looks like we are confronting a fundamental fact about nature, a fact that no human ingenuity or ideological rancor can eradicate: Things are made of atoms.

Realists can therefore say that we find in the history of science just what we expect to find if humans have in fact uncovered some deep truths about the hidden nature of the universe. Realism does not predict that the history of science will have developed in the Whig way. Like antirealists, realists do not expect that science always takes the royal road to truth and never deviates. They expect to see that some ideas have survived in spite of everything, that is, that some ideas time and time again have buried their would-be undertakers. Further, they would expect that these oddly persistent ideas, though perhaps eclipsed for decades or even centuries by opposing ideas, would eventually emerge stronger than ever. Further, they would expect that successive conceptual revolutions, though changing many concepts in fundamental ways, would still have to accommodate those persistent ideas. We do seem to find all these things in the history of science, and realism seems to give a straightforward explanation of why we do.

Laudan's response is that, even if he does concede that realism accounts for certain salient features of the history of science, as a point against antirealism, this argument begs the question. Laudan notes that antirealists have always contended that the realists' argument from a theory's empirical success to its approximate truth rests on a simple fallacy, the fallacy of affirming the consequent. Every introductory logic text tells us that it is fallacious to argue like this:

> If *P*, then *Q*.
> *Q*.
> Therefore, *P*.

For instance, this is not a valid argument:

> If it rained all night, then the streets are wet.
> The streets are wet.
> Therefore, it rained all night.

The conclusion does not follow. The streets could be wet for some other reason; maybe a water main broke. Laudan says that realists argue in the same fallacious way:

> If theory T is approximately true, then it will be empirically successful.
> Theory T is empirically successful.
> Therefore, Theory T is approximately true.

When asked to justify their realism, given that it seems to rest on the fallacy of affirming the consequent, Laudan says that realists respond with just another instance of that same fallacy:

> If realism is true, then it will account for the main features of the history of science.
> Realism does account for the main features of the history of science.
> Therefore, realism is true.

However, says Laudan, if antirealists do not think that a scientific theory is shown approximately true by its true consequences, they surely will not accept that a philosophical theory—realism—is true because *it* has true consequences! To expect otherwise is blatantly to beg the question against antirealists.

SCIENTISTS' OWN REALISM

Do realists argue fallaciously when they defend the approximate truth of past or current theories? Surely at this point some perceptive readers will be asking a deeper question: Do scientists really need the help of *philosophers* in articulating grounds for regarding their theories as (approximately) true? To many, the suggestion that philosophers could offer such help is like saying that philosophers could advise Michael Phelps about swimming. To suggest that scientific claims, about atoms for instance, need an additional defense over and above the sorts of evidence that Perrin and others have given seems to me already to concede half the case to antirealists. My own view is that philosophers have very little to offer in the way of *positive* arguments for the approximate truth of past or currently accepted theories. Arguments from the history of science, however sound, cannot add significantly to the reasons scientists have *right now* for believing in atoms. No such arguments will be as strong as Perrin's evidence that convinced the antiatomic skeptic Ostwald. I think the chief role for realist philosophers is to rebut the skeptical arguments of antirealists, arguments like the pessimistic metainduction. In short, the best reasons for regarding theories as approximately true are not the metascientific arguments of realist philosophers, but the reasons scientists themselves give.

Why then do *scientists* sometimes come to regard particular theories as approximately true? The first thing to note is that scientists do *not* generally argue in the fallacious manner Laudan attributes to philosophical realists. They do not proceed by taking an isolated theory *T*, drawing observational consequences from *T*, and then concluding that *T* is approximately true since its observational consequences are actually observed. A theory must have such empirical success before scientists will accept it, that is, the observational consequences we draw from the theory must be consistent with the actual data. But that is not enough. It is a slogan among some philosophers that "theory is underdetermined by data." What this means is that for any batch of data we might cite, infinitely many possible theories could make predictions consistent with those data. In the abstract realm of sheer logical possibility, where everything not self-contradictory is possible, this thesis of underdetermination is correct. If *all* that matters is that a theory's predictions be compatible with the known data, then infinitely many theories are possible. Another way of expressing underdetermination is this: For any theory that could be the correct explanation for some batch of data, it will always be possible to "cook up" an incompatible theory that also accounts for the data. For instance, any batch of natural phenomena could be explained as the handiwork of a divine creator—or two divine creators or three creators or 137,890,546 creators . . . and so forth.

However, scientists do not find that their challenge is to select from among an infinite number of theories that would account for their data. On the contrary, their biggest problem all too often is coming up with even *one* theory that will do the job! The reason is that it is very hard for a conceivable theory even to qualify as a *candidate* for truth in science. True, a new theory must promise empirical success. For instance, when Wegener first proposed continental drift, he showed that it offered cogent explanations for the "fit" of continents, the distribution of fossils, and geomagnetic data. But again, this is not enough for scientists even to begin to take a possible theory seriously. A candidate theory must be logical, coherent, and consistent with the accepted (or at least rationally *acceptable*) standards and values of scientific theorizing. A "cooked-up" theory will not even make it to the list of candidates. For instance, a theory cannot drag in a lot of arbitrary (what philosophers call *ad hoc*) supporting hypotheses with it, nor should it "explain" things in scientifically unacceptable ways—by appealing to magic, say. Candidate theories must also be consistent with the large body of accepted background knowledge that is presumed true by all qualified practitioners of a field. For instance, a new theory in paleontology that ignored continental drift would not be taken seriously, since paleontologists now presume that drift occurred. So strict are these requirements that at any given time in any field of science there will be a fairly small set of theories that are considered the *only* viable candidates for explaining some batch of phenomena. That is, scientists are confident that the (approximate) truth lies somewhere within the given range of candidate theories.

Once scientists have narrowed the field to a tractable set of candidate theories, the theories are tested against one another. If, in the end, one theory emerges above all the others, scientists then accept that theory as the right one—for the time being and until a better theory comes along. The reasoning here is not fallacious at all. Scientists assemble a small set of candidate theories, which they hold to be the *only* reasonable candidates, and then they eliminate all but one. The logical skeleton of this way of arguing is this:

T_1 or T_2 or T_3. (Suppose these are the only candidate theories.)
Not T_1.
Not T_2.
Therefore, T_3.

Which is a perfectly valid form of argument. So it is not just because a theory is empirically successful that scientists (tentatively) regard it as approximately true but because it is the one that is most successful compared to a field of the only other acceptable candidate theories.

Typically, therefore, theories in science are tested against other theories, and not evaluated on their own merits alone. Let's consider an example illustrating this process of theory selection from a small set of competing candidates. Recall the debates over dinosaur extinction examined in chapter 2. Until 1980, the most respectable theories of mass extinction postulated a series of gradually occurring processes that led eventually to the severe deterioration of the dinosaurs' environment. According to these theories, the late Cretaceous climate slowly became much drier and cooler as shallow seas receded, turning the dinosaurs' lush subtropical environments into savanna or semi-desert. Smarter, faster, warm-blooded, and far more prolific mammals preyed on dinosaur eggs and could thrive in environments hostile to dinosaurs. Freezing winters and smart mammals eventually proved too much for the cold-blooded, dim-witted dinosaurs. In the extinction debates of the 1980s and 1990s, two neocatastrophist theories rose to challenge the old uniformitarian paradigm: the impact and volcanic theories. For many scientists during this period it became clear that whatever had happened to the dinosaurs had occurred at least relatively quickly, so the old gradualist theories could not be right. Further, there was evidence that *some* sort of cataclysm had gripped the earth at the end of the Cretaceous period. For instance, at many sites an abundance of "shocked" quartz grains were found in the rocks marking the Cretaceous boundary (and not in adjacent rock). Only quartz crystals subjected to extreme percussive force exhibit the characteristic deformations that lead geologists to call them "shocked." Impact theorists naturally saw this as evidence for the impact of a massive extraterrestrial body. Other scientists argued that shocked quartz is produced by ultra-violent volcanic explosions, explosions even greater than the ones that destroyed Krakatoa in 1883. No other known natural forces can produce shocked quartz.

The upshot is that by about 1990, there were three, and only three, types of extinction theories vying for supremacy: the older gradualist view (seemingly on its way out), and the neocatastrophist impact and volcanic theories. One complication is that there were different specific theories within these broader types of theory. For instance, David Raup's Nemesis hypothesis was one of the impact theories. Another complication is that some scientists opted for theories combining elements of the different theory types. Thus, some paleontologists said that dinosaurs were stressed by a gradually deteriorating climate and that the final cataclysm, whether volcanic or extraterrestrial, delivered the coup de grâce. Nevertheless, by around 1990 there was a fairly small and reasonably well-defined set of theories accounting for the end-Cretaceous mass extinctions. Why just these theories? After all, innumerable other theories could account for the dinosaurs' extinction. Maybe the dinosaurs were hunted to extinction by extraterrestrial big game hunters. Perhaps dinosaurs were on the ark with Noah, but just could not adapt to the post-flood environment. Why didn't these scenarios get equal billing with the other theories? Well, these two are just too silly for serious scientific consideration. The fact is, though, that there are severe empirical constraints limiting the kinds of plausible extinction theories that can be offered. Mass extinction is hard to achieve. Only a limited number of known natural occurrences *could* do it and not sterilize the earth completely, and only a few of those are consistent with the facts admitted on all sides. As we said before, even coming up with a good *candidate* for an acceptable theory is a hard job.

So scientists tentatively accept a theory as approximately true only if (a) it is empirically successful in an *absolute* sense (it explains the facts it is supposed to, is supported by background knowledge, passes rigorous tests, etc.), and (b) it is *more* empirically successful than every other plausible rival they can think of. "Aha!" antirealist philosophers will exclaim, "What makes scientists so sure that they have *thought* of every possible acceptable rival to the winning theory? How do they know that the true theory is one they never even thought of?" The answer, of course, is that they don't know for sure that they have thought of every possibility (more on this in the next chapter). This is not a trivial problem. Nature is full of surprises, and the experts have been left with red faces on many occasions—even Einstein. When he formulated his general theory of relativity, Einstein was chagrined to find that it implied that the universe is expanding. Einstein was so sure that the universe is not expanding that he introduced a "cosmological constant," a fudge factor, into the theory to offset such expansion. Within a couple of decades, Edwin Hubble had discovered that the universe is indeed expanding. There is a well-known photograph of Einstein peering pensively through Hubble's telescope while Hubble smugly puffs his pipe in the background. You can almost see the thought balloon over Einstein's head reading, "Well, I'll be damned!"

Human mental powers, even the collective powers of the best and the brightest, are anything but foolproof.

Still, there are times when we can feel fairly confident that we have considered every reasonable possibility. How? The answer lies in the *cumulative* nature of scientific knowledge. Scientists take it for granted that we already have some knowledge about how the universe works. We have some acquaintance with natural entities, forces, and processes, and a pretty good idea of the kinds of effects they have. Scientists think that we have grasped some of the laws of nature. Getting back to the example of dinosaur extinction, I think that scientific thinking on this topic goes something like this: Something killed off the dinosaurs. Despite the fondest wishes of many a seven-year-old, they are no more. Was the extinction gradual or sudden? If sudden, as the evidence now indicates, was it caused by something on the earth or something extraterrestrial? What earthly forces could possibly produce a cataclysm sufficiently intense to account for the end-Cretaceous extinctions? Massive volcanism seems to be the only likely candidate; we know enough about volcanoes and their effects to realize the enormous environmental change they can cause. No other earthly force we know of could do the job and not clash with the known facts. We also know that enormous volcanic eruptions did occur at the end of the Cretaceous. If the extinction was caused by something extraterrestrial, what could it be? By far the most likely candidate, given what we know, is a comet or an asteroid striking the earth. We know the laws of physics that tell us the force such a strike would generate, and we can calculate the damage it would cause. Further, other than such an impact, no other cataclysmic cosmic event that we can think of, like a nearby supernova, could possibly cause mass extinctions and be consistent with the evidence. Therefore, the end-Cretaceous extinctions were caused either by gradual deterioration of the environment, extremely violent volcanism, or the catastrophic impact of an object from space (or some combination of these factors).

Of course, we *could* be completely off base and the real reason for the mass extinctions could be something we never thought of. This is why scientific knowledge is always tentative and scientific claims always deserve some (sometimes very small) degree of skepticism. However, earth scientists would be much more skeptical of the suggestion that there might have been an unknown natural force sufficiently powerful to cause mass extinction but that acted in a way such that its effects are indistinguishable from the effects of known causes (Just what did produce those shocked quartz grains?). At the very least, they will demand a plausible suggestion about what that unknown extinction agency might be before they take such a "possibility" seriously. In the meantime, they will feel justified in thinking that the truth lies somewhere in the neighborhood of our current theories.

COULD WE BE WRONG ABOUT EVERYTHING?

Hard-bitten antirealist philosophers are, of course, unimpressed by scientists' confidence in what they think they know. They doubt that scientists have grounds for claiming the approximate truth of the background theories that scientists invoke to narrow down their current candidate theories. For philosophers, unlike scientists, absolute skepticism is a genuine fear. Ever since Descartes, at least, philosophy has been haunted by the idea that, in the words of the title of the classic Firesign Theater album, *Everything You Know Is Wrong!* (Let me hasten to add that, since if we *know* something it must be true, it is incorrect ever to say that what we *know* is wrong—only what we *think* we know, but Firesign Theater's way of saying it is a lot funnier.) Maybe the whole edifice of science is wrong and conceivably an entirely different set of theories could be produced that would be as empirically successful as our current science. Perhaps we could encounter a race of extraterrestrials whose science was even more successful than ours, yet their theories would be incompatibly different from ours.

Still, let us pause to consider what a breathtaking claim it is to suggest that not just a particular theory but perhaps the *whole* fabric of scientific theory could be wrong. What makes this a breathtaking claim is not just the enormous empirical success of individual theories but also the fact that diverse theories fit together so well in a mutually supporting network to give us a highly coherent picture of the world. In every field of intellectual inquiry, even the most mundane, *coherence* is taken as an indicator of the probable truth of a set of beliefs. A highly coherent account will be one where the different pieces of the story all fit together so that each bit reinforces the other to create a compelling overall picture. Readers of mystery stories are familiar with how this works. At first the inspector entertains a number of different hypotheses about whom the murderer could be. Eventually, though, all the pieces of evidence and conjecture come together to constitute a compelling case against one suspect. Similarly in science, different theories often reinforce each other so that each becomes far more credible than it would have been alone. We shall see below that this was the case with Darwinism and Mendelian genetics, which initially seemed opposed but were later shown to be mutually supporting.

W. V. O. Quine, one of the leading American philosophers of the twentieth century, liked to say that hypotheses face the tribunal of experience not alone but in batches. What he meant was that beliefs, whether in science or in ordinary life, form an intricately interconnected network so that evidence that supports or undermines one belief also impinges on many other connected beliefs. Thus, when evidence leads us to give up one theory, we also have to rethink other theories and make modifications or adjustments where necessary. For instance, the Cepheid variable stars, which, as we noted earlier in this chapter, Henrietta Swan Leavitt found to be "standard candles," are not bright enough to help astronomers estimate truly vast cosmological distances. Astronomers think that type ia supernovas are also standard candles, that is, they all have about the same brightness. Type ia supernovas are so bright that, for a time, they outshine all the other stars in their galaxy. That's why they can be seen across cosmological distances. Based on the assumption that type ia supernovas are standard candles, astronomers have made estimates of the size and age of the universe. Also, because type ia supernovas in distant galaxies appear dimmer than expected, astrophysicists have just in the past few years postulated the existence of "dark energy," a mysterious force that exerts a repulsive effect contrary to the pull of gravity. Clearly, then, if we came to question our theories about type 1a supernovas, we would also have to question our theories about the age and size of the cosmos since these estimates depend in part on our understanding of these supernovas. Likewise, much of the discussion about "dark energy" would have to be reexamined.

Our theories therefore do form an intricately interconnected web. Evidence that challenges one theory also challenges connected theories. It works the other way, too. Evidence that supports one theory will indirectly support others. For instance, further evidence that our theories about type 1a supernovas are correct will strengthen the other theories that depend on our beliefs about such supernovas. The point is that the evidence for theories is *holistic*. The support for a theory comes not just from the empirical success of that one theory but also from the whole network of theories and evidence to which it connects.

Philosopher Susan Haack suggests another helpful metaphor for understanding how scientific theories can mutually support one another. She compares the quest for scientific knowledge to solving a crossword puzzle. Nature provides the clues, and theories are the answers we write in. But our crossword-puzzle answers are constrained not only by the clues but also by what we have already written in. For instance, suppose that clue 7 across of the daily crossword is "literary land of little people" and the answer is eight letters. This could either be "the Shire," home of Tolkien's hobbits, or it could be "Lilliput," land of Swift's Lilliputians. The clue, by itself, does not allow us to decide between these two; the clue underdetermines the answer. Suppose, though, that an answer we have already written in gives us an *r* for letter seven. The answer now seems to be "the Shire." Of course, the answer could still be "Lilliput," and the previous answer that gave us the *r* at letter seven is wrong. We won't be sure until we have filled in more answers and have seen that they are mutually supporting.

What the crossword analogy shows is that our confidence in a particular answer (theory) depends not just on how well it fits the clue (the data) but also on how well it fits in with the answers already written in (background theory). In fact, often we become really confident in an answer only after we have filled in a lot more answers and see that they all fit together. In other words, it is not only the fact that our theories are true to the data that justifies our confidence in them but also that they cohere so well in a mutually supporting relationship with other theories that we also accept. Maybe an adamant skeptic could argue that it is possible that an alternative science could have theories just as empirically successful and just as mutually supportive. However, to take the suggestion of a complete, alternative science seriously, we have to be given more than the bare suggestion that it is possible. The skeptic would at least have to start providing us with the nuts and bolts of such an alternative science. Needless to say, no skeptic has yet achieved this. In other words, given the enormous empirical success of our best theories, *AND* that they cohere in a mutually supporting network, surely the burden of proof is on the skeptic.

Looking at the history of science, we do find a net increase in the empirical success of our theories and unquestionably also an increase in the degree to which they constitute a unified, integrated, and mutually supporting network. Many of the most successful theories enjoy their status because they integrated fruitfully with other highly successful theories. Consider the historical relationship between Darwinism and Mendelian genetics. At first, there was no relationship. Darwin was already famous when the *Origin* was published in 1859; it made him a scientific superstar. Gregor Mendel was an unknown monk who published his results in an obscure journal. Darwin apparently never heard of Mendel's results and proceeded to develop his own, completely wrong, theory of genetics. Mendel's work remained unknown until 1900 when three geneticists independently rediscovered it. It quickly rose to prominence, and, because it seemed inconsistent with Darwinian natural selection, soon sent Darwinism into a decline. It is not that scientists stopped believing in evolution *per se*, but many no longer accepted the Darwinian explanation of evolution in terms of the gradual accumulation of small changes as postulated by natural selection. In fact, historians often refer to the "eclipse of Darwinism" in the early decades of the twentieth century. Then, in the 1930s and 1940s, brilliant work by mathematicians and scientists such as R. A. Fisher, Sewell Wright, Theodosius Dobzhansky, Ernst Mayr, and G. G. Simpson established the neo-Darwinian synthesis. The synthetic theory showed that instead of being opposed, Darwinism and Mendelian theory needed and mutually supported each other. In fact, researchers found that the way to understand the genetics of populations was to incorporate models of Darwinian selection. This synthesis was a triumph for both theories. United into an integrated synthetic theory, both Darwinian evolutionary theory and Mendelian genetics were made stronger than ever. In fact, evolutionary theory connects at such a fundamental level with so much in biology that Dobzhansky famously said, "nothing in biology makes sense except in the light of evolution."

The history of science contains many incidents where seemingly disparate theories, even theories in very different fields of science, are found to be mutually supportive. Perhaps, then, we can view such episodes as supporting, at the very least, a minimal form of realism, what philosopher Robert Almeder calls "blind realism" (1991). That is, even if we are not confident that any *particular* theory is an approximation of reality, it seems simply inconceivable that the *whole* edifice of scientific theory could be completely wrong. Likewise, if we have already filled in a good bit of a crossword, it could well be that some of our answers are wrong, but it seems highly unlikely that *every* answer is wrong! Therefore, given the history of greatly increasing empirical success and the vast increase in the cohesiveness and mutual supportiveness of our theories, it just does not seem feasible that we could be completely on the wrong track. If anyone insists that, nevertheless, we might be completely wrong, we can rightly demand that they give us a reasonably detailed picture of what a total alternative science looks like.

At this point many antirealists would complain that the above argument is directed at a straw man (I do not think so. I think that global antirealism is quite common, even faddish, these days.). They would object that they do not assert that the *whole* body of scientific theory could be wrong, merely that particular kinds of theoretical truth-claims cannot be known to be true. Perhaps they would be perfectly willing to admit that theories in evolutionary biology, paleontology, geology, and other fields that postulate measurable, detectable events and processes are perfectly fine with them. Their problem is with theories—like those of fundamental physics—that postulate entities that are *in principle* impossible to directly see or measure. Things and occurrences that are merely unobservable *in practice*, like an asteroid strike sixty-five million years ago, give them no problems (such an event would have been eminently observable to any sentient creature unlucky enough to have been around when it occurred). Their objection is that science so often postulates entities, like electrons, that cannot possibly be directly observed. In cases like this, they say, we go farther than we should in asserting that such things really exist. In the next chapter we examine the claims of this more modest antirealism.

FURTHER READINGS FOR CHAPTER FOUR

Herbert Butterfield coined the term "Whig history" in *The Whig Interpretation of History* (New York: Charles Scribner's Sons, 1951). In another work he gives a very succinct statement of the problems with such historical writing:

> It is not sufficient to read Galileo with the eyes of the twentieth century or to interpret him in modern terms—we can only understand his work if we know something of the system which he was attacking, and we must know something of that system apart from the things which were said about it by its enemies. In any case, it is necessary not merely to describe and expound discoveries, but to probe more deeply into historical processes and learn something of the interconnectedness of events, as well as to exert all our endeavours for the understanding of men who were not like-minded with ourselves. Little progress can be made if we think of the older studies as merely a case of bad science or if we imagine that only the achieve-

> ments of the scientist in very recent times are worthy of serious attention at the present day.

This quote is from pages 9–10 of Butterfield's *The Origins of Modern Science 1300–1800*, revised edition (New York: Free Press, 1957). A good discussion of the problems with Whig history is in chapter 9 of *An Introduction to the Historiography of Science* by Helge Kragh (Cambridge: Cambridge University Press, 1987).

As I note in "Further Readings for Chapter Two," the work that set the records straight on Owen was Adrian Desmond's *Archetypes and Ancestors: Palaeontology in Victorian London 1850–1875* (Chicago: University of Chicago Press, 1982). Desmond shows that Huxley was the vanguard of a movement to completely secularize science and rule out the appeal to metaphysical principles like archetypes. Owen, on the other hand, represented the more conservative elements of Victorian society that were shocked by the materialism and agnosticism of the young scientific radicals.

Steven Shapin and Simon Schaffer's *Leviathan and the Air-Pump* (Princeton: Princeton University Press, 1985) retrieves a rather-obscure controversy from the history of the science and turns a seemingly dry subject into a fascinating narrative. The authors certainly succeed in showing, as did Desmond in the work mentioned above, that seemingly arcane scientific debates are about a lot more than they seem. The fact that scientific issues (like climate change in our day) can become footballs for conflicting political doctrines does create a severe challenge for the autonomy and objectivity of science. For a dissenting view on the Hobbes/Boyle controversy, see the third chapter of Paul R. Gross and Normal Levitt's *Higher Superstition: The Academic Left and Its Quarrels with Science* (Baltimore: Johns Hopkins University Press, 1994). See also the eleventh chapter of Christopher Norris's *Against Relativism: Philosophy of Science, Deconstruction, and Critical Theory* (Oxford: Blackwell Publishers, 1997).

Paul Feyerabend long relished his role as "the worst enemy of science." His book *Against Method* (London: NLB, 1975), was a strident attack on "scientific method." Feyerabend is best characterized as an extreme libertarian; he insisted that no perspective or ideology should enjoy any sort of hegemony. Science should therefore be demoted from its dominant position in Western society and should have no higher status (or more funding) than any way of looking at the world. Astronomy, for instance, should enjoy no higher prestige or support than astrology. The idea that science is a more "objective" or "rational" way of interpreting reality than, say, voodoo, was loudly and contemptuously derided by Feyerabend. A sympathetic but critical study of Feyerabend is John Preston's *Feyerabend* (Cambridge: Polity Press, 1997).

Larry Laudan's insightful remarks about scientific progress are found in his essay "The Philosophy of Science and the History of Science" in *A Companion to the History of Modern Science*, edited by R. C. Olby, G. N. Cantor, J. R. R. Christie, and M. J. S. Hodge (London: Routledge, 1990), pp. 47–59. Laudan's case against convergent realism is given in his essay "A Confutation of Convergent Realism" in *Scientific Realism*, edited by Jarrett Leplin (Berkeley: University of California Press, 1984), pp. 218–49, and in his book *Science and Values: The Aims of Science and Their Role in Scientific Debate* (Berkeley: University of California Press, 1984). Laudan's arguments against scientific realism are clear and challenging. It is an especially effective critique because it is grounded in deep knowledge of the history of science. A powerful riposte to Laudan is given by Stathis Psillos in his book *Scientific Realism: How Science Tracks Truth* (London: Routledge, 1999). Psillos's book is perhaps the most comprehensive and effective defense of scientific realism currently available. Jarret Leplin's *A Novel Defense of Scientific Realism* (Oxford: Oxford University Press, 1997), is also a vigorous and carefully argued statement of scientific realism and critique of antirealist arguments.

Among the writers of science for the general public, some, like Carl Sagan, could compose beautiful, resonant prose that conveyed the wonder and majesty of the cosmos. Others, like Stephen Jay Gould, produced writings that were masterpieces of the essayist's art. But nobody could match Isaac Asimov in sheer expository skill. Asimov had the very rare, perhaps unique, ability to take the most difficult ideas of science and present them so clearly and engagingly that any interested person could read about them with pleasure and profit. Asimov seldom made comments of philosophical significance, but his delightful essay "The Relativity of Wrong" is an exception. It is printed in one of the many collections of Asimov's essays, *The Relativity of Wrong* (New York: Kensington Books, 1988). Asimov makes the simple point—one that seems to have eluded many philosophers—that while there may be only one absolutely correct way to answer a question, there are infinitely many ways to answer it falsely, and some false answers reveal far more insight about the topic of the question than others. Maybe philosophers would do better to spend less time trying to develop theories of approximate truth and more time exploring the cognitive value of insightful falsehoods.

An authoritative and highly readable history of atomism is Bernard Pullman's *The Atom in the History of Human Thought*, translated by Axel Reisinger (Oxford: Oxford University Press, 1997). Pullman traces the ups and downs of atomic theory through the centuries and shows how it finally triumphed only at the beginning of the twentieth century.

W. V. O. Quine expresses his insights into the interconnectedness of all our assertions, from the most theoretical to the most mundane, in many places in his writings. A very clear and succinct statement is in the introduction to his *Methods of Logic*, fourth edition (Cambridge: Harvard University Press, 1982). For an elementary exposition of his view of rational belief, see *The Web of Belief*, second edition, coauthored with J. S. Ullian (New York: Random House, 1978). The "web of belief" is Quine's main metaphor for the complex connections between our beliefs and the way, like an extremely intricate spider web, that different strands support, and are supported by, many other strands. Susan Haack make a similar point with her crossword-puzzle analogy. She develops the analogy and offers many other insights in a book I recommend to the reader with no reservations, *Defending Science—Within Reason: Between Scientism and Cynicism* (Amherst, NY: Prometheus Books, 2003). In several of her works, Haack defines herself as a "passionate moderate" who rejects both the immoderate claims of scientism and the cynicism of the radical science critics.

A work I failed to mention earlier that gives an excellent overview of the history of evolution, including evolutionary synthesis, is Peter J. Bowler's *Evolution: The History of an Idea*, third edition (Berkeley: University of California Press, 2003). Bowler is one of the most prolific and knowledgeable historians of Darwinism and evolutionary ideas. An excellent history of Gregor Mendel's discoveries is Robert Olby's *Origins of Mendelism* (Chicago: University of Chicago Press, 1985).

Robert Almeder's theory of "blind realism" is presented in his *Blind Realism: An Essay on Human Knowledge and Natural Science* (Lanham, MD: Rowman and Littlefield, 1991). Almeder's arguments are sharp and crystal clear. His views on realism have not received the attention they deserve.

The full citation for W. H. Newton-Smith's *Rationality and Science* is given in the "Further Readings" section of chapter 2.

CHAPTER FIVE

TRUTH OR CONSEQUENCES?

Seeing is believing, we are told. About the best evidence we can have for anything is to see it. What would clinch the case for UFOs? If one landed on the Washington Mall, as in the science-fiction classic *The Day the Earth Stood Still*, that would clinch it. What would prove that the Loch Ness monster is real? If the body washed up tomorrow. What would fill churches, synagogues, and mosques with former unbelievers? Atheist philosopher Norwood Russell Hanson imagines the scenario that would do it for him:

> Suppose . . . that on next Tuesday morning just after our breakfast all of us in this one world are knocked to our knees by a percussive and ear-shattering thunderclap. Snow swirls; leaves drop from the trees; the earth heaves and buckles; buildings topple and towers tumble; the sky is ablaze with an eerie, silvery light. Just then, as all the peoples of this world look up, the heavens open—the clouds pull apart—revealing an unbelievably immense and radiant Zeus-like figure, towering up above us like a hundred Everests. He frowns darkly as lightning plays across his Michaelangeloid face. He then points down—at me!—and exclaims for every man, woman, and child to hear: "I have had quite enough of your too-clever logic-chopping and word-watching in matters of theology. Be assured, N. R. Hanson, that I do most certainly exist." (1971, 309)

Yes, it would be hard to remain a skeptic in circumstances like these, but what if the evidence for God's existence is not this good? When is it reasonable to believe in God even without such spectacular Steven Spielberg–type revelations? In general, when should we believe in something we cannot see? Well, maybe if we could hear, taste, smell, or feel it. But what if something cannot be detected by any of the senses? As a matter of fact, people believe in all sorts of things that cannot be detected by any of the senses. Many people believe that gods, souls, evil spirits, luck, fate, numbers, *mana*, *ch'i*, and so on, exist even though they are not objects that can register on the senses. It is very important to ask how we can have confidence that things exist even if we cannot sense them.

A very common misconception about science is that it deals only with the observable. A common antievolutionary charge is that since we cannot see creatures evolve from one major group to another (reptiles to birds, say), it is not scientific to claim that they did. On the contrary, scientists constantly talk about things that are too big, too small, too fast, too slow, too inaccessible, or too distant in space or time to see. If we had time machines, we could do paleontology by traveling back to the past to directly observe prehistoric life. But since we have no time machines, we cannot observe past life directly and have to draw our inferences from the traces we have now. Also, many fields of science talk about things that cannot be directly sensed—extrasolar planets, magnetic fields, electrons, and so on—but are known only through their effects on various sorts of detection devices (spectrographs, electron microscopes, Geiger counters, seismometers, etc.).

Here we need to make an important distinction. There are some things we cannot see, but this is only a *practical* problem. We cannot travel to the earth's inner core, so our knowledge of the core is inferential, not based on direct observation. However, if we *could* devise a way to burrow through thousands of miles of crust, mantle, and outer core, we could directly sample the earth's inner core. The problem is that no conceivable technology would ever permit us to do that. So, the earth's inner core is unobservable—but only *in practice* not *in principle*.

Contrast this with the case of the electron. Even if (to cite another 1950s science-fiction classic) like *The Incredible Shrinking Man* you could be reduced to atom size, you still would not be able to see electrons. An electron is far smaller than the wavelength of light, so electrons are *in principle* invisible (The in practice/in principle distinction is not really very clear, but let's go with it for now). Yet, for over a hundred years, physicists have accepted the existence of electrons. To see why they did, and still do, regard electrons as genuine entities, we need to review the history of their discovery.

ELECTRONS: REAL PARTICLES OR CONVENIENT FICTIONS?

Electrical phenomena have long fascinated people. Since ancient times it has been known that rubbing certain substances like amber (the word for *amber* in Greek is *elektron*) would give them the strange power to attract small bits of paper and other materials. It was also known that this strange attractive power could be passed by direct contact from one body to another. In the eighteenth century, scientists demonstrated that the electrical effect could be transmitted long distances if one used the proper conductive medium. So, it looked like *something* is transferred from electrically charged bodies to uncharged ones. Scientists postulated a subtle electrical fluid (others favored two fluids) that could be agitated by friction and could flow along wires and from one body to another. However, two of the leading researchers of electricity in the nineteenth century, Michael Faraday and James Clerk Maxwell, disagreed with the subtle fluid theory and maintained that electrical charge is a "field" phenomenon, a tension in the medium surrounding electrified bodies.

Another hypothesis favored by some early researchers was that electricity was the effect of the interaction of electrical corpuscles—tiny, irreducible particles bearing the smallest units of electrical charge. Some physicists regarded such electrical particles as fundamental constituents of matter. With the rise of Maxwell's field theory, these hypotheses were largely abandoned, but they soon came back in different guise. In 1891, Irish physicist George Johnstone Stoney introduced the term *electron* as a measure of unit electrical charge. Also in the early 1890s, the noted Dutch physicist Hendrick A. Lorentz attempted to reconcile the electrical particle hypothesis with the Maxwellian field theory by proposing that electrons were structures in the continuous medium postulated by Maxwell. Clearly, this was a time of much ferment and little consensus among physicists in their thinking about electricity. In his classic of irreverence *The Devil's Dictionary*, Ambrose Bierce, the leading American satirist of the late nineteenth century, mocked physicists' apparent indecisiveness about electricity. His tongue-in-cheek definition of *electricity* was, "The power that causes natural phenomena not known to be caused by something else." In other words, in Bierce's view, physicists were clueless about the nature of electricity. However, just when it looks to outsiders like scientists are grasping at theoretical straws, in fact they are often on the cusp of a major advance.

Progress often occurs in science when there is a fruitful interaction of theory and experiment. While the theorists were busy proposing and debating hypotheses, a particular problem had grasped the attention of experimentalists: What are cathode rays? If you take a glass tube and pump enough air out to make a reasonably good vacuum, attach a negative electrode (the cathode) and a positive electrode (the anode) to opposite ends of the tube, and send a strong electrical current into the tube, an electrical discharge will flow from the cathode, through the tube, to the anode. So *something* flows from the cathode to the anode, but what is it? Is it a "ray"—an electromagnetic wave, or a stream of tiny charged particles?

In 1897, English physicist J. J. Thomson, head of the famous Cavendish Laboratory at Cambridge University, performed an experiment that strongly supported the theory that cathode "rays" are particles. Thomson found that an electrical field could deflect the "rays," something that would be expected of charged particles but not electromagnetic waves. Further, he was able to determine the ratio of mass to charge the hypothesized particles would have to possess. In an earlier experiment he had shown that the velocity of the "rays" was far less than the speed of light, which also undermined the electromagnetic-wave theory. Putting all of the pieces of evidence together, Thomson declared that the cathode "rays" are in fact minute particles bearing a negative electrical charge. The name *electron* was eventually given to the postulated particle, and Thomson is usually given the main credit for its discovery.

It is remarkable that at about the same time Thomson was doing his experiment, German physicist Walter Kaufmann was doing the same kind of experiment in Berlin, only, according to some historians, he did it *better*. Yet Kaufmann is not credited with the discovery of the electron because he did not claim to have discovered a new particle. Kaufmann was not less familiar with the physics than was Thomson; his reluctance to proclaim the existence of a new particle was due to a *philosophical* conviction. The philosophy called positivism was very influential among German scientists at the time, largely due to the influence of scientist/philosopher Ernst Mach (1838–1916). The basic principle of positivism is that scientists have the job of systematically describing only what can be measured and should not speculate on the existence of things that cannot be directly detected. For Mach, a scientific theory does not function to reveal to us the hidden nature of reality; it is an *instrument* to enhance our powers of prediction and control over what we can see. Mach therefore had what is called an "instrumentalist" view of theories. Of course, Mach realized that scientists sometimes find it useful to speak of invisible things like atoms or fields, but he held that such terms are used only as figures of speech and are not meant literally. Concepts like "atom" and "field" are nothing more than convenient fictions, says Mach; they are placeholders for batches of observations or measurements that we could perform but are too numerous to mention individually. For instance, when we say that a magnetic field exists in a certain area of space, this is just a shorthand way of saying that, for instance, iron filings scattered in that space would arrange themselves in a certain way, or that a compass needle would be deflected in a certain direction, and so forth.

The appeal of positivism is obvious. To many working scientists of the day it seemed that the chief advantage science possesses over metaphysics is that science sticks to observable facts and refuses to engage in airy speculation about inscrutable entities. Yet positivism arguably impeded the progress of science, and was abandoned, even by some of its most ardent proponents, when the evidence for the existence of unobservable entities became overwhelming. In his 1913 book *Les atomes*, physicist Jean Perrin gave the strongest evidence yet obtained for the reality of the world of atoms and molecules. Every beginning chemistry student is introduced to Avogadro's number (named after Italian chemist Amedeo Avogadro). This is the number of atoms in the gram molecular weight of any substance. To get the molecular weight of any substance, first add up the atomic weights of the atoms that make up the substance (for instance, a molecule of water has a molecular weight of about eighteen since each molecule of water is two atoms of hydrogen and one atom of oxygen, and each atom of hydrogen has an atomic weight of about 1 while the oxygen atom has a weight of around 16). Now take the number of grams of the substance equal to its molecular weight—eighteen grams in the case of water. Eighteen grams of water will contain Avogadro's number of water molecules. Avogadro's number is very large—6.022045 × 1023—and this gives you some idea of how tiny water molecules are since eighteen grams is not much water. Perrin cited thirteen independent ways of determining Avogadro's number, and they all gave very similar values. When thirteen independent methods arrive at pretty much the same answer, it is very hard for even the most skeptical scientist to doubt that there is something real underlying these results. In other words, when we use thirteen completely different procedures to count the number of atoms in a given amount of stuff, and we get close to the same number each time, it is hard to believe that we are not counting *something*.

It surely is much harder to doubt the existence of molecular reality today than it was when Perrin wrote *Les atomes* in 1913. After all, we now have the rapidly growing field of nanotechnology, where we *build* and *use* devices on a molecular scale. We now make micromachines, like gears smaller in diameter than a human hair. Scanning electron microscopes can apparently see *and manipulate* individual atoms and molecules (more on this later in the chapter). Such devices can also be used for ultrafine etching. You could engrave the entire *Encyclopædia Britannica* on the head of a pin. Scientists and engineers engaged in such work would likely feel little patience for a skeptic who still doubted the existence of atoms and molecules. Their response to such skepticism would likely be to point to the images on the screens of their electron microscopes and exclaim with exasperation, "But, dammit, there they are!" But skeptics do still exist, and since they are among the most prominent figures in contemporary philosophy of science, we need to look at their arguments.

VAN FRAASSEN'S CONSTRUCTIVE EMPIRICISM

In 1980, noted philosopher of science Bas van Fraassen published his book *The Scientific Image*. Although this book is not as famous among nonphilosophers as Kuhn's *The Structure of Scientific Revolutions*, it has had a strong impact on the philosophy of science. Van Fraassen defends a form of antirealism he calls "Constructive Empiricism" (or "CE" for short). CE is a more moderate form of antirealism than the instrumentalist view of Mach. As we saw, Mach held that scientific theories are not even *about* the deep, hidden structure of the universe; all they are about is the prediction and control of observable things. Van Fraassen, on the other hand, takes theories literally. In his view, theories about atoms *really do* postulate the existence of such invisible entities and do not merely use the term *atom* as a figure of speech. The same holds for all theories that propose the existence of "theoretical entities," things that are unobservable but that are supposed to explain the phenomena we do observe. Therefore, for van Fraassen, theories that postulate theoretical entities are intended to be literal descriptions of the hidden structure of the universe and what they assert is either true or false. Van Fraassen opposes "scientific realism," which he understands as follows: "[According to scientific realism,] science aims to give us, in its theories, a literally true story of what the world is like; and acceptance of a scientific theory involves the belief that it is true" (1980, 8).

A number of scientific realists have objected that van Fraassen has given a misleading and simplistic characterization of their position. In particular they object that he has construed their view as a thesis about the *truth* of *theories* when scientific realism is really an assertion about the *objective existence* of the kinds of entities postulated by our physical theories (Devitt, 1991). However, for the sake of convenience, and to avoid muddying the waters with too many caveats, qualifications, and distinctions, let's go with van Fraassen's definition for now. A more adequate characterization of scientific realism will be developed at the close of this chapter.

Van Fraassen proposes an alternative to the realist view of the aim of science. He holds that science would be just as worthwhile and valuable an activity if we took the more-modest view that the aim of theory is merely to achieve empirical adequacy. An empirically adequate theory is one that makes predictions consistent with all past, present, and future observations. Van Fraassen certainly thinks that we should *accept* the best-supported current theories, but he holds that accepting a theory is not the same thing as believing it. For instance, you can accept the reigning Standard Model of particle physics without believing that the particles it postulates (quarks, electrons, etc.) are real. You can accept the Standard Model while remaining agnostic about the actual existence of the particles it postulates. Van Fraassen recommends that, with respect to all well-confirmed theories that postulate unobservable entities, we should accept those theories without believing them. He says that by accepting (but not believing) such a theory you commit yourself *only* to the belief that its observable predictions have been and will continue to be reliable. In other words, Van Fraassen is reviving the view that the goal of science is not to discover truth but merely to save the appearances.

So what *do* we want from scientific theories—that they be true depictions of reality or merely that their observational consequences be accurate? Let's begin by looking at van Fraassen's argument in some detail. First we need to get clear on just how strong his claim is. Is he saying that with respect to theories that postulate unobservable theoretical entities, we can never rationally believe that the theory is true? That is, is he asserting that no matter how well such a theory is confirmed, or how completely it meets our theory-choice standards, it is never reasonable to believe that the theory is true? His claim is milder than this, at least as he expresses it in some of his writings. He merely claims that realism is not rationally *compelling*, that is, that an antirealist attitude toward theories, and concomitant agnosticism about the existence of theoretical entities, is just as reasonable as the realist attitude. In other words, antirealism is no less rational a view of scientific theories than realism.

Actually, put this way, van Fraassen's antirealism is so mild a doctrine that it hardly seems to merit the vast amount of commentary and argument it has evoked. Realism need not be seen as a form of intellectual imperialism, though some of the radical science critics mentioned in chapter 3 might think that it is. Even the most hard-bitten realist need not insist that it is downright *irrational* not to believe in electrons, quarks, or whatever. After all, some of the leading physicists have entertained doubts about theoretical entities. Most famously, Niels Bohr, one of the leading physicists of the twentieth century and one of the founding figures of quantum mechanics, took the frankly antirealist view that the goal of physics was to correctly predict experimental results, not to penetrate to an invisible reality behind the appearances. He said, "In our description of nature, the purpose is not to disclose the real essence of phenomena, but only to track down, as far as possible, the relations between the manifold aspects of our experience" (quoted in Kragh, 1999, 210). Werner Heisenberg, another founder of quantum mechanics, agreed with Bohr. It would be bold indeed to say that physicists of the caliber of Bohr and Heisenberg held an irrational view about the nature and goal of physics. So *if* all that van Fraassen is claiming is that his antirealism is not an irrational position—while also admitting that realism is a fully reasonable view—there seems to be no reason to belabor the issue and the point can just be conceded.

Yet there are places where van Fraassen gives arguments that apparently support a somewhat-stronger claim. That is, some of his arguments seem to uphold the view that it is *less* reasonable to believe a well-confirmed theory (i.e., regard it as true) than merely to accept it as empirically adequate. For instance, while he admits that the unobservable entities scientists postulate *may* exist, van Fraassen contends that we can never *know* that they do. In other words, he is not claiming that there *is no* reality below the level that we can observe. Rather, he agrees with scientific realists that there is such a reality, he merely contends that we cannot be sure what that reality is like (In philosophical jargon, van Fraassen's claim is epistemological, not metaphysical.). No matter how much evidence we might have supporting the existence of, for example, electrons, it is always possible, since we cannot see the electrons, that there is something else there that is not an electron but just mimics the way we think electrons ought to act.

Also, van Fraassen seems to make a rather-strong claim about the kind of observational knowledge science can give us. He puts a great deal of stock in what we can observe with the unaided senses. He does not deny that our observations are "theory laden," that is, that the content of our observational knowledge is pervasively and fundamentally dependent on the general theories we accept. Even the most mundane observation, like "this is a glass of water," is not just a report of raw experience but involves the ideas "glass" and "water," which are general concepts that we use to *interpret* our experience. Nor does van Fraassen think that there is a difference in kind between statements that assert observational claims and those that assert theoretical claims (earlier philosophers of science made a strict distinction between "observation language" and "theory language"). However, he does think that, as a matter of fact, some objects are observable, while some hypothesized objects are in principle unobservable. That is, some of the objects postulated by theories are such that organisms of our biological sort could not observe them no matter where we were located or when we might exist. Van Fraassen counts as "observable" things (like Charon, the moon of Pluto) that are just too far away for us to see or events (like the impact that supposedly eradicated the dinosaurs) that happened long before there were any humans. Such objects and events still count as observable for van Fraassen because *could* we be transported to Pluto, or *had* we existed at the end of the Cretaceous, we could have observed those things. Only things that are observable, in this broad sense, are objects of empirical knowledge for van Fraassen. He is skeptical about the claims scientists make about things that are in principle unobservable, even if instruments like microscopes supposedly detect them. It is not that he denies the existence of things too small to see, only that he thinks that we cannot *know* the microworld as we know the world we can observe with the unaided senses.

I therefore take van Fraassen to be making two claims about scientific knowledge: (1) Our observational knowledge extends only so far as our unaided senses can, in principle, go. A considerable degree of skepticism is warranted with respect to what supposedly can be detected only by instruments. (2) When we accept a scientific theory that postulates unobservable entities, we should commit ourselves *only* to its empirical adequacy and not make the stronger and riskier claim that it is *true*. As we noted in the last chapter, false theories can make true predictions, so a false theory might prove as empirically adequate, even in the long run, as a true theory. When we accept a theory, therefore, it looks like a safer bet if we commit ourselves only to its empirical adequacy and suspend belief about its truth. Why stick your neck out any farther than you have to? In the next subsection I shall examine van Fraassen's claims about observational knowledge, and in the following subsection move on to examine his skepticism about theories that postulate unobservable entities.

DO WE OBSERVE THROUGH MICROSCOPES?

How much trust should we place in our unaided senses? Is there any reason to think that the operations of our sensory faculties are, in principle, more reliable than other ways of getting information about the world? We have to be very careful here. Just because we are more familiar with using our eyes than, for instance, an electron microscope does not mean that our eyes are less likely to convey false or misleading information. The fact that we use our eyes naturally and generally without any special training makes us feel that seeing is a simple and unproblematic process. But seeing is neither simple nor unproblematic. For one thing, the human visual apparatus, consisting of eyes, the optic nerve, and the portions of the brain that process visual input, constitute a staggeringly complex system (or, rather, a system of systems of systems of . . .). Compared to the human visual apparatus, a machine like an electron microscope is really a pretty simple device. In fact, it is fair to say that neuroscientists have only in very recent years begun to put together anything like an adequate account of how the brain processes visual input (this story is engagingly told in the late, great Francis Crick's *The Astonishing Hypothesis*). Much, of course, is still unknown. By contrast, the physical principles on which the electron microscope operates are as well-known as any in physics.

More to the point, a vast scientific literature exists detailing how sense perception is *not* reliable in many circumstances. Consider UFO sightings. Psychologist Terence Hines describes some of the visual illusions that have contributed to many UFO "sightings":

> A number of well-known visual illusions play a role in what witnesses report in UFO sightings, especially those that take place at night. One is the *autokinetic effect*. This effect refers to the fact that, if one views a small source of light in a dark room, the light will appear to move, even though it is stationary, and even though the observer's head is stationary. . . . Another illusion is that of *apparent motion*. Consider two positions in a dark room, A and B. A small light is turned on at A, then turned off. Moments later, a second light is turned on at B. . . . What is perceived is a single light appearing at point A, moving to point B, and then going off. (1988, 170–71)

Even more interesting are the ways that people can "see" what they expect to see—or what they want to see or what they think they ought to see—rather than what actually occurred. Numerous psychological experiments have demonstrated how personal expectations and social pressures bias perception; no doubt this is the source of many reports of "miraculous" and "paranormal" happenings. Some local news programs tout the reliability of their reporting by naming their broadcasts "Eyewitness News." Any competent trial lawyer can quickly show how fallible eyewitness testimony can be.

The point is not that we should never trust our senses (after all, what choice do we have?). If we could never trust our senses, we would have no basis for believing those sciences that tell us that our senses are not always to be trusted! The point is that observation is highly fallible, easily misled, and often subject to revision or correction. Indeed, observation is often far more reliable when we supplement our unaided senses with the use of instruments. *Prima facie*, then, there is no reason to think that the unaided senses are in general more trustworthy than machines, like an electron microscope, specifically designed to show things beyond the reach of human senses.

Why do scientists trust what they see through microscopes? How do we know that what microscopes show us is real and not just an artifact, a false image produced by the microscope itself rather than the actual features of the specimen we are supposedly observing? Well, for one thing, many of the things we can see only through microscopes look really real! If you collect a tiny bit of detritus from your bedroom carpet and put it under a good microscope at about 1,000 power, you will see—or, in deference to van Fraassen, you will *seem* to see—fearsome-looking creatures. Biologists, who believe that these creatures are real and not figments of microscopy, tell us that these creatures are dust mites, tiny arachnids (related to spiders, that is) that live in your carpets and eat the skin flakes you constantly shed. The microscope will apparently reveal a great deal of detail about this creature. It seems to have eight legs, serrated front claws, protective plates of body armor, and holes for breathing, eating, excretion, and copulation. Again, it looks very real. I would expect an artifact, a false image produced accidentally by the operation of the instrument, to look like a blur or a blob, not something that the computer animators at Pixar Animation Studios would have been proud to create. In fact, microscopes seem to reveal an entire microworld, one as rich in detail and as diverse in population as our everyday world. To take seriously the suggestion that all of this could be an artifact, we would have to be given a very convincing explanation of just how microscope makers could have created this world within the world.

Another reason to trust some of the images microscopes give is that some imaging processes can display a continuous range of magnifications, from just below what the unaided eye can see to very high powers. For instance, there is a well-known series of micrographs that begins with an image of the tip of a pin that is magnified slightly. You can easily see the smoothly tapering tip. Also, there are tiny yellow clusters on the tip, but you can't quite make out what they are. In the next image you can still see the very tip of the pin and the yellow clusters have begun to resolve into piles of tiny, individual shapes. Closer views show that these piles are composed of fuzzy, yellow, lozenge-shaped bacteria that find a home in the niches of the pin. What you have then is a series of pictures that connect at one end with what the unaided eye can see. Successive magnifications resolve and bring into focus features that were visible but less clearly seen in the previous images. In short, you have a series of pictures, continuous with ordinary vision, that display just what you would expect to see if you were getting closer and closer views of the subject.

Whether we actually see through a microscope is a question philosopher Ian Hacking raises in one of the most acute criticisms of van Fraassen's skepticism about the reliability of instruments designed to see what the human senses cannot detect. Hacking's essay "Do We See through a Microscope?" begins by reviewing the history of microscopy from Leeuwenhoek's first primitive device in the seventeenth century, little more than a toy, to today's sophisticated instruments. Hacking shows how refinements in lenses removed the distortions that made early microscopes practically worthless for serious research. In the late nineteenth century, Ernst Abbe, part-owner of the now world-famous Zeiss optical firm, succeeded in mating theory and practice to produce light microscopes of superior resolution. Finally, in the twentieth century, microscopes based on various different physical properties of light as well as different kinds of electron microscopes were developed. As microscopes advanced, they became of greater and greater use, and finally they were indispensable to the practice of many branches of science.

Hacking says that one reason to think that what microscopes show us is real is that the physical principles, the laws of optics, used in the construction of microscopes are very well understood. However, Hacking says that this is only a small part of the reason why we can trust the images of microscopes. A better reason is that we can physically interact with the structures that we see through the microscope. For instance, we can watch as we insert tiny needles into cells and see the liquid ooze out as we inject it. Thus, we do not just passively look through a microscope, we can use it to do work in the microworld, moving around and intervening in that world. We noted earlier that there are now electron microscopes so powerful that they apparently allow us to manipulate individual molecules. Surely, if we can push things around, build things with them, or inject stuff into them, they must be real.

Perhaps the most persuasive evidence of the reality of the denizens of the microworld is that different microscopes, designed according to work on entirely different physical principles, show us pretty much the same thing. In fact, we can manufacture grids, label each grid with a letter, and shrink the grids to microscopic size. When we look through a microscope, any kind of microscope, we can still see the grid and the little letters in them. Hacking therefore challenges the antirealist:

> Can we entertain the possibility that, all the same, this is some gigantic coincidence? Is it false that the disc is, in fine, in the shape of a labeled grid? Is it a gigantic conspiracy of 13 totally unrelated physical processes that the large scale grid was shrunk to some non-grid which when viewed using 12 different types of microscopes still looks like a grid? To be anti-realist about the grid you would have to invoke a malign Cartesian demon of the microscope. (1985, 147)

René Descartes imagined that there could be an all-powerful demon that so deceived us that we would go wrong no matter how obviously something seemed to be so. Hacking says that in order to remain skeptical about the reality of the things we seem to see through microscopes we would have to postulate such a demon.

Hacking's essay was written nearly three decades ago. Subsequent developments in nanoscience have greatly strengthened his case. Nanoscience studies phenomena at the "nano" scale; a nanometer is one billionth of a meter, far smaller than anything the unaided eye can see. Various devices have been developed to study reality at these scales. Two leading nanoscientists, Mark Ratner and Daniel Ratner, describe the use of "scanning probe" instruments, comparing their operation to the way that we find out about surface textures by sliding a finger over them:

> In scanning probe instruments, the probe, also called a tip, slides along a surface the same way your finger does. The probe is of nanoscale dimensions, often only a single atom in size where it scans the target. As the probe slides it can measure several different properties, each of which corresponds to a different scanning probe measurement. For example, in *atomic force microscopy* (AFM), electronics are used to measure the force exerted on the probe tip as it moves along the surface. This is exactly the measurement made by your sliding finger, reduced to the nanoscale. (2003, 39–40)

In principle, "feeling" at the nanoscale is analogous to feeling with the finger. In practice, of course, computers must be used to enhance the signal from the tiny probe, and this is a fairly complex process. But again, feeling with the human finger is a far more complex process and is not as well understood as the physical principles on which atomic force microscopy operates. From the perspective of our best-confirmed scientific theories, which is the perspective van Fraassen endorses, there seems to be no reason to think that atomic force microscopy is any less reliable in conveying information about the things it "feels" than our fingers feel when we detect the roughness of sandpaper or the smoothness of velvet.

Further, nanoscience has evolved into nanotechnology, that is, we now have the tools to build structures at the nano scale that have the properties we design into them, properties with effects detectable at the "macro" scale. The most obvious examples are computer microchips. As of February 2002, when the Ratners were writing, the smallest features commonly etched onto commercial microchips were about 130 nanometers across. The diameter of a human hair is 50,000 nanometers, and the smallest things the unaided eye can see are 10,000 nanometers wide. Given our ability to build things at the nanoscale that reliably operate the way they are designed to operate, it looks like we have a pretty good idea of what is going on down there. If we don't know what we're doing in making a microchip, then our efforts to design and manufacture them are just a shot in the dark, and it is amazing not just that we can make them but also that we are rapidly getting better and better at it!

Let me pause to emphasize that van Fraassen does *not* maintain the patently absurd view that there is no microworld or that all of our images of that world *must* be artifacts or fabrications. However, he does hold that our beliefs about things that, if they exist, are detectable only through instruments cannot be nearly as solidly grounded as our beliefs about things we can directly sense. He also opposes arguments like Hacking's that seem to be intended to show that skepticism about microscopic reality is simply perverse. In his reply to Hacking, van Fraassen focuses on the charge that so many different kinds of microscopes, operating on very different physical principles, would not show the same thing unless what they showed was real:

> Imagine I have several processes that produce very different visual images when set in motion under similar circumstances. I study them, note certain similarities; as I repeat this, I discard similarities that do not persist and also build machines to process the visual output in a way that emphasizes and brings out the noticed persistent similarities. Eventually the refined products of these processes are strikingly similar when initiated under similar circumstances. Now I point to the similarities and say that they are too striking to be there by coin-

> cidence, though, of course, the discarded dissimilarities were mere idiosyncrasies of the individual processes. What is the status of my assertion? What principle of reasoning could support it? Since I have carefully selected against nonpersistent similarities in what I allow to survive in the in the visual output processing, it is not at all surprising that I have persistent similarities to display to you. (1985, 297–98)

In other words, it could be the microscope makers, not the supposedly objective features of the examined specimen, that account for the agreement in what is seen. That is, a microscope is deemed a reliable instrument only when it has been adjusted to show images like those of other microscopes. But perhaps building microscopes is just a case of "the blind leading the blind," to use the biblical metaphor, and microscopes have been unwittingly designed to display the same false images.

However, this is a weak argument that does very little to undercut Hacking's point. Hacking's point is that it is very unlikely that different microscopes, designed to operate on very different physical principles, would show closely similar images unless those images were real. Van Fraassen's reply would be cogent only if microscope designers, wittingly or not, *constructed* the image to look like the images from another microscope. Merely *selecting* among various images will not do; you can only select from what is *there*. Neither can you *refine* or *emphasize* the similarities between images unless the similarities are *there*. Van Fraassen offers no argument for thinking that microscope makers construct factitious images rather than select and refine striking similarities that are already there.

It looks like a good cumulative case can be made that we do *observe* things with microscopes that are really there but are too small to be seen with the unaided eye: Many of the things we seem to see look very real, some with legs and other organs very much like the ones we see on large-scale creatures. We have no idea how the microscopes themselves could have created such detailed, lifelike images. Also, microscopic images offer a continuum of magnifications that connect at the low end with things that we can see with the unaided eye. Increasing magnifications show just what we would expect to see from closer-up views, with features indistinctly seen from farther away appearing clearer and more detailed from closer up. Further, the physical theories we use in the design of microscopes, like the laws of optics, are very well understood. Also, microscopes not only permit us to look at the microworld but also to move around in it and manipulate the things we seem to find there. The things we build at the nanoscale work—they have reliable effects in our big-scale world—so we seem to know what we're doing down there. Finally, there are now many different kinds of microscopes and other devices for detecting micro and nano reality. These instruments operate on the basis of very different physical principles, but the images they give us are often in agreement. Van Fraassen's attempt to explain away this fact is a failure.

Does the cumulative weight of these arguments suffice to *prove* that van Fraassen's skepticism is profoundly wrongheaded and that we most definitely do observe micro and nano reality as it is? No, skepticism can never be entirely banished, not even skepticism about ordinary, unaided vision. Philosophers can never eradicate skepticism; they can only hold it at arm's length. However, the upshot of these arguments seems to be that if there is genuine observational knowledge of the physical world, it is arbitrary to say that such knowledge extends only as far as our unaided senses can go.

BUT WHAT ABOUT THINGS THAT ARE *REALLY* UNOBSERVABLE?

Another reason for thinking that van Fraassen is wrong to give such priority to the unaided senses is that the distinction between the observable and the unobservable is not clear. Science writer David Bodanis is sensitive to the thought that something like the salmonella bacterium, which is supposed to exist beyond the range of direct observation, might not be real. He considers and rejects such skepticism:

> Because these creatures are so small it is tempting to think they are not really there, merely some sort of scientific construction. This is false. If you have good eyes you should be able to make out a good range of dust flecks caught in a beam of light in a darkened room. These can be as small as 20 microns (20/1000 millimeter) long. Salmonella are about a tenth of that, so if you had only slightly better vision you could imagine seeing those hairy wriggling submarine life forms all about you. They exist not in a distant, unreal realm, but just a little beyond the normally visible. (Bodanis, 1986, 85)

As Bodanis notes, it is easy to imagine a slight enhancement of our vision, perhaps a genetic mutation would give some people the power to see bacteria. For a person with such enhanced vision, seeing bacteria would soon become as ordinary as seeing cats. The difference between the observable and the unobservable therefore seems to be an accident of evolution, and it is hard to see how a strict epistemological distinction can be based on that.

Van Fraassen might be willing to admit that the boundaries between the observable and the unobservable are fuzzy and not fixed. However, he could insist that many scientific theories do clearly postulate entities that are not and will never be observable. For instance, physicists have postulated entities like quarks and superstrings that cannot and probably never will be observed in even the loosest sense of the term. Astrophysicists talk about mysterious dark matter, which, they say, constitutes the bulk of the universe's mass. This dark matter supposedly exerts gravitational influence but does not otherwise interact with "ordinary" matter. Astrophysicists also postulate the existence of black holes, putative objects that *by definition* cannot be directly observed. Surely, van Fraassen will contend, there is need for skepticism about such hypothetical entities. After all, the *only* evidence we could have for the existence of such indubitably unobservable things is that they constitute the *best explanation* for the things we can observe. Van Fraassen argues that inference to the best explanation is not a reliable means of establishing the existence of things we cannot observe.

Inference to the best explanation (IBE for short) is a form of reasoning that we use all the time. Perhaps the most famous instance of IBE in literature is in Daniel Defoe's *Robinson Crusoe*. In the story, Crusoe has been shipwrecked for many years on an island that he thinks is utterly deserted. One day while walking on the beach he sees the clear outline of a human footprint. Crusoe does not *see* anybody; he only sees the footprint. Crusoe immediately knows that he is no longer alone on the island. This realization comes so quickly that it is not a process of conscious reasoning, but it is an inference nonetheless. If we were to spell it out, the reasoning would probably go like this: Here is a very distinctive shape in the sand. The only thing that could explain the existence of such a shape is a human foot. Other hypotheses, like, for instance, that the shape was a random effect of wind and surf, are just too implausible to consider. The only reasonable hypothesis is that a human foot made this shape, and where there is a functioning human foot, there is a human being. Again, reasoning of this sort is used every day, and in those mundane contexts van Fraassen has no objection to it. For instance, he is happy to admit that if we hear certain suspicious patterings and scratchings behind the wainscoting, we could rightly conclude that we have mice. After all, a mouse is a readily observable creature, and we can learn firsthand the sorts of noises an active animal of that size is likely to make behind the walls. Likewise, van Fraassen would have no quarrel with those paleontologists who look at dinosaur footprints and infer things about dinosaur behavior. We *Homo sapiens* came on the scene about sixty-five million years too late to see a dinosaur, but *had* we been around in the Mesozoic, dinosaurs would have been most readily observable. So even a dinosaur counts as an observable in van Fraassen's sense.

The case is entirely different, says van Fraassen, when we try to use IBE to justify belief in unobservables. The problem, he says, is that for IBE to do the job here, to give us adequate grounds for thinking that a theory postulating unobservables is likely true, we would have to have a sort of privileged access. That is, we would have to have some *a priori* way of knowing that the *correct* explanation of whatever we are trying to explain is probably among the hypotheses we have so far considered. Van Fraassen puts it this way:

> Inference to the Best Explanation is not what it pretends to be, if it pretends to fulfill the ideal of induction. As such its purport is to be a rule to form warranted new beliefs on the basis of the evidence, the evidence alone, in a purely objective manner. . . . It cannot be *that* for it is a rule that only selects the best among the historically given hypotheses. We can watch no contest of the theories we have so painfully struggled to formulate, with those no one has proposed. (1989, 142–43)

In other words, IBE may be great for choosing among the hypotheses we have already thought up, but how do we know that we are not just selecting the best of a bad lot? That is, what authorizes us to think that one of the hypotheses we already have is the right one? Maybe all of our hypotheses are way off base and the true explanation is one that never has and maybe never will occur to us. This isn't a problem when we use IBE in most everyday situations. We have independent reasons for believing that (in most circumstances) a human foot *and nothing else* is likely to make the distinctive-looking footprint. But things seem different when we postulate an unobservable entity—let's call it *e*—as the cause of some puzzling fact *f*. In this case, because *e* is unobservable, and we have no prior or independent access to *e*, there seemingly are no independent grounds for saying that *e*, *and only* e, could explain *f*. Maybe some other unobservable entity, *e**, could be the correct explanation for *f*, and it simply never occurred to us that *e** might be the cause. Hence, IBE apparently requires privileged access when it is used to argue for the reality of unobservables. We apparently must have some sort of privileged, prior assurance that the true explanation probably lies among the ones we have so far considered. What could give us such assurance? Could we argue that God, or evolution, has designed the human mind so that it is likely to light upon true hypotheses?

Does IBE require privileged access when it is used as evidence for the reality of unobservables? Let's consider the case of black holes. Black holes, if they exist, are among the strangest things in the universe. The best way to begin to think about black holes is to consider what happens to stars as they age and die. Here's the account astrophysicists give: A star's fate is determined by its mass. A star with the mass of our sun will have a rather-quiet ending: After about ten billion years of existence, its hydrogen fuel starts to run out and the star swells to become a red giant. Eventually, the outer layers of the star will be shed into interstellar space, creating a beautiful planetary nebula and leaving behind a compact object known as a white dwarf. A white dwarf is a very odd object. It can pack the mass of the sun into a planet-sized sphere. This means that a white dwarf is extremely dense. A typical average density would be thirty-five thousand grams per cubic centimeter, thirty-five thousand times the density of water and three thousand times the density of the matter at the earth's core. Since a white dwarf no longer generates its own energy through thermonuclear reactions, it slowly cools over billions of years until it becomes a "black dwarf," a cold, giant crystal harder than diamond.

The fate of massive stars, those significantly larger than our sun, is much more spectacular. They end their lives in cataclysmic supernova explosions. When a supernova occurs, a star in its death throes can briefly outshine the total luminosity of the hundreds of billions of other stars in its galaxy. If the remnant left behind after a supernova explosion has a mass between about 1.4 and 3 solar masses, that residual star matter will become a neutron star. A neutron star is a very bizarre object with the inconceivable density of one hundred million metric tons per cubic centimeter. Neutron stars spin very rapidly, as fast as hundreds of times per second, thus creating magnetic fields a trillion times stronger than Earth's. These ultrastrong magnetic fields generate beams of microwaves that rotate with the star. When such a beam sweeps past an earthly observer, it is detected as a microwave pulse, so rapidly spinning neutron stars are called *pulsars*.

The most extreme fate befalls those stars so massive that the supernova remnant has a mass more than three times that of the sun. With such great masses, the gravitational collapse is complete, and, according to the theory of general relativity, everything collapses into a dimensionless point called a *singularity*. Within a singularity, according to general relativity, the density becomes infinite and space/time itself is infinitely distorted. Under such conditions the very laws of physics break down. The singularity is shielded from the rest of the universe by an "event horizon" (that nature always decently forbids "naked" singularities by hiding them behind an event horizon is called the "cosmic censorship hypothesis"). The event horizon marks an absolute boundary. Once anything—*anything*—passes inside the event horizon, it is trapped forever. No force in the universe can extricate it. Not even light can escape from inside the event horizon since the gravitational field is so intense that not even the speed of light, the fastest possible speed, achieves escape velocity. Essentially, the event horizon marks a hole in space, and since nothing, not even light, can escape the hole, it is absolutely black.

Because black holes cannot emit or reflect light or any other kind of signal, they cannot be directly observed. Even indirectly detecting a solitary black hole, perhaps as a spot of utter emptiness, would be very unlikely because they are quite small and very far away. Why, then, are many astrophysicists, who are almost always hardheaded empirical types, nevertheless convinced that black holes really exist? General relativity predicts that such things could exist, but why do so many scientists think that they really do exist? They appeal to IBE, that is, black holes seem to be the best explanation for some things that we do observe. Do astrophysicists, as van Fraassen charges, have to claim privileged access? Do they illicitly assume that the truth is probably to be found in the hypotheses they have so far considered? Let's look in some detail at their reasons for believing that there really are black holes (for the sake of brevity, I'll only consider the evidence for stellar-mass black holes and will not examine the very intriguing evidence for supermassive black holes in the centers of galaxies).

Cygnus the Swan is a magnificent constellation that is high overhead on midsummer nights in the Northern Hemisphere. At a point within the constellation, astronomers have identified an intense x-ray source they unimaginatively call Cygnus X-1. This source is about eight thousand light-years away and emits x-rays as powerful as the total luminosity of ten thousand suns. Telescopic observation reveals that

Cygnus X-1 is a supergiant star in a binary system with an unseen companion. Astrophysicists have inferred that the best explanation of such intense x-ray emission is the existence of a black hole as the unseen companion of the star. The gravitational acceleration into the black hole of material stripped from the companion star could create the energies needed. The intense gravitational field would pull material from the companion star that would accumulate in a rotating accretion disk around the black hole. Compression and friction within the disk would raise the temperature of the accreted material to one million degrees Centigrade, which would generate the observed x-rays.

The problem is that the energy necessary to create the x-rays could also be created if the unseen companion were a white dwarf or a neutron star. However, variations in the x-ray emissions on the order of a millisecond indicate that the unseen component of Cygnus X-1 must be less than 300 km in diameter. Light travels only 300 km in a millisecond, so any variation in x-ray emissions that takes only a millisecond to occur cannot come from a body larger than 300 km. Since a white dwarf is about earth-sized, this shows that the unseen component cannot be a white dwarf. Also, a law discovered by Johannes Kepler in the early seventeenth century allows astronomers to place lower limits on the masses of bodies in orbit about a common center of gravity. These calculations indicate that the compact object in the binary system has a mass of at least 3.4 solar masses, well above the upper limit of 3.0 solar masses for a neutron star. Most astrophysicists appear to think that the invisible companion in the Cygnus X-1 binary system is probably a black hole. Some disagree, but they dissent because they doubt some of the calculations of distances and masses, not because they reject the underlying reasoning.

Now van Fraassen would probably agree that of the available hypotheses—white dwarf, neutron star, or black hole—the black-hole hypothesis is the best of the lot. But he would deny that astronomers have any grounds for further asserting that the black hole hypothesis is *probably* true. After all, how do we know that the invisible component of Cygnus X-1 is not some sort of object never imagined by theorists?

Van Fraassen is certainly right that the invisible object in Cygnus X-1 could be something we never imagined. Maybe the whole edifice of theory and observation that leads astrophysicists to believe in black holes, neutron stars, and white dwarfs is wrong. However, to say this is to say nothing more than that scientific knowledge is *always* tentative and fallible—something everybody admits. Unless we have some reason to think that our most careful observations and the predictions of our best-confirmed theories *are not* trustworthy, and not just that they *might not* be, there cannot be anything objectionable about basing our probability judgments on what those theories predict and what those observations seem to show. *Observation* shows that one component of the Cygnus X-1 binary system is invisible, so it cannot be an ordinary star. Further observations and measurements appear to show that the object is very small and very massive. General relativity, one of the best-confirmed physical theories, predicts straightforwardly and in considerable detail what will happen to matter when it achieves such densities. What it predicts is that a singularity will form surrounded by an event horizon of predictable radius. In other words, a black hole will form.

Put simply, astrophysicists think that the invisible component of Cygnus X-1 is a black hole because, given everything they think they know about the universe, it just couldn't be anything else! It can't be a white dwarf or a neutron star, so a black hole is all that's left. Do astrophysicists implicitly appeal to special privilege to limit the candidate hypotheses to just these three—white dwarf, neutron star, or black hole? No, these three, and only these three, are the candidates permitted by the total state of our presumed astrophysical knowledge at this time. Astrophysicists think they have good reasons for holding that all stars below a certain mass will eventually form white dwarfs or neutron stars and not anything else. Further, according to well-confirmed theory, anything denser than a neutron star will collapse into a singularity to form a black hole. There is no implicit appeal to privileged access.

We can summarize the reasons an astrophysicist might give for believing that black holes probably exist: We observe cases like Cygnus X-1 (other binary systems that are *better* black-hole candidates have been more recently discovered) where we can see an ordinary star orbiting a common center of mass with an unseen companion. Further observation and inference (where the inferences are based on indisputably correct premises, like the value of the speed of light or the reliability of Kepler's third law) show that the unseen companion is too large to be a white dwarf and has a mass beyond the theoretical limit for neutron stars. We think that when matter achieves certain densities, a singularity will form. We think this because it is a straightforward prediction of the theory of general relativity, one of the best-confirmed theories in science. Therefore, of the three possible invisible causes of the massive x-ray emissions detected in Cygnus X-1, the best explanation is a black hole.

We see that IBE enters in only after a great deal of theory and observation have given us reasons to delimit our candidate hypotheses to three and only three—white dwarf, neutron star, or black hole. The decision to limit our candidates to these three is therefore apparently based on *scientific* grounds and does not appeal to any sort of privileged or *a priori* knowledge. It therefore seems that van Fraassen is wrong to think that IBE must always appeal to privileged access when it concludes that an unobservable entity probably exists. There are often good *scientific* reasons for thinking that the right explanation lies within the batch of hypotheses under consideration. Maybe van Fraassen was right that we risk less if we only expect our theories to be empirically adequate and refrain from asserting their truth. But when it looks like we *can* get some truth, why not go for it?

Having taken a rather-lengthy detour through astrophysical details, let's pause to remind ourselves of van Fraassen's initial objection to IBE: IBE can only select from among the hypotheses we happen to have thought of, and there is no way, short of claiming privileged access, that we can claim to know that the true explanation probably lies within the batch of hypotheses we actually have. Privileged access, once again, would mean that we have some sort of prior assurance that true explanations are likely to be among the ones we just happen to think up. The only way to argue for such privileged access would be to claim that our minds are somehow designed (by God or evolution) so that the hypotheses that we think up are likely to contain true explanations. Having placed this heavy burden of proof on realists, defenders of CE can enjoy the smug confidence that realists are unlikely ever to establish such recondite theses. But if, as we have seen, it is the straightforward work of science, and not privileged access, that assures scientists that the true explanation probably lies within a small set of candidates, then the burden of proof shifts to the critics of IBE. Such critics would now have to show why the use of IBE in astrophysics is really in principle different from its use when we say that there is probably is a mouse behind the wainscoting. Whether it is mice or black holes that best account for our data, we seem to have good reasons to think that we have gotten the *right* explanation.

Of course, we have just scratched the surface of the realism/antirealism debate here, and antirealist philosophers would stoutly contest the conclusion that I have just drawn. There are many other arguments and issues relevant to the debate that we do not have room to examine. At the present time among professional philosophers of science this debate is at a stalemate. Neither side seems to be capable of gaining a decisive advantage over the other. This does not mean that the realism/antirealism debate is going to go away. It has been around for thousands of years, and it is an issue that spontaneously arises from the nature of scientific inquiry. Indeed, any attempt by a scientist, philosopher, or theologian to penetrate to the reality behind the appearances is bound to invite a skeptical riposte.

SO WHAT REALLY IS THE GOAL OF SCIENCE?

If we do reject the claim that the goal of science is merely to save the appearances, should we say that science aims to generate theories that tell us the truth about the natural world—the view van Fraassen attributes to scientific realists? Well, it is no doubt misleading to speak of "the" goal of science, since science has many goals, but it is intuitively plausible to say that science seeks truth. Surely scientists would not insist that theories be tested so rigorously if they did not care for truth.

Truth is important in science, but the pursuit of truth is subordinate to the goal of *understanding*. Sometimes we understand something by learning the literal, exact truth about it—but not always. Sometimes we cannot know what something *is*; the best we can do is to say what it is *like*. Often in understanding some aspect of the natural world we have to use models, analogies, and metaphors rather than literal descriptions. Sometimes scientists make physical scale models—like the plastic-and-metal model that Francis Crick and James Watson built to guide them in uncovering the double-helix structure of DNA—but these are not the kinds of models meant here. Here the kind of model we are talking about is an abstract, usually mathematical representation that might not give any sort of *picture* of what it represents but only symbolic tools for talking about that hidden reality. Other models will offer only a very simplified or idealized depiction of some part of the natural world. Such a model is not a portrait of reality, but it is more like a sketch, or even a caricature that emphasizes only the most scientifically crucial aspects while omitting all unnecessary detail—including all those crucial details that often make those objects so fascinating for us. As far as gravitational theory goes, it does not matter at all whether a mass in free fall is a lead sphere, a brick of gold bullion, or an NBA superstar. So, when our theories employ such abstract and simplified models, the theory does not aim to give a literally true description of the world.

In fact, science could hardly work if we always insisted that theories tell the exact truth. Consider the law of free fall discovered by Galileo. Mathematically, the formula is $d = gt2/2$, that is, the distance traveled by a freely falling body, d, over a time t, is equal to the acceleration due to gravity, g, times the square of the elapsed time, t, with the product divided by two. Since the acceleration due to gravity is 32 feet per second squared, if a skydiver jumps from a plane and is in free fall for four seconds, he or she will fall 256 feet, according to this law. The discovery of this law was a great landmark in the history of science. It is still found in all elementary physics textbooks. The problem is that it is not true. Strictly speaking, the law holds only for objects falling in a total vacuum, and total vacuums do not exist naturally (not even in interstellar space) and cannot be created in laboratories. In real-life situations, Galileo's law is only an approximation (often a very good one). In fact, if you jump from a plane, you probably will not fall *exactly* 256 feet in four seconds. If we insisted on an absolutely true law for real-life falling bodies, we would have to take into account the very complex influence of factors like the effects of air resistance on bodies of various shapes, and the influence of ambient temperature and humidity. Such "laws" would be far too complex to use.

The upshot is that the laws of nature as scientists formulate them are a compromise between truth and usefulness. The complexity of exact truth has to be sacrificed for the virtues of simplicity and universality. This is why the laws of physics so often postulate ideal entities like total vacuums, frictionless surfaces, and point-masses. It is not that the laws of physics lie; they do not even fib. Rather, they constitute what philosopher Stephen Toulmin calls "ideals of natural order"—representations of what nature would be like under ideal circumstances (recognizing, of course, that circumstances are never actually ideal). Such ideals function as baselines for explaining the behavior of real entities. To the extent that things behave like the ideal, that behavior is considered natural and in need of no further explanation. Behavior that departs *too* far from the ideal requires further explanation (*slight* departures from the ideal are expected). This is not just true of the laws of physics, by the way, but also of those of biology. Mendel's genetic theory predicts ratios of the various phenotypes that are ideally expected in a given generation, but the ratios actually found are very unlikely to match precisely with the ideal value. In fact, if the actual experimental value matched the predicted value *too* closely, we would suspect that the experimenter had fudged the data!

Realists often defend their views by asserting that scientists spontaneously take a realist view of theoretical entities. This is a plausible claim. Generally, physicists seem to think that there really are quarks; chemists think there really are atoms; biologists think that

DNA really does have a double-helix structure, and so on. However, it does not follow that scientists automatically hold that getting literal truth is THE goal of science. Scientists take a more pragmatic view and are often quite willing to say, as did Newton, "*hypotheses non fingo!*" when a theory works but nobody knows why it does. A case in point is quantum mechanics. There is no more successful theory in the history of science than quantum mechanics. Hundreds, if not thousands, of tests have never turned up any evidence against it, and its use is so basic to contemporary physics that the field would not exist without it. Yet quantum mechanics, for all its usefulness, presents an impenetrable mystery.

As represented in the mathematical language of quantum mechanics, there are certain properties of quanta—the tiniest bits of matter—that are given no definite values until they are actually measured. The so-called "dynamic" properties of quanta—such as position and momentum—are represented as being in no definite state prior to measurement. The values of dynamic properties for unmeasured quanta are represented as a "superposition of states," a mixture of possible states rather than a definite value. Only after a detecting device actually performs a measurement does the property take on a definite value. Now *nobody* knows what is going on here. The standard interpretation, called the *Copenhagen interpretation*, says that it is not just that we *don't know* the values of those dynamic properties prior to measurement; there *are* no values before we test for them! When we interact with those quanta in order to measure them, then—and only then—do they take on definite values. Nobody understands how the act of measuring quanta could cause indefinite properties to become definite. This is the famous "measurement problem" of quantum mechanics.

Does it bother most physicists that nobody knows what is really going on here? Not at all. Of course, many other *interpretations* have been offered that try to make sense out of all of this, but most physicists seem to think that such interpretations are more philosophy than science since they have nothing to do with the use or application of quantum theory. The observed quantum facts are the same on all such interpretations. As one wag put it, the average quantum mechanic has no more interest in philosophy than the average auto mechanic. The working scientist accepts quantum mechanics because it *works*; it tells us what we want to know and lets us do what we want to do. The fact that it doesn't tell us what is *really* going on when we measure a particle's dynamic properties is not held against it any more than Newton's failure to explain why gravitational force exists kept people from using his law of universal gravitation. Maybe we will never know the *whole* story about what is really going on in the quantum world, and we will have to be satisfied with a theory that leaves such gaps.

Clearly, therefore, the statement that truth is THE goal of science is simplistic and misleading. We often have to use models, metaphors, and analogies to understand things that we cannot describe literally; sometimes these give us our *only* tools for grasping things. Other accepted theories represent idealizations or simplifications that are not literally true of the actual world. In particular, the laws of nature as we formulate them generally strictly apply only to idealized situations that are never perfectly realized. Further, theories like quantum mechanics often fail to tell us some of the things we would really like to know, but they do provide invaluable mathematical tools for explanation, prediction, and control. A good scientific theory enhances our understanding of the world, and understanding sometimes is and sometimes is not the same thing as knowing the literal, exact truth about things.

In fact, all of the talk about metaphors, analogies, and models may have left you wondering whether there is much difference between scientific realism and constructive empiricism if the best we can often do is to present an abstract model of unobservable entities. Hasn't the realist really largely given up the game and sided with the antirealist? The realist has to admit that we cannot have a literal description of the dynamic properties of unobserved electrons. The best we can presently do, and maybe the best we will ever do, is to make an abstract mathematical model (the "wave function") of the electron that we use to make reliable predictions about what we will probably find when we do perform a measurement. As we note above, what *really* is going on with unobserved electrons is anybody's guess (The realist can adopt a realistic *interpretation* of what is going on at the quantum level, but, again, this is an interpretation added to the theory, not an element of the theory itself.).

So the scientific realist has to admit that even our best theories in even our most "mature" sciences often represent aspects of nature in ways that are not even truth-like. On the other hand, constructive empiricists do not deny that there *is* an unobservable reality, and they have to admit that in case after case the physical world acts exactly *as if* that unobservable reality were constituted of unobservable particles of specific types. In other words, constructive empiricists admit that the best models of the unobservable structure of the cosmos are precisely the ones that postulate electrons, protons, quarks, and so on. So, at least in dealing with the postulated particles of fundamental physics, both camps seem to be saying something tantamount to this: "There exists a level of reality far too small to be directly observed. We cannot say exactly what that reality *is*, but we can construct abstract mathematical models of that reality that reliably predict what we will experimentally observe."

It would be nice to accept the irenic suggestion that at rock bottom not much separates the scientific realist from the constructive empiricist, but the proposed reconciliation won't wash. Though scientific realists might admit that we do not and maybe cannot know everything about electrons, they do hold that "electron" names a natural kind, that is, that our concept "electron" corresponds to a real category of thing in the natural world. In other words, the realist thinks that when we classify some things as electrons, we are classifying them as they *are*. Further, they hold that there are very many particular instances of the electron kind. As philosophers like to say, there exist many "tokens" (individual electrons), of that "type" (the kind of particle we name by the term *electron*). They also hold that there are some properties of electrons, the so-called static properties such as charge and mass, that we can say that electrons always have under all circumstances, whether or not they are being observed. Constructive empiricists do not think we can make any such positive assertions about what sorts of natural kinds exist in the unobservable realm. *All* we can say is that things can be modeled in certain ways and that those models make reliable predictions about observable consequences, so there is an irreducible difference between scientific realism and constructive empiricism.

Taking into consideration the entire discussion of rationality, realism, and antirealism through this whole book, I offer the following definition of scientific realism:

Scientific realism is the doctrine that makes the following three claims: (1) There is a physical world, with determinate properties, that exists independently of our perceptions or understanding of it. (2) Science aims to give us, in some *of its theories, a literally true picture of aspects of the physical world. This aim extends to giving a literally true picture of some of the unobservable parts of the world. It is eminently reasonable to regard well-confirmed theories that have that aim as truth-like. (3) Further, science has* actually succeeded *in identifying and accurately describing some of the natural kinds of observable and unobservable things in the world.*

Scientific realism first makes a *metaphysical* claim that there exists a real physical world "out there" that is the way it is independently of how we perceive it or think about it. This may seem patently obvious to many and not worth stating. Recall, however, that some of the people discussed in earlier chapters did seem to question this basic assertion. For instance, in places Kuhn seemed to be saying that the world that exists changes from paradigm to paradigm. Also, Bruno Latour and Steve Woolgar in *Laboratory Life* and elsewhere really do sometimes seem to question the existence of a determinate, independent physical reality. Scientific realists affirm that there is such an objective physical reality.

Scientific realism also makes an *epistemological* claim, namely, that it is entirely reasonable to regard some well-confirmed theories as truth-like. Scientific realists affirm that when a theory aims to tell us the truth about some aspects (observable or not) of the world, and when that theory is supported by sufficient evidence or passes sufficiently stringent tests, we are warranted in regarding those theories as approximately true. Concomitantly, it is a legitimate goal of science to try to discover as much truth as we can get about the universe, including truths about its deep, unobservable structure. Modestly asserting that seeking theories that are literally true is only *one* goal of science is consistent with the possibility that theories that are merely empirically adequate may sometimes give us our best possible understanding of some natural phenomena. A modest scientific realism therefore rejects the stark choice offered at the beginning of the chapter—whether we want our theories to be true or merely to have true consequences. We now see that "truth or consequences" is a false dilemma. In a nutshell, the scientific realist thinks that science should strive to maximize our understanding of the world, and he or she holds that this project will often involve the discovery of truths about unobservable aspects of the natural world. For the scientific realist, the problem with all forms of antirealism is that, by refusing to believe what science tells us about the unobservable structure of the world, these doctrines settle for *less* understanding than we can get.

Scientific realism as I formulate it also claims that science really has succeeded in correctly identifying and describing some of the constituents of the natural world, even some that are unobservable. Some scientific realists shy away from tying their position to any particular body of scientific claims. If we do, and those claims are later shown wrong, what happens to scientific realism? However, it seems to me that if we do not insist that science *really has* succeeded to *some* extent in identifying and accurately describing (i.e., describing with a high degree of truth-likeness) some of the actual constituents of the world, we concede the whole case to the pessimistic metainduction. If science hasn't gotten anything right so far, then it looks like the main epistemological claim of scientific realism, that we are warranted in regarding some theories as truth-like, is pretty doubtful to say the least. Finally, to distinguish scientific realism from the "blind realism" mentioned in the last chapter—where we affirm that many scientific theories must be right but we are noncommittal about which ones actually are—many scientific realists may wish to add to the above definition that we *do* have good grounds for identifying some *particular* theories as truth-like.

Having formulated and defended my definition of scientific realism, it follows only to recommend it to the reader as the best philosophical perspective on science. As I see it, modest scientific realism of the sort defined above is an attractive *via media*, a middle course, between philosophical extremes. It does not make the hyper-realistic claim that truth is THE goal of science; it concedes that our best theories sometimes might be merely empirically adequate or employ models, metaphors, or analogies that cannot give us a literal picture of things. On the other hand, modest scientific realism thinks that antirealists go too far in recommending that we *never* believe theories about the unobservable parts of the universe. Such a view leaves us with an impoverished understanding of the world. Finally, by affirming the actuality of an independent physical reality, scientific realism resists the claims of social constructivists, postmodernists, and relativists that, in various ways, make physical reality an artifact of our concepts. Even if it is coherent to claim that reality is an artifact, and I'm not sure that it is, it is an extremely counterintuitive notion for many people. I think that John Searle is right that the "default setting" of the human mind is to think that there is a world "out there" that exists independently of us. Those who are not persuaded by the arguments for changing that default setting will favorably regard scientific realism's affirmation of the independence of physical reality.

When you stop and think about it, it really was incredibly brash for people to start thinking that they could come to understand the hidden inner workings and structure of the amazing and beautiful but extremely complex and mysterious cosmos that we are a small part of. It is a brashness that can only be captured by that wonderful Yiddish word *chutzpah*. Maybe the biggest gamblers in history were the pre-Socratic (prior to Socrates) Greek philosophers who had the wild idea that the human mind could penetrate to the inner essence, the *archē*, of things—the primordial reality underlying our world of shifting appearances. Of course, each philosopher gave a different answer about what that underlying reality was supposed to be, but they agreed that there was such a reality and that the human mind could discover it.

Now, 2,500 years later, the pre-Socratics' big gamble seems to have paid off—spectacularly. Natural science now offers a rich set of powerful theories that purport to tell us about unobservable parts of the world. These theories enjoy the support of meticulously gathered evidence and have survived the most rigorous attempts to refute them. But have they earned the right to be *believed*, or does it still require *chutzpah* to believe that there are (for instance) electrons and that we know some literal truths about them? In these last two chapters I have given some reasons why I think that we are not going too far out on a limb to think that science does sometimes tell us the real story about the unobservable things in the universe. But issues such as this have been debated *ad nauseam* by philosophers of science, and, as I said earlier, the realism/antirealism debate is deadlocked.

Maybe some progress could be made if we cast our net somewhat wider, that is, if we placed the realism/antirealism debate in the context of the broader epistemological question about how, in general, we could have warranted beliefs about unobservable things. At the beginning of this chapter I noted that humans often believe in many things, besides the theoretical entities postulated by science, that they cannot detect with their senses, and that these claims raise their own questions about what we can rationally believe when we move beyond what we can observe. For instance, I asked what it would take to make it reasonable to believe in God even if He doesn't make spectacular public appearances. Perhaps then we could make progress in the philosophy of science *and* the philosophy of religion if we could get clearer on some basic epistemological points, like how, in general, claims about unobservables (whether gods, ghosts, or gravitons) might be warranted. It seems to me that the most promising line of inquiry here would be to examine more deeply the nature of IBE. Philosopher Peter Lipton (1991) and others have written engaging and enlightening works on this subject. We see at the end of the next chapter that perhaps IBE can ride to the rescue and solve a problem that has long plagued scientific methodology.

FURTHER READINGS FOR CHAPTER FIVE

Norwood Russell Hanson was a brilliant young philosopher of science who was killed in a flying accident at the age of forty-three. His tremendously amusing essay "What I Don't Believe" is found in *What I Do Not Believe, and Other Essays*, edited by Stephen Toulmin and Harry Wolf (Dordrecht, Holland: D. Reidel, 1971), pp. 309–31.

The story of the discovery of the electron is related clearly and authoritatively in *The Discovery of Subatomic Particles*, revised edition, by Nobel Prize–winning physicist Steven Weinberg (Cambridge: Cambridge University Press, 2003). Weinberg is a top-notch scientist who also is an outstanding writer of science for the general public. He tells the story of Walter Kaufmann, the German physicist who did the same sort of experiment Thomson did, only better, in his *Dreams of a Final Theory* (London: Vintage, 1993).

Jean Perrin's 1913 book *Les atomes* was published as *Atoms*, translated by D. Hammick (Woodbridge, CT: Ox Bow, 1991). Perrin's work was the subject of a distinguished study by Mary Jo Nye titled *Molecular Reality: A Perspective on the Scientific Work of Jean Perrin* (New York: American Elsevier, 1972).

Bas van Fraassen's best-known work is *The Scientific Image* (Oxford: Clarendon, 1980). Van Fraassen stated his constructive empiricism in this book and thereby kicked off what was probably the liveliest debate in the philosophy of science since the publication of Kuhn's SSR. By 1980, philosophy of science had finished a long struggle to throw off the dominance of logical positivism, a view that had been strongly influenced by Mach's instrumentalism. Then van Fraassen came along offering a new form of antirealism, constructive empiricism, supported by a very challenging set of arguments. Van Fraassen offered additional arguments against realism in *Laws and Symmetry* (Oxford: Clarendon Press, 1989). Van Fraassen is a clear writer, but he often gets into quite-technical issues. An introduction to his views for beginners, written from a sympathetic perspective, is *Understanding Philosophy of Science* by James Ladyman (London: Routledge, 2002). The arguments in van

Fraassen's critics, with a reply by van Fraassen (which is the van Fraassen citation from 1985 in this chapter), were brought together in *Images of Science: Essays on Realism and Empiricism*, edited by Paul M. Churchland and Clifford A. Hooker (Chicago: University of Chicago Press, 1985). A pointed critique of van Fraassen's and other forms of antirealism is Michael Devitt's *Realism and Truth*, second edition (Oxford: Blackwell Publishers, 1991).

Niels Bohr's antirealism is examined in Helge Kragh's *Quantum Generations: A History of Physics in the Twentieth Century* (Princeton: Princeton University Press, 1999). This is an excellent, nontechnical history of physics in the twentieth century.

Francis Crick tells the story of neuroscience in elucidating the process of seeing in his *The Astonish Hypothesis: The Scientific Search for the Soul* (New York: Simon & Schuster, 1994). Crick's enthusiasm is unbounded and infectious. Seeing comes so automatically for us that it comes as a shock to realize just how astonishingly complex the process is and how much we still have to learn. Even more shocking are the ways that "seeing" can go so wrong. When people hear that large numbers of sane, intelligent, and honest people have "seen" miracles, UFOs, Bigfoot, ghosts, and the like, it is almost irresistible to think that there must be something to all these things. However, Terence Hines in *Pseudoscience and the Paranormal: A Critical Examination of the Evidence* (Amherst, NY: Prometheus Books, 1988), shows how easily we can "see" what isn't there.

A fascinating, and disturbing, look at the microworld in your own house is found in David Bodanis's *The Secret House* (New York: Simon & Schuster, 1986). The pictures of the dust mites and other tiny creature that live with (and on and in) us are inevitably somewhat disquieting. Bodanis also has the picture of the pin with the bacteria clustered on the tip. It makes you thankful to have a good immune system.

Ian Hacking's essay "Do We See through a Microscope?" is published from pages 132 to 152 in the book *Images of Science*, mentioned above. In my view, Hacking is one of the most interesting of the current philosophers of science, and his arguments for realism are among the most effective. Another essay by Hacking defending realism is "Experimentation and Scientific Realism," in *Scientific Realism*, edited by Jarret Leplin (Berkeley: University of California Press, 1984), pp. 154–72. If Hacking were writing these essays today, he would probably make use of the recent developments in nanotechnology. A very readable introduction to these amazing developments is *Nanotechnology: A Gentle Introduction to the Next Big Idea* by Mark Ratner and Daniel Ratner (Upper Saddle River, NJ: Prentice Hall, 2003). Van Fraassen's reply to Hacking is in *Images of Science*.

The inimitable Isaac Asimov wrote about black holes in his usual stimulating and crystal clear manner in *The Collapsing Universe* (New York: Walker, 1977). A good book by a black-hole specialist is *Black Holes* by Jean-Pierre Lumient, translated by Allison Bullough and Andrew King (Cambridge: Cambridge University Press, 1992). If black holes don't give you the creeps, not much in the natural world will. They are fascinating objects: you just have to be glad that, so far as we know, they are very far away.

Stephen Toulmin's *Foresight and Understanding: An Enquiry into the Aims of Science* (New York: Harper Torchbooks, 1961), is a succinct (115-page-long) work that, unlike so much other writing about science from the early 1960s, remains fresh, relevant, and insightful. Toulmin really knows the history of science, and this knowledge adds substance to his philosophical conclusions.

Probably the best book for general audiences on the various interpretations of quantum theory is Nick Herbert's *Quantum Reality: Beyond the New Physics* (Garden City, NY: Anchor, 1985). There are loads of books on quantum mechanics written by authors with an ideological ax to grind who want to use quantum mechanics to justify mysticism, idealism, or occultism. Quite a few of these know nothing about quantum mechanics; others understand the physics but put a highly tendentious spin on it. A good corrective for these is Victor J. Stenger's *The Unconscious Quantum: Metaphysics in Modern Physics and Cosmology* (Amherst, NY: Prometheus Books, 1995). A very readable and interesting introduction to the philosophy of physics that defends a realistic view is

Peter Kosso's *Appearance and Reality: An Introduction to the Philosophy of Physics* (Oxford: Oxford University Press, 1998).

The best current book-length study of IBE is Peter Lipton's *Inference to the Best Explanation* (London: Routledge, 1991). Lipton specifically addresses van Fraassen's and other criticisms of IBE.

CHAPTER SIX

MYSTERIES OF METHOD

INDUCTION AND DEDUCTION

Science textbooks frequently begin with a statement (often oversimplified or misleading) of the author's understanding of the scientific method. This seems right since everybody knows that what makes science special is that scientists follow something called "the scientific method," right? Actually, whether there is such a thing as THE scientific method is highly questionable. Science clearly has methods. Different sciences employ different procedures, techniques, and practices, often with little apparent overlap. Astrophysics and archaeology, for instance, are done in very different ways. Astrophysicists check their theories by the analysis of the tenuous traces of electromagnetic radiation collected from distant quasars, supernovae, pulsars, and so forth. Archaeologists meticulously recover and reconstruct artifacts from which they read the records of ancient lives. Each science does things its own way, and the methods used in one would often be impossible to apply in another.

However, does this apparent diversity of techniques and procedures really rest on a deeper methodological unity? Are these various methods just different ways of realizing the same epistemological goals and logical processes? Can we offer a generalized description of scientific method that is broad enough to apply to all of the natural sciences yet not so vague as to be uninformative? Can we, in fact, develop demarcation criteria that permit us to distinguish genuinely scientific enterprises from those that, though they might present themselves as scientific, fail to meet the standard? That is, can we make a clear distinction between science and pseudoscience?

Debate over such questions is practically as old as science is. It would be impossible in this chapter to summarize the history of debates about scientific method. Whole books can and have been devoted to that. I will not even attempt to summarize the modern debates about the confirmation of theories. As noted in the preface, the purpose of this book is not to summarize but to immerse the reader in a set of problems, controversies, and debates. The problem I address in this chapter is an overarching theme of the whole history of the debate about scientific method: How can we substantiate universal claims—claims that have the form "All *a*'s are *b*'s"? Our experience is of individual *a*'s and *b*'s. For instance, we see only individual cats. How do we come to know anything about cats in general? Science is full of universal claims; for instance: all copper wire conducts electricity; all snakes are carnivorous; all hydrogen atoms absorb and emit energy of given wavelengths; the period between the peaks of maximum brightness of a Cepheid variable star is directly proportional to its absolute magnitude. How do we know about all copper wire, all snakes, all hydrogen atoms, and all Cepheid variable stars when, in all of human history, we can only have examined a tiny portion of the ones that exist?

The answer traditionally given is that we can employ a kind of reasoning called *induction*. Inductive reasoning supposedly permits us to draw universal conclusions from the observation of individuals. From observation of individual cats, atoms, or stars, we can know about all cats, atoms, or stars if we properly employ inductive reasoning. This chapter will trace the ups and downs of inductive reasoning, starting with its first great exponent, Aristotle (384–322, BCE). Aristotle saw induction as culminating in an act of infallible intellectual vision. From the critical examination of individuals, we can discern a universal nature shared by all things of that type, which we call the "form" or "essence" of the thing being examined—its defining characteristic. I will then move on to Francis Bacon (1561–1626), who severely criticized Aristotle's theory of induction and attempted to put this sort of reasoning on a much more secure basis. Bacon saw induction as needed for both the discovery and the confirmation of scientific hypotheses. I will then turn to the twentieth century, when the hypothetico-deductive (HD) method became the standard account of scientific method and put induction firmly in place as the means whereby a theory's successful predictions can indicate the probable truth of the theory.

The serpent in the garden of scientific methodology was David Hume (1711–1776). In his classic *Enquiry concerning Human Understanding*, Hume offered a simple but powerful argument that seemingly undermined all forms of inductive reasoning. For Hume, there is no inductive *reasoning*, only habitual *expectation* conditioned and reinforced by repeated experience—what he called "constant conjunction." The twentieth-century philosopher Sir Karl Popper (1902–1994) agreed with Hume's critique of induction and proposed a method of science intended to turn only on deductive reasoning.

So the questions I will ask in this chapter are these: Is inductive reasoning sound? Can it be legitimately employed in scientific method? Can science be rational if it dispenses with induction? Are Hume and Popper the final word on this topic? Is there a third kind of scientific reasoning—inference to the best explanation—that has the strengths of inductive reasoning but lacks its alleged weaknesses?

ARISTOTLE: THE FIRST METHODOLOGIST

Aristotle was the first to offer a comprehensive set of logical and methodological prescriptions for the practice of science. Aristotle was himself a scientist, perhaps the greatest biological theorist prior to Darwin. He was also a logician; indeed, he devised the first system of formal logic. Logic addresses the issue of correct inference. If you know certain things, what else can you legitimately conclude? That is the central concern of logic. For instance, if you know that all whales are cetaceans and that some whales have teeth, can you conclude that some cetaceans have teeth? Yes, you can, and the kind of reasoning you do here is called *deductive reasoning*.

In philosophy, reasoning takes the form of arguments, and an argument is a discourse that moves from certain assumptions, which are called premises, to conclusions that are propositions taken to be supported by the premises. The strongest kind of support that the premises could give to the conclusion would be where the truth of the premises *guarantees* the truth of the conclusion, that is, where it must be the case that the conclusion is true when the premises are true. The above reasoning is of this sort. If we assume, that is, accept as true, the following premises:

(1) All whales are cetaceans.
(2) Some whales have teeth.
Then it must be true that:
(3) Some cetaceans have teeth.

If we accept the two premises (1) and (2), we have to accept (3) that some cetaceans have teeth. Why must we accept it? Because the most basic law of logic, the law of noncontradiction demands it. The law of noncontradiction forbids you to assert a contradiction; a contradiction is any proposition that takes the form "*p* and not-*p*." For instance, you are not allowed to say, referring to the same time and place, "It is raining and it is not raining." We can use the rules of deductive logic to derive a contradiction from the assumption that certain propositions (premises) are true and others (conclusions) are false. But when certain propositions are inconsistent with the falsehood of another, this is the same as saying that where the former are true, the latter *must* be true. The rules of deductive logic therefore permit you to start with assumed premises and derive other propositions that *must* be true when the premises are.

For example, if we accept (2) above—that some whales have teeth—

then (using recognized rules of deductive logic throughout) there must be some whale, let's name him "Willy," that has teeth. Further, if we accept (1)—that all whales are cetaceans—then Willy, being a whale, is a cetacean. Willy therefore has teeth and is a cetacean. However, if we deny (3), our conclusion, then we assert, "It is not the case that some cetaceans have teeth," or, in other words, "No cetaceans have teeth." If no cetaceans have teeth, then Willy, being a cetacean, does not have teeth. Using the rules of deductive logic we arrive at the conclusion that Willy has teeth and that Willy does not have teeth, and this is a manifest self-contradiction. We see, therefore, that we cannot, on pain of self-contradiction, affirm our premises (1) and (2) while denying our conclusion (3). In other words, if (1) and (2) are true, and (3) *cannot* be false, then (3) *must* be true.

Arguments that have the property we have just illustrated, where the truth of the premises guarantees the truth of the conclusion, are known as valid deductive arguments. A theory of deductive logic is therefore a theory of valid argumentation. A theory of valid argumentation proposes certain rules of inference that, when applied to certain types of premises, will always generate a true conclusion.

The first such theory was proposed by Aristotle and is called the theory of the syllogism. A syllogism is a form of argument consisting of two premises and a conclusion. Only a particular kind of proposition, called a categorical proposition, can constitute a premise or conclusion of a syllogism. Categorical propositions express relations of inclusion or exclusion between classes of things. One class is mentioned in the subject of the categorical proposition, and the other is mentioned in the predicate. Aristotle recognized four types of categorical propositions, which are called A, E, I, and O, where *S* and *P* stand respectively for the subject and predicate terms of the proposition:

A: All *S* are *P*.
Example: All whales are cetaceans.
E: No *S* are *P*.
Example: No whales are reptiles.

I: Some *S* are *P*.
Example: Some whales are toothed animals.
O: Some *S* are not *P*.
Example: Some whales are not toothed animals.

In short, A propositions state that all members of a given class belong to some other class; E propositions state that no member of a given class is a member of another; I propositions state that some members of a class are members of another; O propositions state that some members of a class are not members of another. A syllogism, then, consists of three categorical propositions that are so related that the truth of the first two guarantees the truth of the third.

Some syllogisms are valid arguments:

No whales are reptiles.
Some whales are toothed animals.
Therefore: Some toothed animals are not reptiles.

The first premise excludes the class of whales from the class of reptiles. The second premise says that some members of the class of whales are toothed animals. If both of these premises are true, then it has to be the case that some toothed animals, namely toothed whales, are not reptiles. Since the truth of the premises guarantees the truth of the conclusion, this is a valid argument.

Some syllogisms are invalid; for instance:

All pinnipeds are mammals.
Some endotherms are mammals.
Therefore: Some endotherms are pinnipeds.

The first premise includes the class of pinnipeds in the class of mammals. The second premise tells us that some members of the class of endotherms are mammals but tells us nothing about whether any of them are also included in the class of pinnipeds. They might be or they might not be. If they are not, then the premises can be true and the conclusion false, which means that the argument is invalid.

For Aristotle, the structure of scientific knowledge must take the form of a set of valid syllogisms with true premises. Valid arguments with true premises will give us all true conclusions; indeed, the truth of the conclusions will be guaranteed. If our body of scientific knowledge therefore consists of the conclusions of valid syllogisms with true premises, then our scientific knowledge will consist of proven truths. The theory of the syllogism can give us rules for determining which syllogisms are valid and which are not. However, this still leaves us with the problem of coming up with true premises. Where do our premises come from?

Some of them can be gotten from the conclusions of prior syllogisms, and the premises of those other syllogisms from the conclusions of yet other syllogisms, and so on. However, there has to be a starting point; this proving process cannot stretch back *ad infinitum*, or nothing can be proven. The conclusion of a valid argument is only as certain as its premises, which means that the proof is only provisional until the premises themselves are established. Yet if one set of premises depends on a prior set of premises, and those in turn on prior ones, and so on forever, then *all* of our conclusions will be merely provisional and nothing can ever be proven! So, moving backward from conclusions to premises to yet earlier premises and so on, we eventually have to hit something that is certain *without having to be proven*. That is, we have to come back to premises that state truths that are just *self-evident* and hence not in need of being proven from any earlier premise or set of premises. In other words, at some point we will have to have the first syllogisms of our science, which, as first, cannot derive their premises from any prior syllogisms. To get our initial premises to start building our science, we therefore need a different form of reasoning. We need to used induction and not just deduction.

Sometimes you hear people explain the difference between deduction and induction by saying that deduction moves from the general to the particular while induction moves from the particular to the general. While this characterization may be true of a number of individual deductive and inductive arguments, it is not precise enough to serve as a satisfactory definition. A better way of explaining the difference begins by noting that deductive arguments are attempts to make the premises guarantee the truth of the conclusion. If the premises of a valid deductive argument are true, the conclusion must be true. Inductive arguments are different. The premises of an inductive argument can never guarantee the truth of the premises. The law of noncontradiction is not violated if you affirm the truth of an inductive argument's premises and then deny the conclusion. Instead of guaranteeing the conclusion, the premises of an inductive argument are meant to make that conclusion more probable.

Here is an example of inductive argument:

> A poll of five hundred voters indicated that 47 percent favored candidate Wingnut.
> Therefore: Approximately 47 percent (within a margin of error) of all voters favor candidate Wingnut.

The premise makes the conclusion probable, not certain; there is still some degree of probability that the percentage of voters preferring candidate Wingnut falls outside of the margin of error for the sample. An even more familiar form of induction is called *induction by simple enumeration*. An example would be as follows:

> Of one hundred crows observed, all have been black.
> Therefore: All crows are black.

Induction by simple enumeration notices that all observed things of a given type share a particular property and concludes that all things of that type (observed or unobserved) have that property. If each of one hundred crows so far observed is black, this gives us some reason to think that all crows are black, but there is no guarantee. There might be white crows someplace we have not looked.

Another way of contrasting inductive and deductive reasoning is to say that the former is an "ampliative," and the latter is not. Induction is called "ampliative" because its conclusions add to, or amplify, the information contained in the premises. From the fact that the sun has risen every day for the last several billion years, we conclude that it will rise tomorrow. However, that it will rise tomorrow is new information, information not contained in the statement of past risings. In a deductive argument, however, the conclusion only makes explicit the information that was already in the premises. From the premises (1) that all humans are mortal and (2) that Socrates is human, we already have the information that Socrates is mortal. The conclusion "Socrates is mortal" only spells out that information. Because deduction does not really tell us anything new, we can be as sure of the conclusion of a valid deductive argument as we are sure of its premises. Because inductive arguments tell us something new, something not contained in the premises, the conclusion might be wrong, even if the premises are all true. The dilemma of human reasoning seems to be this: You can be certain, or you can learn something new. You cannot do both.

For Aristotle, the scientific method proceeds by a two-step process that begins with induction and ends with deduction in the form of syllogisms. We begin the process of science by noting the characteristics of many individual things. For instance, we notice that humans come in a great variety, with differences in sex, height, weight, age, skin color, hairiness, intelligence, and personality. We also notice that they have some things in common. For instance, all humans are bipedal and featherless. Hence, some ancient philosophers defined humans as "featherless bipeds." We arrive at the conclusion that all humans are featherless bipeds by a process of induction by simple enumeration, that is, by noticing that in our experience each individual human is a featherless biped and generalizing to the conclusion that all are featherless bipeds.

The conclusions of inductive inference give us the initial premises of the syllogisms that constitute the body of our scientific knowledge. But not just any inductions will do. Sometimes, though a property may be shared by all members of a class, that property might be just an accident and have nothing to do with the nature of that class of entities. Only universal characters that are *essential* to that sort of thing—part of what Aristotle calls the *form* of the thing—can serve as the starting premises of science. The essential features of a kind of thing are those properties such that if something lacks those features it is just not *that* kind of thing. Accidental properties, on the other hand, are those that a thing can have or lack without ceasing to be *that* kind of thing. For instance, at one time all emus lived on the continent of Australia. Yet there seems to be nothing essential about emus that they must live in Australia. They do not stop being emus as soon as they are removed from Australia. We therefore need a way of telling which universal characteristics are essential and which are merely accidental.

Unfortunately, Aristotle is not sufficiently clear on this point. In his work *Posterior Analytics*, one of a set of methodological treatises collectively referred to as the "*Organon*" ("instrument"), Aristotle asserts that we *do* have a natural, inborn, intuitive faculty for apprehending the universal natures of things. By a seemingly mysterious but powerful act of the mind, we perceive that certain features constitute the inner essence or true natures of certain kinds of things. Thus, I can recognize that rationality is an essential human feature and not, say, being a featherless (though humans are indeed the only extant featherless bipeds). Put another way, if I were asked to define *human*, I might reasonably leave out that humans are featherless, but I could not leave out that humans are rational creatures.

Earlier I said that the dilemma of human reasoning seems to be that we can either learn new things or be certain, not both. Aristotle attempts to escape this dilemma by making the final stage of induction a kind of intellectual seeing; we mentally perceive the universal natures of things. Our intellectual vision here has the immediacy and certainty of visual perception, though it is intellectual and not visual. We learn something new, the essences or forms of things, but our knowledge is as secure and certain as the immediate perceptions of the senses.

Once we have an understanding of the essential natures of things, we can express this understanding in the form of universal categorical sentences, and these can then form the initial premises of the first syllogisms that will constitute the body of scientific knowledge. Further syllogisms will take as their premises the conclusions of those first syllogisms, and thus extend our knowledge. For Aristotle, then, scientific knowledge is a body of necessary truths. Each scientific truth is necessary because either it is immediately intuited as an essential characteristic, or because it is proven, ultimately, from premises that state such essential characters.

An additional, and essential, feature of our scientific knowledge is that it tells us the *causes* of things. Merely knowing facts is not enough; you must understand *why* the facts are so. As Aristotle puts it, there is "knowledge of the fact" and then there is "knowledge of the reasoned fact." The latter kind of knowledge is better since it tells us not only *that* something is but also *why* it is. Aristotle's examples here may seem a bit contrived or opaque for modern readers, so let me offer some of my own.

Consider the following syllogism:

All flying birds are creatures with hollow bones.
Falcons are flying birds.
Therefore: Falcons are creatures with hollow bones.

This syllogism not only informs us of the fact that falcons have hollow bones but also indicates *why* they do. Flight requires structures that are light yet strong. This is why balloons are not made of lead. Flying birds need skeletons that are light but strong, and this is why falcons have hollow bones that are lighter than solid bone but are still remarkably strong.

Now consider this syllogism:

All creatures with hollow bones are flying birds.
Falcons are creatures with hollow bones.
Therefore: Falcons are flying birds.

Presuming that the premises are true, we have demonstrated the fact that falcons are flying birds. However, this syllogism gives us no indication of *why* falcons are flying birds. After all, birds do not fly because they have hollow bones; they have hollow bones because they fly. This is why syllogisms of the first type are preferable to syllogisms of the second type. The first type gives us knowledge not just of the fact but also of the reasoned fact, that is, *why* the conclusion is true.

In more familiar terms, Aristotle's distinction between knowledge of the fact and knowledge of the reasoned fact is his way of indicating the difference between correlation and cause. Some things consistently go together, so that where you have one you can infer that the other is present also, but there is no causal or explanatory relationship. For instance, suppose, as was true for a while in the late 1950s, every kid has a Hula-Hoop. In that case, where you would find kids, you would find Hula-Hoops; and where there were Hula-Hoops, you would find kids. So there was a very strong correlation between kids and Hula-Hoops, but kids don't cause Hula-Hoops, and Hula-Hoops don't cause kids. On other occasions, when two things are found constantly conjoined, one will be causally connected to the other and therefore helps explain the other. For instance, lightning and thunder are not only consistently conjoined; lightning causes thunder by explosively heating the air and thereby creating an intense shock wave that we perceive as thunder. For Aristotle, scientific knowledge is knowledge of causes and not merely of correlations.

Having proven above that falcons have hollow bones, we might extend our scientific knowledge as follows:

> All creatures with hollow bones are creatures with a high basal metabolism.
> Falcons are creatures with hollow bones.
> Therefore: Falcons are creatures with a high basal metabolism.

We have observed many creatures with hollow bones and arrived by induction at the conclusion that all hollow-boned creatures have a high basal metabolism. We also know, from the conclusion of our earlier syllogism, that falcons are creatures with hollow bones. These two true statements therefore form the premises of our above syllogism, which is a valid syllogism so we know that the conclusion is true. Further, flying is a very energy-intensive activity, so we would expect creatures that have adaptations for flight, like hollow bones, also to have a high rate of basal metabolism since high energy budgets also go with flying (if you "eat like a bird" you consume twice your body weight in food daily). The above syllogism therefore not only tells us *that* falcons have a high rate of basal metabolism, but, once more, indicates *why* this should be so.

To sum up, for Aristotle, our body of scientific knowledge is expressed in the form of syllogisms. The premises of these syllogisms may be derived from the conclusions of prior syllogisms, but this process cannot stretch back *ad infinitum*. At some point, the premises of our syllogisms must be inferred directly from the perceived natures of things. The scientific truths expressed in the conclusions of our syllogisms not only are true, but they *must* be true since they are inferred from true premises by valid arguments. The conclusions of our science therefore possess certainty. Not only are those conclusions certain, we see *why* they are so. That is, the premises from which we infer those conclusions also indicate the *causes* of those conclusions. Falcons do not just happen to have hollow bones, they have hollow bones *because* they are flying birds. The truths of our science therefore are certain and their causes are known.

The Aristotelian concept of scientific knowledge sounds authoritative. It promises a lot. It so impressed the leading medieval writers on scientific method, such as Roger Bacon and Robert Grosseteste, that they retained the basic structure and content of the Aristotelian method while adding some original developments. Basically, though, for nearly two thousand years Aristotle's version of scientific method was viewed as, if not the final word, at least its most authoritative statement.

BACON: SCIENTIFIC METHOD RENOVATED?

The veneration of Aristotle came crashing down with the beginning of modern science. Whereas Aristotle had been perhaps overrated before, he came to be the whipping boy for the new scientists and their apologists. The most famous apologist and propagandist for the new science was Francis Bacon (no relation to the earlier Roger). Bacon, while not himself a scientist of note, was the most articulate defender of an allegedly new method for science, one that scorned Aristotle and saw itself as, literally, making a whole new start. Bacon even titled his writings to show his aim to reject the Aristotelian method and begin anew. His *Great Instauration* (*instauration* means "renovation") proposed an allegedly new set of principles to guide scientific inquiry. Bacon's *Novum Organum*, offered a new set of "instruments" for the conduct of science to replace Aristotle's old *Organon*.

Did Bacon propose a whole new way of doing science? His successors, until well into the nineteenth century, hailed him as the true prophet of the modern science. In the twentieth century, some revisionist historians of science depicted him as overrated and minimized his contribution to the Scientific Revolution. Still, there is no doubt that Bacon strongly influenced succeeding generations of scientists and philosophers of science.

How did he achieve such influence? For one thing, Bacon was trained as a lawyer and possessed an advocate's rhetorical and debating skills. He was also a literary stylist whose famous *Essays* are part of English literature. Such was Bacon's literary flair that some of those who claim that Shakespeare did not write the plays published under his name attribute the plays to Bacon! Most scholars regard this claim as almost certainly untrue, but that it was entertained at all is a testament to Bacon's literary talent. Bacon's ability to express recondite points of methodology in a vivid and cogent way no doubt made his writing engaging and persuasive. Further, Bacon may have ironically suffered the same fate as Aristotle—overrating followed by underrating. The upshot will, I think, be a more balanced evaluation of the contributions of both Aristotle and Bacon.

One of Bacon's lasting and important contributions was the recognition that there are major impediments to objective thinking that must first be cleared away before we can set foot on the path to truth. Some of these impediments are intrinsic to human thinking, others are idiosyncratic and particular to one's individual cognitive quirks, others arise from the obfuscation generated by ideology and dogma, and others are inherent in the nature of language itself. Bacon called these impediments "idols" because, like golden idols, they have a specious glitter that makes them deceptively appealing. Bacon confronts these false idols of the intellect in the *Novum Organum*.

The "idols of the tribe" are the first sort:

> The *Idols of the Tribe* lie deep in human nature and in the very tribe or race of mankind. For it is wrongly asserted that the human sense is the measure of things. It is rather the case that all of our perceptions, both of our sense and of our minds, are reflections of man, not the universe, and the human understanding is like an uneven mirror that cannot reflect truly the rays from objects, but distorts and corrupts the nature of things by mingling its own nature with it. (Bacon, 1994, 54)

This quote illustrates Bacon's aptitude for making deep points vividly. The human understanding does not reflect truly, but like a carnival funhouse mirror, presents a distorted image.

This is a deep insight and one that current cognitive science supports and elaborates. All too often the spontaneous working of the human mind serves to bias or distort rather than truly represent things. A particularly pernicious example mentioned by Bacon is what we today call "confirmation bias," that is, the tendency to form beliefs first and then seek confirming instances while disregarding any contrary evidence. Bacon explains insightfully:

> The human understanding, once it has adopted opinions, either because they were already accepted or believed, or because it likes them, draws everything else to support and agree with them. And though it may meet a greater number and weight of contrary

> instances, it will, with great and harmful prejudice, ignore or condemn or exclude them by introducing some distinction, in order that those earlier assumptions may remain intact and unharmed. (57)

He illustrates confirmation bias with a story from the Roman author Cicero: A visitor to a temple was shown a picture of those who had called on the gods and were rescued from a shipwreck. Does this not illustrate the power of the gods? But the skeptical visitor wanted to know where there were pictures of those who had called on the gods and drowned anyway. People want very much to believe that the gods will save them when they are in desperate straits, so they willingly look for evidence that prayers are effective and ignore the evidence that they are not.

We see confirmation bias at work everywhere. The bigot sees everything bad about those he despises and ignores all that is positive. Confirmation bias is particularly pernicious because it often works unconsciously. The reason that strict protocols, like requiring that studies be "double-blind," are so important that without such controls, even trained researchers and scientists will unconsciously project their own expectations or hopes onto the perceived results. Companies will often tout their many satisfied customers, but you have to go to the Better Business Bureau to find out about their many unhappy ones. Bacon sums up:

> The human understanding is not a dry [i.e., unbiased] light, but is infused by desire and emotion, which give rise to "wishful science." For man prefers to believe what he wants to be true. . . . In short, emotion in numerous, often imperceptible ways pervades and infects the understanding. (59–60)

The second sorts of idols are the "idols of the cave":

> The *Idols of the Cave* arise from the individual's particular nature, both of mind and body, and come also from education, habits, and chance. (61)

Each of us has individual quirks of personality, background, or habit that can distort our understanding. For instance, as Bacon notes, some people have a partiality for old, settled answers and are automatically suspicious of anything new, while others are immediately attracted to any new idea and are hostile to old ones (62–63). Scientists and scholars, being only human, have as many personal crotchets, obsessions, and animosities as anybody else, and these can sometimes constitute major distractions.

Even Einstein had his blind spots. Einstein, though impressed by the techniques of quantum mechanics, could not accept that it described fundamental reality. Quantum mechanics appeared to imply that nature, at bottom, follows statistical rather than deterministic laws, and Einstein famously replied that God does not play dice with the universe. Worse, it violated Einstein's basic intuitions about reality when quantum mechanics implied that there could be what Einstein called "spooky action at a distance." That is, quantum mechanics predicts circumstances in which the measurement of the properties of one particle instantly determines the properties of another particle, however distant. Einstein devised ingenious thought experiments that, he held, revealed the weakness of the conceptual foundations of quantum mechanics. Yet since Einstein's death, actual experiments have confirmed even the "spookiest" predictions of quantum mechanics.

Bacon continues:

> There are also idols arising from the dealings and association of men with one another, which I call *Idols of the Market-place*, because of the commerce and meeting of men there. For speech is the means of association among men; but words are applied according to common understanding. And in consequence, a wrong and inappropriate application of words obstructs the mind to a remarkable extent. . . . Indeed words plainly do violence to the understanding and throw everything into confusion, and lead men into innumerable controversies and fictions. (55)

We must use words to communicate, and, indeed, to think, but words derive their meaning from common usage, and common usage might reflect common misunderstanding. Thus, words themselves can play an obscurantist role when their received meaning reinforces erroneous concepts.

An example not mentioned by Bacon but much discussed in recent decades is the word *race*. For nineteenth-century writers on race, the word was held to designate essential biological kinds that clearly distinguish and categorize all humans. Each individual was a member of some particular race, and certain bodily characteristics, such as skin color or hair texture, were taken as definitive of one's racial status. Your race was supposed to largely determine your essential nature—who you are.

Over time, though, biologists came to realize that *race* is a scientifically useless and misleading term. Geneticists have pointed out that the genetic diversity *within* a "race" is higher than the average diversity *between* "races." There is, for instance, more genetic diversity among Sub-Saharan Africans than there is on average between Sub-Saharan Africans and northern Europeans. Indeed, it now looks like the whole concept of "race" is a social construct, that is, a conventional category arbitrarily imposed for sinister purposes. However, rethinking the concept of "race" is hampered by the fact that the word retains all of its old connotations for many people and these associations channel thought in particular directions despite the most diligent efforts of debunkers. If racist assumptions are built into the very meaning of a word, then any use of the word tends to reinforce those assumptions. Small wonder that Bacon considered the idols of the market-place to be the most troublesome of the idols! (64)

Last are the "idols of the theatre":

> Finally, there are idols which have crept into human minds from the various dogmas of philosophies. . . . These I call the *Idols of the Theatre*, because I regard all of the philosophies that have been received or invented as so many stage plays creating fictitious and imaginary worlds. (55–56)

Bacon sweepingly condemns all previous philosophies as fiction! When fiction is taken for fact, confusion and error follow.

We do not have to be as comprehensive in our condemnation of systems of philosophy as Bacon to see the cogency of his point. I think Bacon is best understood here as cautioning against the pernicious influence of ideology. I understand "ideology" to be a fixed system of all-encompassing dogma. The ideologue is one whose concepts have become rigid pigeonholes into which all experience must be forced. A good example would be the Young Earth Creationist, mentioned in chapter 2, who MUST, however implausibly, interpret all the features of the earth's crust in terms of the Noachian deluge. The absurdity of holding that the fossil record can be explained in terms of a single, geologically recent, universal deluge shows the danger of ideological distortion.

For Bacon, Aristotelianism was a pernicious and obscurantist ideology, an idol of the theatre. He accuses Aristotle of reaching conclusions beforehand, like the true ideologue, and then twisting the facts to fit the foregone conclusion:

> Nor should it count for much that in his book *On Animals* and in his *Problems* and other treatises he [Aristotle] often cites experiments. For he had come to his decision beforehand, without taking proper account of experience in setting up his decisions and axioms; but after laying down the law according to his own judgment, he then brings in experience, twisted to fit in with his own ideas, and leads it about like a captive. (70)

This is a damning judgment and one that Bacon does not adequately substantiate. However, as a characterization of how, in general, ideology distorts experience, it is right on target.

The deleterious effects of dogma are plain, but we must pause to ask an important question: What, really, is the difference between ideology and method? To adopt a scientific method is already, at least in a preliminary way, to make certain judgments about nature. When we prescribe a method for science, we already indicate the kinds of questions that we intend to consider and, at least in a general way, the kinds of answers we expect to find. Does this not bias our inquiry in the same way, if not as blatantly, as an ideology? For instance, as we saw in chapter 3, science famously (or infamously, to the creationist) accepts methodological naturalism. That is, the answers we look for in science are presumed to be natural causes. Such putative supernatural causes as gods, ghosts, and souls are simply dismissed from the beginning because science is not even *looking* for such supposed causes. Isn't this biased?

To reiterate and clarify some of the points made in chapter 3, it is important to recall that all inquiry must start from somewhere, some set of assumptions about what there is or might be and how best to proceed. The idea that we could adopt a perspectiveless perspective, what philosophers ridicule as "the view from nowhere," is an illusion. Indeed, can there even be coherent experience unless we employ a basic toolkit of interpretive principles, concepts, and categories to impose intelligible order on the manifold of raw sensory input? Without basic concepts to categorize, evaluate, and relate the flood of incoming sense data, we would be as helpless as newborns.

While a legitimate method may tell us where to start our inquiry and what kind of answers to seek, it does *not* dictate where inquiry must conclude. Method is a guiding hand, not like the dead hand of dogma. We begin with working hypotheses and tentative assumptions, but we might end up with a big surprise. As we noted in chapter 3, scientific methods could quite conceivably confirm religious hypotheses, as in the case of the crucial experiment whereby Elijah triumphed over the prophets of Baal (1 Kings, chapter 18). As Bacon accuses Aristotle of doing, the ideologue has his answers in hand before he starts. Such methods as the ideologue employs are ones guaranteed to produce the desired result. Surprises are ruled out.

In a sense, then, a method must sit lightly. It must umpire without rigging the game. For one thing, if a method ossifies into dogma, it will become obscurantist, because it is important to be able to reject old methods if better ones come along. As we saw in the debate between Hobbes and the experimentalists such as Boyle, Hobbes played the role of the obscurantist because he was unwilling to relinquish old methods when scientists developed better ones.

Bacon held, then, that Aristotle's method had become dogma, but what were his specific complaints? First, Bacon takes issue with the syllogism itself:

> The syllogism does not apply to the principles of the sciences and is applied in vain to the intermediate axioms, being quite unequal to the subtlety of nature. Therefore, while it commands assent, it fails to take hold of things. (46)

Further:

> The syllogism consists of propositions, propositions of words, and words are tokens of notions. Therefore—and this is the heart of the matter—if the notions themselves are muddled and carelessly derived from things, the whole superstructure is shaky. (36)

Bacon charges, first, that, though the syllogism can indeed prove conclusions from propositions that are already known, they provide no means for getting the initial premises. As we saw, Aristotle held that the first premises are arrived at by induction, but Aristotle understood this as a power of the mind that can distill pure essences from experience of a multiplicity of individuals. Bacon is contemptuous of this claim and contrasts it with true induction:

> There are and can only be two ways of inquiry to the discovery of truth. The one flies from the senses and particular axioms to their [supposed] immutable truth, proceeds to judgment and the discovery of intermediate axioms. And this is the message that is now in use. The other calls forth axioms from the senses and particulars by a gradual and continuous ascent, to arrive at the most general axioms last of all. (47–48)

What Aristotle thinks is a deep, infallible power of the mind, Bacon dismisses as a hasty leap of speculation. The broadest and deepest generalities about nature don't just jump out at us; they have to be won by hard work and careful experimental investigation. Surely, Bacon seems to charge, it is arrogant or naïve to think that the deepest secrets of nature can be won so cheaply. Further, what guides an Aristotelian in deciding which instances to gather as the data for inductive inference? The actual practice seems to be haphazard and unguided: You observe a few randomly collected or encountered instances and leap to a wide-ranging conjecture about essential natures.

The basic problem, Bacon charges, is that Aristotelians depend on induction by simple enumeration, which is a woefully inadequate because it only collects positive instances and ignores negative ones (57–58). Therefore, simple enumeration supports the sort of confirmation bias that is one of Bacon's idols of the tribe. If I notice, for instance, that ten teenagers text while driving, I might conclude that all teenagers text while driving if I am not careful to note also the ones who do not.

The correct procedure, as spelled out by Bacon, is to examine phenomena systematically, making sure that we have encompassed many diverse instances and not just a few examples of the phenomenon of interest. Further, we have to consider relevant cases where the phenomenon is *not* present and others where it is present but in varying degrees. For instance, suppose that we are interested in the nature and causes of heat. To study heat, we first draw up a table of "Instances Agreeing in the Nature of Heat," which is a list of natural things that generate heat. For instance, he lists the sun's rays, flame, natural hot springs, thunderbolts, and so on. He next lists ostensibly similar things that generate no heat, for instance, the light of the moon and the stars, St. Elmo's fire, *ignis fatuus* (will-o'-the-wisp), and the like. Finally, he lists those things that have various degrees of heat. For instance, decaying matter produces some small degree of heat and bodily exercise generates heat as well, the more you exercise, the more heat you generate (145–68).

Once we have a large and diverse body of data pertaining to a given phenomenon, then, and only then, do we begin our inductive ascent.

We use induction to frame hypotheses, and then to refine and revise them in the light of further inductions. Instead of immediately leaping to the broadest principles, as Bacon accuses the Aristotelians of doing, we begin with low-level generalizations that apply only to specific sciences. When we have gained a substantial body of these lower-level generalizations, which are the common facts of a given science, we begin to look for generalities between the facts of the various sciences. We avoid listing accidental correlations by consulting all of our tables, not just the ones that list positive instances. For instance, we might observe that many hot things glow brightly and leap to the conclusion that heat is caused by a bright glow. However, by consulting our table of things that do not generate heat, we see that many things with a bright glow, like the moon or fireflies, do not generate heat. We therefore are not misled into thinking that a bright glow is the source of heat.

Ultimately, as our generalizations grow broader and more comprehensive, we hope to arrive finally at what Bacon calls the "forms," by which he means the basic laws of nature. Once we have the basic laws of nature, the broadest generalities and regularities that hold within all the sciences, the inductive part of our inquiry comes to an end. We then can employ deduction in which we explain particular phenomena in terms of the discovered laws of nature. For instance, Bacon held that if we pursue the study of heat to its conclusion, we will discover that heat of a body is nothing more than the motion of its components, and that motion is therefore the cause of heat (174).

Remarkably, our modern understanding of thermodynamics confirms this theory. Hotter bodies are those with more energetic molecular motions. The difference between hot water and cool water is that in the hot water the mean kinetic energy of the water molecules is higher. Once we know the true nature of heat, we can use that knowledge to explain particular instances, and so our scientific knowledge will advance. To take a modern example, the most volcanically active body in the solar system is Io, the moon closest to Jupiter. What makes Io so hot? Being close to Jupiter's enormous mass, the interior of Io is exposed to massive gravitational tidal forces as it orbits Jupiter. The internal back-and-forth flexing of the moon due to those tidal forces generates the enormous heat that powers Io's volcanoes.

This, in a nutshell, is Bacon's methodology for the sciences. How should we evaluate it? Bacon's methodology has been roundly criticized. Contemporary philosopher of science Peter Urbach summarizes the charges of Bacon's critics:

> First of all, they [the critics] say that Bacon exaggerated the contribution that facts alone can make to science, and, in particular, they criticize his claim that science could be completed, if only a sufficiently comprehensive stock of facts were gathered. Some also object that it would be impossible to collect *all* the facts about any phenomenon, for there are simply too many of them. (Some even go so far as to say that their number is infinite.) Next, it has been brought against Bacon that separating fact-collection and the subsequent use of those facts in generating axioms, while not impossible (though some hold that it is), is not the way that science has actually progressed. In practice, they claim, an investigation proceeds in the inverse order, starting out with a hypothesis, which then determines which facts are relevant, these in turn leading to the confirmation or perhaps the disconfirmation of the original hypothesis. Finally, Bacon is sometimes taken to task for imagining that the factual basis for science is incorrigible, it being pointed out that observational reports inevitably include some hypothetical, and hence fallible, component. (1987, 151)

Urbach defends Bacon against his detractors, arguing that, when carefully read, we see that Bacon's view of science was much sounder and more "modern" than recent commentators have thought. For instance, Urbach argues that Bacon did not recommend an effort to collect *every* fact relating to a phenomenon, a task he clearly realized to be impossible (154). His interest was not in compiling a comprehensive set of facts but one that emphasizes the greatest *variety* of instances evincing the phenomenon (154). Urbach also argues that Bacon recognized the need for a guiding hypothesis in the search for facts, recommending that we collect facts in the light of working hypotheses, and then revise our hypotheses in the light of discovered facts, and then search for more facts guided by our revised hypotheses, and so on (155–56).

Another recent commentator expresses what now is the standard judgment on Bacon:

> It is a historical fact that Bacon's philosophy is contemporaneous with the birth of modern thought. But it is also a fact that modern thought has developed in a way which does not accord with the idea that Bacon gave of the new science. Of course, we can praise the Chancellor's [Bacon's] sense of reformation, his critique of false science, his comments on academic institutions and the politics of science; we can even say as the French Encyclopedists did, that he was the herald of experimental philosophy. But the fact remains that the Baconian concept of science, as an inductive science, has nothing to do with and even contradicts today's form of science. (Malherbe, 1996, 75)

But this is a bit harsh. For one thing, Bacon's tables of the presence, absence, and variation of a phenomenon clearly anticipate the canonical statement of such methods by John Stuart Mill. Mill's Methods, as they are called, have indeed proven useful in the conduct of science. Journals such as *Nature* and *Science*, and popular magazines like *New Scientist* and *Scientific American* regularly report studies based on versions of these methods. In the end, though Bacon claimed that his methodology was a new broom that swept away the Aristotelian cobwebs, his "instauration" was much less than revolutionary. In fact, Bacon's methods can be seen more as a revision or correction of Aristotle's basic inductive methodology.

Aristotle and Bacon can therefore be seen respectively as the grandfather and the father of the modern branch of the philosophy of science called "confirmation theory," that is, the study of how scientific hypotheses and theories are confirmed by evidence. Evidence clearly has a very significant bearing on the decisions of scientific communities to accept or reject certain theories, but spelling out the precise nature of that relationship is difficult. It is well known by philosophers of science that no body of data uniquely determines a theory. That is, no matter how much evidence we have for a theory, it is at least conceivable that some other theory would be consistent with all that data. For instance, as we shall see in a later chapter, all of the evidence that we think with our brains is construed by Cartesian dualists as evidence for an alternative theory, namely that brain states closely correlate with but do not cause mental states, which are the activities of a nonphysical soul. Clearly, then, we need a more precise understanding of how evidence relates to theory and supports it without determining it.

THE HYPOTHETICO-DEDUCTIVE METHOD

A survey of modern confirmation theory is impossible in a short space. Here I can present only what for a long time was the received or canonical account, the hypothetico-deductive (HD) theory of confirmation. Unfortunately, there will be no room to develop Bayesian confirmation theory, perhaps the most prominent contemporary theory of confirmation. I offer the account of the HD method as lucidly expounded by one of the leading philosophers of science of the twentieth century, Carl G. Hempel. Note, though, that the description of the HD method immediately below is an oversimplification (it omits discussion of the essential role of auxiliary hypotheses; this omission will be corrected later on).

The essence of the HD method can be expressed very simply: Legitimate scientific theories have empirical content, that is, they tell us something about the physical universe. If a theory tells us something about the physical universe, then from a statement of that theory, *T*, we should be able to deduce a statement of a test implication, *I*. This statement *I* has an "if . . . then" form; it tells us that if certain conditions are met, you can deduce the prediction that some occurrence, in principle observable by us, will take place at some specific point in space and time. We then check to see if that prediction is true. If the prediction is seen to occur, that confirms the theory, that is, it provides grounds for affirming the truth of the theory. If the prediction does not occur, this disconfirms the theory, that is, it provides grounds for denying the truth of the theory. Put another way, a positive instance (a true prediction) confirms the theory and a negative instance (a false prediction) disconfirms a theory.

Consider, for instance, the Copernican theory and the prediction of parallax mentioned in the first chapter. Recall that parallax is the shift from our point of view of foreground objects with respect to more distant background objects. You observe parallax when driving and nearby objects appear to shift with respect to more distant objects. The Copernican theory seemingly generates the test implication that nearby stars will appear to shift with respect to more-distant background stars when viewed from opposite points of the earth's orbit. In July, the earth is on one side of the sun. In January, it is on the opposite side and is approximately 186,000,000 miles from its position six months earlier. Such a large shift in position should, seemingly, result in an apparent shift of the closer stars with respect to the more distant ones. The older, geocentric theory, does not predict that we will see stellar parallax.

The Copernican theory, like any good theory, can provide multiple predictions. Another test implication we can derive from that theory is that the planet Venus should show phases just as the moon does. That is, sometimes when we view Venus we should see its disc fully illuminated, and at other times only partially so—like the moon when it is between new and full. The geocentric theory did not predict that Venus would show phases.

Let's set out the HD method a little more formally. The structure of a scientific test would be like this, where *T* states our theory; *I* is a test implication of that theory; *C* is a statement of what are called "initial conditions," the particular relevant conditions at the time of the test; and *O* is the statement of the predicted observation:

If *T*, then *I*.
If *I* and *C*, then *O*.

We then check to see if *O* in fact occurs when condition *C* is met. If *O* does not occur, we reason as follows:

> Not-*O*.
> Therefore not-*I* or not-*C*.
> But *C* is true (i.e., the conditions specified in *C* are actually met).
> Therefore, not-*I* (that is, the test implication of *T* is false).
> Therefore, not-*T* (if *T* implies *I*, but *I* is false, then *T* has to be false too.).

Here a negative instance, the nonoccurrence of the predicted event, disconfirms the theory *T*. With respect to our example, if *T* is a statement of Copernican theory, *T* implies *I*, that if stars are observed at six-month intervals (when, according to *T*, the earth would be at opposite sides of its orbit), then they will appear to shift in position. Suppose that we do observe the stars at six-month intervals, say July 15 and January 15, but observe no apparent shift in position. In that case we would have to say that it is false that if the stars are observed at six-month intervals, then they will appear to shift in position. We would therefore have to conclude that Copernican theory is false since its test implication is false. Note that our reasoning in the above schema is strictly deductive.

But what if *O* is true when the condition specified by *C* is met. For instance, what if we *do* observe parallax when we observe certain stars at six-month intervals? What can we conclude? Well, we cannot *deduce* anything. From "If *I* and *C*, then *O*," and "*O*," we cannot conclude "*I* and *C*." Neither can we conclude "*T*" from "*I*." Even if *all* of the test implications of Copernican theory are true, we cannot deductively infer that Copernican theory is true. To do so would commit the well-known logical fallacy of "affirming the consequent." However, if a theory has a variety of test implications, and they are independently verified to be true, and no rival theory has a similar record of generating true test implications, we inductively infer that the theory has a good chance of being true.

It is especially favorable evidence for a theory if the test implication predicts something surprising, something that we would regard as highly improbable if the theory is not true. The most famously surprising prediction of Copernican theory, as augmented and extended by Newtonian celestial mechanics, was the prediction by astronomer Edmund Halley, a contemporary of Newton's, of a comet that would appear many years later, in 1758. In December of that year, long after Halley's death, the comet did reappear as predicted. Comets were seemingly capricious at the time, and the prediction that one would appear in a given year seemed highly improbable. Therefore, the appearance of Comet Halley (as it has since been called) in 1758 was hailed as spectacular confirmation of astronomical theory. Practically speaking, if a theory, like the Copernican theory, enjoys success after success over an extended period, correctly making diverse and surprising predictions, it comes to be regarded as very probably true.

The HD method therefore captures our strong intuitions that theories must be testable *vis-à-vis* the evidence, and that positive instances, particularly when diverse and surprising, confirm a theory and negative instances disconfirm the theory. The HD method recognizes other factors besides empirical testing that have a bearing on whether or not we accept a theory. One is theoretical support. When we test a theory, we do not forget everything that we have learned before. We assess a theory in part on its own merits, that is, on its own success in making true predictions, but we also have to consider it in the light of what we already know, or think we know. A degree of epistemic conservatism has to be part of our methodology. Of course, new ideas may force us to change our old ideas, but, if the old ideas still appear solid, we have to consider whether the new idea conflicts with those established ideas. Ideally, the new idea and the old ones should be mutually supporting. Sometimes it takes a while to sort out the relations between new ideas and old ones. When Mendelian genetics was established at the beginning of the twentieth century, it appeared to conflict with Darwinian natural selection, so, though evolution was not abandoned, the mechanism of natural selection faded into the background. Later, with the rise of the neo-Darwinian synthesis of the 1930s, it was realized that Mendelian genetics and Darwinian natural selection, instead of conflicting, were mutually supportive, and the resulting synthetic theory was stronger than any previous theory.

Also, other things being equal, that is, given equal predictive success between two theories, we prefer the simpler of the two. When a theory becomes extremely complicated, as the old Ptolemaic theory did, we start to suspect that its predictive successes are less due to nature than to human cleverness in manipulating the theory to make it square with observation. Some philosophers and scientists have expressed the conviction that nature is, at bottom, simple. We hope that it is. As Stephen Hawking once supposedly said, it would be great to have a final theory, a Theory of Everything, that would be simple enough to put on a T-shirt. However, there seems to be no way to know, *a priori*, whether or not the universe, at rock bottom, is simple. We just have to wait and see if we can find out.

Ironically, the notion of simplicity seems to be quite complex. There are a number of different senses in which a theory could be called "simple," some of which might conflict so that a theory could be simple in some ways but not in others. Perhaps the clearest sense in which a theory could be simpler than a rival is that it requires fewer assumptions for its predictive success. The predictive success of Ptolemaic theory required epicycles, equants, and eccentrics. There was no independent evidence for these things; they just had to be assumed to make the theory square with observation. The developed Copernican theory, on the other hand, needed no such assumptions to make accurate predictions.

Clearly, the HD method is much more sophisticated than anything explicitly developed by Bacon. Further, the role that induction plays in the HD method is different from its role for either Aristotle or Bacon. Proponents of the HD method do not regard induction as encompassing a sort of intuitive intellectual vision as Aristotle did. Neither, unlike Bacon, did they hold that induction is useful both for framing and testing hypotheses. For HD theorists, there is no "logic of discovery"; the formulation of theories is a mysterious process involving imagination and creativity, and perhaps even hunches or flights of fancy. In HD theory, induction is useful strictly in the confirmation of theories. If we have a large and diverse number of positive instances, including some that were new and surprising—and the theory is not disconfirmed—we use induction (with considerations of theoretical support and simplicity) to say that the theory is probably true, or at least approximately so.

However qualified and circumscribed inductive reasoning is in HD theory, it is nevertheless an essential component. Yet all inductive methodologies must eventually face up to the simple but powerful arguments of David Hume. Hume, often regarded as the leading philosopher ever to write in the English language, delivered a potent attack on the whole basis of inductive reasoning in his philosophical classic *Enquiry concerning Human Understanding*. So powerful was this attack that many philosophers regard it as unanswered to this day.

HUME ON INDUCTION

Hume begins by making a crucial distinction between "matters of fact" and "relations of ideas." All truths known by human beings, he says, either express matters of fact or relations of ideas. Truths that express relations of ideas are "intuitively or demonstrably certain," that is, their truth either is self-evident (as soon as we understand a self-evident truth, we see that it must be true) or can be strictly and rigorously proven. Examples of such truths are: "1 + 1 = 2," "Bachelors are unmarried males," and "If something is square it has a shape." Here are some truths that can be strictly proven: "The square of the length of the side of the hypotenuse of a right triangle is equal to the sum of the squares of the lengths of the other two sides," "There is no highest prime number," and "If two angles of a triangle are 60 degrees and 90 degrees, the third angle must be 30 degrees."

All relations of ideas are known *a priori*, that is, as Hume says, "they are discoverable by the mere operation of thought, without dependence on what is anywhere existent in the universe" (1988, 71). For instance, we prove truths about triangles purely from the axioms and theorems of geometry and without measuring or inspecting any actual triangles. Indeed, the provable truths about triangles hold even if there are no triangular objects in the world. All truths based on relations of ideas are *necessary*, that is, they cannot be untrue. The reason is that truths expressing relations of ideas are true because their denial entails a self-contradiction.

Truths expressing matters of fact, on the other hand, are those that we discover to be in fact true but that do not *have* to be true; you can deny their truth without contradicting yourself.

Here are some matters of fact: "The moon's average distance from the earth is 238,900 miles," "Da Vinci painted the *Mona Lisa*," and "Charlemagne was crowned emperor on Christmas Day in the year 800." All matters of fact are known *a posteriori*, that is, by seeing what the world is actually like. No matter of fact can be known *a priori*. All matters of fact are *contingent*, not *necessary*—which means that they are true but they do not *have* to be. Further, and crucially, all matters of fact that are not immediately known by the senses are known through the relation of *cause and effect*.

For Hume, causal reasoning involves observing consistent conjunctions between types of events and then projecting that relationship into the future. In other words, from having observed that two events *A* and *B* are consistently spatially and temporally contiguous, we extrapolate and say that this relationship of contiguity holds universally so that wherever we find *A*, we will find *B*, and *vice versa*. For instance, we have always observed lightning to be followed by thunder, unless we are too far away to hear the thunder. We therefore project that whenever we are in the proximity to a lightning bolt, we will immediately after hear the crash of thunder. The projection of consistent regularities, the spatial and temporal contiguity of certain types of events, is all that cause-and-effect reasoning can achieve. Causal reasoning is a form of inductive reasoning. According to Hume, all inductive reasoning works by taking note of past patterns and projecting them into the future. For instance, I have eaten oatmeal many times in the past and it has not poisoned me. Therefore, if I eat oatmeal for breakfast tomorrow, it will not poison me.

Hume says that all inductive reasoning rests on an unstated assumption, the assumption of the uniformity of nature. To say that nature is uniform means that patterns observed to hold consistently in the past may be projected into the future and expected to hold consistently there. Now consider the claim "Nature is uniform." If this is a true statement, then, like all other true statements, it must express either a relation of ideas or a matter of fact. Let's consider these in turn.

(1) Does the statement "Nature is uniform" express a relation of ideas? If it does, then denying that statement will lead to a contradiction. Recall that all true statements expressing relations of ideas are true because if you deny them, you contradict yourself. For instance, if you deny that all bachelors are unmarried males, you must hold that there is at least one bachelor who is not an unmarried male. However, if you speak the English language, you know that *bachelor* means "unmarried male," and so all bachelors must be unmarried males. Therefore, you are committed to the contradictory proposition "At least one bachelor is an unmarried male, and all bachelors are unmarried males." However, if someone denies the uniformity of nature and says "It is not the case that nature is uniform," there is no contradiction. It is perfectly consistent to suppose that patterns observed in the past will not be observed in the future. So the claim that nature is uniform, if true, does not express a relation of ideas.

(2) Does the statement "Nature is uniform" express a matter of fact? If it does, then it must be something we immediately perceive or something we know by the relation of cause and effect. We do not immediately see that nature is uniform, that is, it is not self-evident, so if it is a true matter of fact, it must be known to be true by arguments based on cause and effect. But cause and effect arguments *presuppose* the uniformity of nature. Without rational confidence that nature is uniform, we cannot reasonably observe past causal regularities and

project those into the future. Obviously, you cannot *prove* that nature is uniform by appealing to a kind of argument that must *presuppose* that it is uniform. To do so would be to argue in a circle, like saying that God exists because the Bible says so, and the Bible should be believed because it is the word of God.

So if the uniformity of nature is not something that we can know by analyzing our ideas or establish by cause-and-effect reasoning, there is no reasonable basis for saying that nature is uniform. But if we cannot know or even reasonably assume that nature is uniform, we cannot know that the future will be like the past. We cannot even know that the sun will rise tomorrow. It follows that no inductive argument can make its conclusion certain, or even probable. However many times we have observed a pattern to hold in the past, there is no rational basis for projecting that pattern into the future. Such a projection assumes that nature is uniform, but apparently there can be no reason for thinking that it is.

Hume's argument has the cogency and simplicity that is the mark of a true work of genius. Philosophers have found his conclusion to be highly antithetical. Surely, as some philosophers have argued, part of what it means to be rational is to expect that the future will resemble the past in key ways and to base our reasoning and behavior on that expectation. If, for instance, doing *A* has always resulted in *B*, only a very irrational person would do *A* and not expect *B*. So strong is this expectation that you frequently hear the quip sometimes attributed to Albert Einstein: "Insanity is when you keep doing the same thing but expect different results." Hume therefore seems to be attacking a fundamental element of human rationality by his critique of induction.

Hume cheerfully accepted the consequence; in fact, he insisted upon it. Why, when *A* occurs, do we expect *B* when *B* has always followed *A*? Hume's answer anticipates the later discovery of the conditioned reflex by Ivan Pavlov. Pavlov famously showed that dogs could be trained to salivate at the sound of a bell by ringing a bell at the time the dogs were fed. If Hume is right, we are a lot like Pavlov's slobbering dogs. When we consistently experience *A*s followed by *B*s, we develop a reflexive propensity to expect *B*s when we encounter *A*s. That's all there is to it. Inductive "reasoning" is nothing more than the projection of a conditioned response, the sense of anticipation arising from a habitual association. As Hume put it, "Reason is and must be the slave of the passions" (1975, 415). So much for human rationality.

POPPER AND THE REJECTION OF INDUCTION

The twentieth-century philosopher of science Sir Karl Popper held that human rationality can be preserved even in the face of Hume's devastating debunking. In fact, Popper claims to have solved the problem of induction (1972, 1). Popper's reply to Hume's problem of induction is the key to his whole view of scientific methodology, to which we now turn.

Popper begins by rephrasing the Humean problem in what he considers a more rigorous and accurate way:

> Can the claim that a universal explanatory theory is true be justified by "empirical reasons"; that is, by assuming the truth of certain test statements or observation statements? (7)

Hume probably would not have recognized this "restatement," so let's be clear about what Popper is saying. For Popper, the goal of science is to discover true universal explanatory theories (191ff). A classic example of a universal explanatory theory would be Newton's law of universal gravitation. Newton's theory was that the force of gravity between any two masses equals the product of their masses, divided by the square of the distance between them, times the universal gravitational constant:

$$Fg = G\frac{M_1 M_2}{R^2}$$

This law was presented as universal, that is, as applying to all gravitational phenomena everywhere. Einstein later showed that it does not have universal application, but let's pretend that it does. In that case, Newton's law holds universally. It applies to earthly bodies, like the famous falling apple, as well as celestial ones, like comets and planets. It is also explanatory. Most famously, Newton's law explains the laws of planetary motion earlier discovered by Johannes Kepler. Kepler had discovered that (1) the planets revolve about the sun in an elliptical orbit with the sun at one focus of the ellipse, (2) a line drawn from the planet to the sun would sweep out equal areas in equal times, and (3) a simple mathematical relation exists between the time it takes planet to complete one orbit around the sun and the mean distance of the planet from the sun. Kepler discovered *that* these laws hold; Newton's theory explains *why* they do.

If universal explanatory laws are the Holy Grail of science, the big question is, how we know when we have found one? The traditional answer, as we saw in discussion of the H-D method, is that laws are confirmed by their positive instances. That is, we confirm universal generalizations with statements—called test or observation statements—that report particular instances that occurred in conformity with those universal statements. Using the language of predicate logic, we confirm a universal statement of the form

$$(\forall x)\ (Px \rightarrow Qx)$$

by statements reporting individual instances:

$$Pa \ \&\ Qa,\ Pb \ \&\ Qb,\ Pc \ \&\ Qc,\ Pd \ \&\ Qd \ldots$$

In English: We confirm generalizations of the form "All *P*s are *Q*s" by observation statements reporting individual instances of *P*s that are *Q*s. That is, the statements reporting that *a* is both *P* and *Q*, *b* is both *P* and *Q*, *c* is both *P* and *Q*, *d* is both *P* and *Q*, and so on, serve to confirm the universal generalization that all *P*s are *Q*s.

Can any number of positive instances confirm a universal generalization? Note that by "confirm," we do not mean "prove." Defenders of inductive inference do not claim that positive instances can conclusively establish a universal generalization, but they do hold that a sufficient number and variety of such instances can provide evidence for the truth of such a generalization. The more such instances we consider, and if there are no contrary instances, nothing that is *P* and not *Q*, the more confidence we should have that all *P*s are *Q*s. If, in many cases and in many different circumstances, we have observed many things that are *P* and have always observed that they are also *Q*, we are justified, it seems, in having confidence that all *P*s are *Q*s.

Not so, says Popper. Popper, like Hume, holds that NO number of positive instances can give us even the slightest reason for thinking that a universal generalization is true (7). A million observations of *P*s consistently being *Q*s gives us NO reason at all to think that all *P*s are *Q*s. Actually, Popper's formulation of the problem understates the strength of Hume's argument. If Hume is right, and there is no rational basis for our trust in the uniformity of nature, then, not only is there no reason to think that *all* future *P*s will be *Q*s; there is no reason to think that *any* future *P* will be *Q*. Starting with the million-and-first thing that is *P*, all future *P*s could be non-*Q*. A lifetime of observing that lead balls dropped near the surface of the earth will fall down gives no rational support at all to the claim that all unsupported lead balls in proximity with the earth's surface will fall down, or even that any of them will. Starting right now, things could start to fall up rather than down.

This seems preposterous. Are we to conclude that any empirical universal generalization is as good as any other? Surely, only a lunatic would walk off a cliff, drink cyanide, fire a loaded gun at his head, or walk into a Texas country & western bar with a T-shirt reading "Willie Nelson Sucks"—and not expect to get killed. Are Popper and Hume saying that lunacy is as rational as sanity?

No, says Popper. We can still make rational judgments about universal generalizations. Positive instances do not confirm, but negative instances conclusively disconfirm. How many negative instances does it take to conclusively show that not all *P*s are *Q*s? Just one. We have to discover only one thing that is *P* and not *Q* to refute the universal generalization that all *P*s are *Q*s. The reasoning here is deductive, not inductive. From "There exists a *P* that is not a *Q*," it is a very simple and straightforward proof that concludes "It is not the case that all *P*s are *Q*s." In other words, it is easy to show that "All *P*s are *Q*s" is false if there is a *P* that is not a *Q*. Here is the proof for those who know some predicate logic:

(1) $(\exists x)\ (Px\ \&\ {-Qx})$	premise
(2) $(\exists x)\ ({--Px}\ \&\ {-Qx})$	1, double negation
(3) $(\exists x)\ {-({-Px}\ \mathrm{v}\ Qx)}$	2, De Morgan's rule
(4) $(\exists x)\ {-(Px \rightarrow Qx)}$	3, implication
(5) ${-(\forall x)}\ (Px \rightarrow Qx)$	4, quantifier negation

Since the refutation of a universal generalization by counterexample is thus an operation of deductive logic, the problem of induction simply does not apply. We therefore *can* rationally compare universal generalizations. We can reject those that have been proven false by counterexample and retain those that have yet to be proven false.

For Popper, the above simple logical point is the key to scientific method. He calls it the method of "conjectures and refutations." A scientific theory is a conjecture, a guess, and it can never be anything more than that. Scientific method consists in making bold conjectures (theories) and subjecting our theoretical guesses to the most rigorous and severe tests we can devise—an all-out effort to refute them. A bold theory is one that makes strong claims; it takes risks. The more you claim, the richer and more scientifically significant is your theory. If I propose a theory that explains the motion of only one planet, I have claimed a lot less than if I offer a theory that explains the motion of all the planets. Of course, the more you claim, the less likely it will be that everything you claim will be true, so bolder theories are less probable than cautious ones. But bolder theories are so much more scientifically interesting because a more-comprehensive theory explains so much more.

In Popper's view, therefore, scientific rationality does not consist in trying to prove our theories right, but, on the contrary, proposing bold, improbable theories and then doing our best to prove them wrong. If a theory survives the most stringent tests, Popper says that it is "corroborated" rather than confirmed (18). To say that a theory is corroborated is NOT to say that it is to any degree confirmed, that is, made more probable. A well-corroborated theory is one that, despite our best efforts, has not been proven false—and it is no more than that. Surviving a million tests without having been shown false is not one iota of evidence that a theory is true. Our theories start as conjectures, and so they remain. In a way, Popper's verdict on theories is like the verdict in a criminal case. The accused is acquitted if the jury rules "not guilty." A "not guilty" ruling is not the same as "innocent"; rather, it is a judgment that the state has failed to meet the burden of proof to establish guilt. Likewise, trial by scientific testing can find a theory not guilty (i.e., not proven false), but cannot prove it innocent (i.e., true).

Logically, then, all we can ever say about a universal explanatory theory is that it has not yet been proven false. Then what do we *do*? What is the rational way to live if all we can ever have are theories that are not yet proven false? We have to act in *some* way, since doing nothing is just as much a way of acting as any other. For practical purposes, which theories do we choose to follow? For instance, if I want to build a spaceship, what scientific principles should I follow, given that I have no grounds for thinking that any one principle is true?

Popper says that since we have to act on the basis of *some* assumptions, the *rational* thing to do is to go with the theories that have been most rigorously tested and have so far have not been shown false.

> Since we *have* to choose, it will be rational to choose the best-tested theory. This will be "rational" in the most obvious sense of the word known to me: the best-tested theory is the one which, in the light of our *critical discussion*, appears to be the best so far, and I do not know of anything more "rational" than a well-conducted critical discussion. (22; emphasis in original)

Of course, the fact that I can then successfully do what I want to do—build a spaceship, say—on the basis of certain assumed generalities, gives us no reason at all to think that those generalities are true. The usefulness of a law in producing a practical result is just one more instance of a test that the law fails to fail.

So important for Popper is the falsification of theories that he makes falsifiability—the possibility of being shown false—the very criterion that demarcates science from nonscience. What makes science special? Why is science a more rational enterprise than its competitors? Why is astronomy more rational than astrology and psychology more rational than parapsychology? Why trust your doctor more than a peddler of snake oil? These questions are especially urgent if one holds, as does Popper, that the empirical success of a science makes its general laws no more probable than any that might be formulated in any scheme of pseudoscience or mumbo jumbo?

Popper answers that genuinely scientific theories are *falsifiable*, that is, they are expressed in such a way that observation or experiment could prove them wrong. For instance, recall Antoine Lavoisier's famous experiment discussed in chapter 4. Lavoisier's claim that combustion is chemically an oxidation reaction was eminently falsifiable. In fact, the competing theory that burning substances released a substance called "phlogiston," was also a scientific theory because it could be shown wrong. Lavoisier devised an ingenious crucial experiment that weighed substances before and after burning. If they weighed more, this would falsify the phlogiston hypothesis since something should weigh less after releasing phlogiston into the air. If they weighed less, this would falsify the oxygen hypothesis since something should weigh more after having something, oxygen, added to it. Lavoisier's experiment found that burned substances weigh more. Popper would interpret this result not as showing that the oxygen hypothesis is confirmed (nothing can do that), but as showing that the competing phlogiston hypothesis is false. Both the oxygen and phlogiston hypotheses count as scientific for Popper. The phlogiston hypothesis was falsifiable and was, in fact, falsified; the oxygen hypothesis was falsifiable, and, though not falsified, it *could* have been had the results of the experiment been different.

On the other hand, consider the following hypothesis:

God made the world in six literal, twenty-four-hour days approximately six thousand years ago. He created the world with an appearance of great age, that is, he created the world with fossils already in the ground and geological formations that appear to be of vast antiquity. In other words, the world, though young, has the appearance of being billions of years old.

Now such a "theory" could hardly be described as scientific. Nothing could possibly falsify it. No evidence of the antiquity of the world could possibly count against it because the hypothesis is that the world was created by God with all such apparent evidence already in place. Trilobites never really existed. God put their fossils in the ground to make it appear that there were once such creatures when, in fact, the world was created six thousand years ago much as it is now. Of course, we might ask why God would set out, apparently, to deceive us, but that is a different question. Clearly, the above theory is not falsifiable and so, for Popper, cannot count as scientific.

Such, then, are the basics of Popper's methodology for science: You make bold guesses and try your best to show that they are wrong. When the most rigorous tests have failed to falsify a theory, we do not regard it as confirmed, that is, probable, but say that it is *corroborated*. For practical purposes, the most rational thing to do is to act on those theories that have been corroborated, that is, subjected to the most rigorous critical discussion without being disqualified. Therefore, Hume's aspersions cast on human rationality are wrong. We rationally evaluate scientific claims—and only those claims that can be so evaluated are to count as scientific—in the light of the best evidence and most stringent tests, and choose to follow those that pass. Of course, in meeting Hume's challenge, we do have to concede that he was right in that no degree of evidence can confirm a universal generalization. However, it does not follow that we are slobbering dogs.

Does Popper meet Hume's challenge, and has he given us a workable and rational prescription for the practice of science? We need to ask four questions: (1) Does Popper effectively address Hume's logical argument as he construes it, that is, that no universal explanatory theory is ever confirmed by reports of its positive instances? (2) Has Popper shown, *contra* Hume, that it is reasonable to prefer our best-tested theories for practical purposes? (3) Would Popper's methodological recommendations be a useful and fruitful set of guidelines for the conduct of science? (4) Does falsifiability provide a legitimate criterion for distinguishing science from nonscience?

As for the first issue, if you can "solve" a challenge by conceding it entirely, then Popper is successful. All corroboration amounts to, as Popper freely admits, is a report on how a theory has performed *up to now* (18). From the fact that a theory has so far not been falsified, we can make no predictions at all about its future performance. The very next test might prove it wrong. To say that one theory is better corroborated than another is only to say that it has passed more and stricter tests, but corroboration can no more indicate a theory's truth than confirmation.

For Popper, the real significance of saying that a theory is corroborated is that it is reasonable to adopt it for practical purposes. Does Popper then at least succeed in showing that it is rational to follow our best-tested theories and that we are not making an irrational leap? It is hard to see that Popper would have any advantage over the confirmation theorist here. If Hume is right, then *nobody* can reasonably have *any* confidence in the future reliability of theories, however well tested—not Aristotle, not Bacon, and not Popper. To downplay this conclusion would be to fail to take Hume's argument as seriously as it should be taken. We can "critically discuss" until we are blue in the face, but nothing will change this. Rational decision and action, *by their very nature as rational*, require projection, some reasonable expectation about what *will* happen in the future, not just what has happened up to now. Otherwise, it is just a leap in the dark. Does Popper provide a reasonable basis for any such practical projection?

Bryan Magee, an expositor and defender of Popper, says this:

> If we are rational we will always base our decisions and expectations on "the best of our knowledge" as the popular phrase so rightly has it, and provisionally assume the "truth" of that knowledge *for practical purposes*, because it is the least insecure foundation available. (1973, 27; emphasis in original)

But, if Hume is right, except in a psychological sense, there can be no such thing as a "least insecure" theory. We may come to *feel* that some outcomes are expected, but, logically speaking, all theories that make universal claims have the same level of security: none at all. Corroborated theories—those that have passed all tests up to now—are in precisely the same boat as refuted ones. That is the whole point of Hume's argument. At NO point can we have ANY confidence in the reliability of ANY universal theory, WHATEVER its record of success or failure. Indeed, even the most mundane action assumes that we can make *some* reasonable projections. Yet the Hume/Popper argument seems to imply that there is no rational basis for *any* such projection.

I suspect that the real reason that we are not more bothered by this argument is that nobody—nobody—really takes its conclusion seriously. If you did seriously consider that we can make no reasonable predictions about the future, and so we can make no rational decision about any action, you *would* soon be insane and probably would be involuntarily frozen into total catatonic immobility. As for Popper's recommendations, then, I think we have to say that if this be methodology, there is madness in it.

Popper himself attempted to back away from some of the seemingly absurd implications of his deductivist methodology. For instance, he develops a notion of "verisimilitude" (Popper, 1972, 52–60). For Popper, theory T_1 has more verisimilitude than theory T_2 if and only if T_1 contains more truth—but not more falsehood—than T_2. Further, he holds that strongly corroborated theories—that is, those that have survived more rigorous testing—are reasonably regarded as having more verisimilitude than theories that have not been strongly corroborated. Thus, while we can never have grounds for thinking that any theory is true, rigorous empirical testing can give us reasons to think that some theories are more approximately true than others. However, W. H. Newton-Smith has severely, and in my view decisively, criticized Popper's concept of verisimilitude and his use of that concept (1981, 44–76).

Would science be pursued more fruitfully, efficiently, or successfully if scientists followed Popper's recommendation and did not seek to confirm their theories but only to falsify them? In actual practice, scientists do not conform to any such dictum. They do want to discredit false theories, but they also try very hard to show that their theories are true. Darwin, for instance, sought not only to discredit special creation but also to argue that evolution by natural selection was the *correct* account of the diversity of life. So scientists do not typically act in the way Popper thinks they should. Ought they?

The strength of Popper's methodology is that it underscores two important insights: (a) all scientific knowledge is tentative and subject to revision, and (b) we should subject our theories to the most rigorous tests possible in hopes of proving them false if they are. However, it is not clear that scientists need to be apprised of these insights; they seem to be old news. Aristotle thought that scientific knowledge had to be certain, but scientists have long since abandoned that idea. A quip attributed to T. H. Huxley famously notes that the greatest tragedy of science is a beautiful hypothesis killed by an ugly fact. If even the most beautiful hypothesis can be killed by hateful, recalcitrant facts, then scientific knowledge must be recognized as tentative and only good so long as our theories do not run into those pestiferous facts.

As for the importance of rigorous testing, every inductive methodology more sophisticated than simple enumeration looks at negative instances as well as positive ones, as Bacon did. In fact, Bacon says quite plainly, "every contradictory instance overthrows a conjecture as to form" (1994, 171). That is, just as Popper would insist, a single counterexample suffices to overturn a proposed universal law. The confirmation theorist is just as interested in discarding bad theories (e.g., heat is caused by a bright glow) as Popper.

The way that scientific testing actually proceeds is a good bit more complicated than the above two paragraphs indicate. From the standpoint of formal logic, Popper's point that a single counter instance refutes a universal generalization is impeccable. However, in actual scientific practice, a well-established theory is hardly ever overthrown by a single recalcitrant fact. Often, it is far more reasonable to hang onto the theory and reject the apparent counterexample. The reason is that scientific theories are never tested in isolation but always along with various *auxiliary hypotheses*, that is, additional assumptions that are not part of the theory but that must hold if the test is to be a genuine test. As was noted at the time, the previous presentation of the HD theory was oversimplified. It omitted discussion of the crucial role of auxiliary hypotheses in the generation of test implications.

A scientific theory, by itself, seldom generates test implications. It does so only if we assume certain auxiliary hypotheses. That is, it is the theory and the auxiliary hypotheses combined that entail the test implication. The real schema of our scientific test is therefore this:

> If T (and A_1, A_2, A_3, . . . An) are all true, then I.
> If I and C, then O.

If, then, we test I in condition C and do not observe O, our reasoning goes as follows:

> Not-O.
> Therefore not-I or not-C.
> But C is true (i.e., the conditions specified in C are actually met).
> Therefore, not-I (that is, the test implication of T is false).
> Therefore, not-T or not-A_1 or not-A_2 or not-A_3 . . . or not-An.

Therefore, from the failure of the prediction, we cannot infer that the theory is false. All we can infer is that the theory OR some auxiliary hypothesis is false. The failure of the prediction might not be due to the theory but rather be due to the failure of an auxiliary hypothesis.

For instance, Copernican theory alone did not imply that scientists of the seventeenth century would observe stellar parallax. It only entailed that test implication if the stars, even the nearest ones, are not extremely far away. Opponents of the Copernican theory assumed that the nearest stars were reasonably close, but, as noted in chapter 1, this assumption is false. Even the nearest stars are unreasonably far away. Therefore, the failure to observe parallax with the instruments of the seventeenth century showed only that *something* was wrong. It turned out that what was wrong was not the Copernican theory but the auxiliary assumption that the nearest stars were close enough to have observable parallax.

For another example illustrating the importance of auxiliary hypotheses, recall again the phlogiston theory and Lavoisier's test. Lavoisier's experiment apparently refuted that theory by showing that the products of combustion weighed more, not less, as we would expect if some substance, phlogiston, had escaped during the burning process. Lavoisier's experimental test rested on an unstated auxiliary assumption, however. He assumed that phlogiston, like other substances, has mass and so has weight in the vicinity of a massive body like the earth. What, though, if this auxiliary assumption is wrong? What if phlogiston is a very different kind of substance, one that has *negative* weight? In that case, when phlogiston flows from a burning substance, we would expect the residue to weigh *more*, just as Lavoisier found! Hence, if the auxiliary hypothesis about the positive weight of phlogiston is wrong, Lavoisier's test fails to discredit the phlogiston hypothesis!

Of course, the idea that phlogiston would have "negative weight" is an arbitrary hypothesis that serves no function except to insulate the phlogiston theory from apparently falsifying evidence. Scientists and philosophers of science rightly scorn such sleazy dodges and label them "*ad hoc*" hypotheses. But the logical point remains. Counter instances show that *something* is wrong. When we deduce a prediction from a theory and its auxiliary hypotheses, and that prediction turns out wrong, the theory must be rejected *or* one of its auxiliary hypotheses must be discarded.

The upshot is that Popper's methodology offers no help at all in deciding what to do in (real-life) test situations like these. Obviously, the disconfirmation of theories is far more complex than a simple application of deductive logic. Of course, Popper admonishes scientists to face up to falsification and not resort to *ad hoc* dodges, but this is preaching, not a methodological prescription.

Does falsifiability give us a sound demarcation criterion for distinguishing science from nonscience? For instance, when something like Young Earth Creationism (YEC) comes along that challenges established science and claims scientific credentials for itself, can we deploy the falsifiability challenge to discredit it and consign it to the trash can of pseudoscience? Here it is important to make a distinction between the intransigence of scientists and the unfalsifiability of a theory. It may well be that proponents of YEC will dismiss, distort, or ignore all contrary evidence. They will thereby defy Popper's characterization of the true scientific attitude whereby bold conjectures are made and equally bold attempts are made to refute them. Defenders of YEC seem to be interested in buttressing, by any means possible, a dogma dictated by their fundamentalist convictions. To be fair, though, even the most indisputably genuine scientists can be awfully pigheaded and just as inflexibly attached to their own theories. There is a story often told about Einstein that he was informed by an eager graduate student that his theory of general relativity had been strongly confirmed by observation. Einstein reportedly shrugged. When asked by the student what he would have done had the observations gone against his theory, Einstein supposedly said: "Then I should be sorry for the Dear Lord. The theory is correct."

Falsifiability has to be a logical, not a psychological criterion. It must be a standard that scientific theories, and only scientific theories, can pass. The fundamental problem with using falsifiability as a demarcation criterion is that it is impossible to formulate that criterion so that it is neither too loose nor too strict. A demarcation criterion is supposed to be a gatekeeper, like one of those big guys with the velvet rope at an exclusive gala event. He is to let in only the invited celebrities in and keep out the *hoi polloi* like you and me. A demarcation criterion must admit only genuinely scientific theories and keep all others out. However, if "falsifiable" is construed too loosely, all sorts of pseudosciences might pass the test. A proponent of YEC, for instance, could say, "Sure, I will believe in evolution if you produce a missing link that is precisely in the middle between apes and humans in all of its anatomical features." Evolutionary theory does not predict such half-and-half missing links; transitional creatures are mosaics with some derived features and some ancestral ones. Still, the creationist would meet Popper's criterion, loosely construed, fair and square. He has mentioned a possible discovery that would conclusively falsify his evolution-denying theory, so his theory would have to count as scientific.

On the other hand, if we make the conditions for falsifiability too strict, we will rule out some theories and activities that are unquestionably scientific. Practically every theory clashes with *some* evidence or observation. Newtonian celestial mechanics famously clashed with two well-known observations: The orbit of the planet Uranus was not what would be expected on the basis of Newtonian theory. Also, there was the problem with the precession of the perihelion of Mercury. A planet's "perihelion" is the point of its elliptical orbit where it is closest to the sun. Mercury's perihelion "precesses," that is, it moves from one point along the orbit to another. According to Newtonian theory, this should not happen. These anomalies were eventually understood. The perturbations in Uranus's orbit were due to the presence of another planet, Neptune, which was eventually discovered. The precession of Mercury's perihelion was later explained by Einstein in terms of relativity theory. However, during the many years that these anomalies went unexplained, should Newtonian celestial mechanics have been regarded as unscientific because it was falsified by observation? Surely, if any theory has ever been scientific, Newtonian celestial mechanics was.

A criterion that cannot walk the fine line between being too strict and too lenient is not a useful tool for demarcation. In fact, almost all philosophers of science now agree that there is no single criterion that distinguishes science from nonscience. Science is just too multifaceted and complex an activity to characterize or categorize simply. I think that all we can really say is that a good theory will have many virtues and few vices and a bad theory will be the opposite. Sometimes a theory has such egregious vices and so few and minor virtues that it is rightly stigmatized as "junk science" or "pseudoscience."

Popper's methodology does not seem to be a viable way of meeting Hume's challenge. How should this challenge be answered? Obviously, the problem of induction is too deep and complex a problem to be meaningfully addressed in the small space we have here. Many answers have been given to Hume's problem, some of them formidably technical (*Stanford Encyclopedia*). What I present below are just some speculations and suggestions about possible responses.

What did Hume mean by "the uniformity of nature"? Clearly, there is no *general* expectation that nature will be uniform. Even armed with banks of supercomputers, forecasters cannot reliably predict the weather more than a few days ahead. Further, we recognize that some currently orderly arrangements are only temporarily stable patterns of phenomena that are fundamentally chaotic. At present the celestial mechanics of the solar system is a highly developed and accurate science. We can predict many years ahead of time when and where solar eclipses will occur. But we know that the orbits of solar-system bodies are not absolutely regular and are actually inherently chaotic, so they will change over time (see Peterson, 1993).

In general, many of the natural regularities that we take for granted are known to be temporary or have well-known exceptions. The sun rises approximately every twenty-four hours in nonpolar regions of the earth, but we know that in the deep past the earth rotated on its axis more rapidly, so days used to be considerably shorter than twenty-four hours. In the distant future, a day will be considerably longer than twenty-four hours. Further, some billions of years from now, when the sun swells to a red giant and engulfs the earth, earthly sunrises will be no more.

What these examples appear to indicate is that neither our

mundane nor scientific inductive inferences seem to depend on nature being uniform in any strict sense. The "uniformity of nature" might be only a relative and not an absolute uniformity, but it is stable enough long enough not to matter. Natural patterns might even change rapidly enough for us to notice, but do so in a predictable way that we can follow and adjust our expectations accordingly. The earth's magnetic field is not constant but fluctuates and actually reverses polarity over geological time (recall that the record of these shifts in seafloor rock was important evidence of continental drift). Yet we can follow this change and are not confused by it.

Philosopher Charles Echelbarger (private communication) says that Hume was aware of such real and imagined instances of nonuniformity, and that Hume asserts the uniformity of nature in the sense that we expect constancy in what we call causal laws. On Hume's interpretation, to say that *A*s cause *B*s really amounts to no more than the assertion that an unvarying relationship exists: *A*s are always followed by *B*s. The expectation that nature will be uniform therefore consists in the expectation that, for any occurrences *A* and *B*, when *A*s have consistently preceded *B*s in the past, we may expect *B*s to occur whenever *A*s occur in the future.

However, the ground for our trust in causal regularities is not just that we have experienced them to hold consistently in the past but also that we think that they rest ultimately on the basic laws of physics. We expect that the spark from the sparkplug will ignite the gas-vapor-and-air mixture in the cylinder of an internal-combustion engine. We explain the chemical volatility of air and gas mixtures ultimately in terms of the laws of physics. Our trust in the uniformity of nature therefore seems to come down, at rock bottom, to a trust in the constancy and stability of the laws of fundamental physics. For instance, there are some physical parameters that physicists take to be absolutely stable and universal. It is assumed, for instance, that the charge of the electron, the speed of light in a vacuum, and the gravitational constant are universal constants. Perhaps we should take Hume as claiming that not even these allegedly universal constants are dependable. He might demand to know how we can have confidence that these are constant and, despite past consistency, will not take on different values in the future. Maybe light will speed up or slow down. Who is to say that it cannot?

Professor Echelbarger also suggests (private communication) that Hume regarded the confidence we have in a basic uniformity to be what Hume regarded as a "natural belief." According to Hume scholar J. C. A. Gaskin, (1) natural beliefs are not generated by any process of reasoning and are impervious to skeptical doubt; (2) they are necessary presuppositions for knowing and acting, that is, you cannot really think or do anything unless you take these for granted; and (3) they are held by all humans everywhere (1993, 337). In other words, there are some beliefs that are so fundamental to rational thought and action that we just have to take them for granted, and, besides, we just do not have the power to doubt them, however cleverly skeptics might argue. That nature evinces sufficient stability to ground our mundane projections of past regularities may therefore be a natural belief, an expression of a fundamental trust that is necessary if we are to think and act rationally. As we said above, madness seems to be the only alternative to trusting that we can make *some* reliable projections. Perhaps this is all we really need to say about the expectation of some degree of uniformity in nature: We just cannot do without it.

With respect to the basic laws of physics, perhaps we can say a bit more. It has been realized fairly recently that the basic constants of nature are "finely tuned," that is, that had they not had definite values within very narrow ranges, no universe capable of supporting carbon-based life would have arisen from the big bang. Further, if those basic constants were now to vary by much, the present order of the universe would be so disrupted that life as we know it would no longer be possible. Perhaps the most popular current argument for the existence of God is based on this "fine-tuning." The proportion of life-friendly possible universes, where the constants are right for the existence of complex life, seems to be so vastly outnumbered by life-unfriendly possible universes, where the constants would not permit complex life, that, it is claimed, our universe must have been set up by a life-friendly creator.

The point is that if the values of the fundamental physical parameters were to change very much in a short period of time, we would not be here to notice the change. Life, or at least life as we know it, depends on the maintenance of a quite-precise degree of uniformity and constancy in the basic physical parameters. If gravity were suddenly reversed and became a repulsive rather than attractive force, we would not be around long enough to be chagrined at the change. Perhaps, then, we can say that we may assume that we will continue to see relative uniformity in the values of the basic physical parameters so long as we are here to observe those values. Perhaps, then, an "anthropic" answer might be the best one to give to Hume's challenge. We can count on the basic uniformity of nature, in the sense of the constancy of our constants, as long as we are in a position for such uniformity and the predictions we base on them to matter to us.

INFERENCE TO THE BEST EXPLANATION (IBE) TO THE RESCUE?

Finally, perhaps the Humean problem of induction can be largely sidestepped if we re-emphasize the extent to which, as noted in chapter 5, science relies on a form of reasoning that is neither deductive nor inductive. Inference to the best explanation (IBE) is very widely used in many different branches of science. Like induction and unlike deduction, IBE is an ampliative form of reasoning. The premises of an IBE argument do not guarantee the conclusion, but, like an inductive argument, give it some degree of support. I see what appears to be a footprint in the sand. It might be that the image was caused by the random actions of wind or wave, but, by far, the most reasonable explanation of an apparent footprint in the sand is that it was caused by a human foot.

IBE is like induction in the sense that it supports but does not prove a conclusion, but in other ways it is crucially different. Consider the following inference based on enumerative induction:

I observed crow C_1 and it was black.
I observed crow C_2 and it was black.
I observed crow C_3 and it was black.
. . . I observed crow *Cn* and it was black.
Therefore: All crows, both observed and unobserved, are black.

This argument cites premises stating past observations of crows and, on that basis, projects that crows observed in the future will also be black. This is precisely the kind of argument that Hume criticized. Just because we have observed a past regularity, the blackness of crows, on what basis do we project that past regularity to unobserved cases?

Consider now the following similar but crucially different IBE argument:

I observed crow C_1 and it was black.
I observed crow C_2 and it was black.
I observed crow C_3 and it was black.
. . . I observed crow *Cn* and it was black.
Therefore: All crows are black because this is the best explanation for the fact that all crows ever observed have been black.

This argument does not project a past regularity into the unseen future. Rather, it offers to explain that regularity, that all crows so far observed are black, in the most plausible and rational manner. If we have observed many crows at many different times and places, and without exception they have been black, it certainly seems more reasonable to explain this remarkable regularity by postulating that all crows are black rather than, say, 95 percent of them. Of course, we could be wrong. Maybe 99.9 percent of crows are black and we just have not had the good luck to see the rare nonblack ones. However, the more we experience uniformly black crows—crows that are black whenever and wherever we encounter them—the more plausible it is to postulate that all crows are black than some lesser percentage of crows.

Again, note the crucial difference. The kind of inductive arguments criticized by Hume *projected* future regularities on the basis of past ones. IBE *postulates* a natural regularity as the *best explanation* of the regularities we have observed. In other words, the most plausible explanation of the fact that we have consistently, without exception, and in a great variety of circumstances, observed *A*s to be *B*s is the hypothesis that all *A*s are *B*s. Absolute uniformity of observation is *more to be expected* on the hypothesis of the uniformity of reality. In other words, a universal regularity *in reality* is postulated as the best explanation of a universal regularity *in our experience*. *That* is how we really establish universal laws, the proponent of IBE will say, and not by a projection from the observed to the unobserved.

Naturally, nothing can be settled in a single chapter, and there is much, much more to be said about all of these points. All that this chapter has aimed to do is to introduce some of the main thoughts and problems concerning the perennial methodological problem of how we know universals when we encounter only particulars.

Working scientists might be impatient with this whole discussion. They might well wonder what right philosophers have to tell them their business or to dictate methodology to them. As the scientist often sees it, scientific methods *work*, that is, they suffice to extend our ability to predict natural phenomena and extend our technological reach. Further, those methods work in another sense: They let scientists get on with their jobs. For one thing, the accepted methods can bring about consensus in the community of researchers who have the requisite expertise about the matter at hand. For instance, paleontologists know much more about past life than any philosopher. If method *M* convinces the community of paleontologists that some claim *C* about past life is true, what qualifies philosophers to raise skeptical quibbles about *M* or its application in confirming *C*? Who appointed philosophers to be the justices of the Supreme Court of Reason? What qualifies them to judge the standards in fields where they stand as amateurs? Isn't science *obviously* on far more solid epistemological ground than philosophy? In the next chapter we will see that some of the leading philosophers of the twentieth century raised some of these points and have in fact recommended that science take over much of the investigation of certain topics that philosophy has traditionally regarded as part of its sovereign domain.

FURTHER READINGS FOR CHAPTER SIX

There are several excellent general histories of scientific method. One of the most popular introductory-level books is *A Historical Introduction to the Philosophy of Science*, by John Losee, third edition (Oxford and New York: Oxford University Press, 1993). Also recommended, but at a somewhat more advanced level, are *Scientific Method: An Historical and Philosophical Introduction*, by Barry Gower (London and New York: Routledge, 1997), and *The Arch of Knowledge: An Introductory Study of the History of the Philosophy and Methodology of Science* (New York and London: Methuen, 1986).

Aristotle develops his theory of the syllogism in the *Prior Analytics* and his methodology of science in the *Posterior Analytics*. Frankly, all of Aristotle's methodological works are very tedious to read. Authoritative secondary sources are very helpful here. A classic work of exposition is Sir David Ross's *Aristotle*, which first appeared in 1923. The sixth edition, with an introduction by John L. Ackrill, is now available (London and New York: Routledge, 1995). Ross gives a straightforward, detailed, no-frills explication of Aristotle's works. A very clear and succinct account of Aristotle's logic and methodology is given in G. E. R. Lloyd's *Aristotle: The Growth and Structure of His Thought* (London and New York: Cambridge University Press, 1968). Aristotle's logic is still studied in elementary logic courses to this day. A clear and helpful presentation of syllogistic logic is in *The Elements of Logic*, by Stephen F. Barker, fifth edition (Boston: McGraw-Hill, 2003). THE authoritative, professional-level study of Aristotle's biology is *Philosophical Issues in Aristotle's Biology*, edited by Allan Gotthelf and James G. Lennox (Cambridge: Cambridge University Press, 1987).

Unlike his *bête noir* Aristotle, Francis Bacon wrote highly accessible and readable prose. Even his methodological work *Novum Organum* is a pleasure to read. Perhaps one reason that Bacon had such success as an anti-Aristotelian polemicist is that, even in translation, Aristotle is so much harder to read. The edition of Bacon's *Novum Organum* that I use here is translated and edited by Peter Urbach and John Gibson (Chicago: Open Court, 1994). Urbach also wrote *Francis Bacon's Philosophy of Science* (La Salle, IL: Open Court, 1987), a lively, informative, and persuasive defense of Bacon as a philosopher of science. Ironically, Bacon's reputation traced the same sort of trajectory as Aristotle's. At one time Bacon was hailed as the high authority, and even in the nineteenth century major scientists like Darwin tried to meet what they perceived as the Baconian ideals of scientific method. Revisionist historians then downgraded Bacon almost to a minor figure of marginal significance. Urbach puts him back in his (IMO) rightful and deserved place. The more negative judgment by Michel Malherbe was from his essay "Bacon's Method of Science" in *The Cambridge Companion to Bacon*, edited by Markku Peltonen (Cambridge: Cambridge University Press, 1996), pp. 75–98. A very helpful and elementary-level introduction to inductive logic, including Mill's Methods, is *Choice and Chance*, by Brian Skyrms (Stanford, CT: Wadsworth, 2000).

The classic, elementary exposition of the hypothetico-deductive method is Carl G. Hempel's *Philosophy of Natural Science* (Englewood Cliffs, NJ: Prentice-Hall, 1966). This work, though long outdated in many aspects, is still a must-read for understanding the logical empiricist tradition that long dominated the philosophy of science in the twentieth century. In all of his writings, Hempel showed that philosophical rigor and clarity of exposition can go together.

Hume's critique of induction is most clearly presented in section 4, "Skeptical Doubts concerning the Operations of the Understanding," in *An Enquiry concerning Human Understanding*. I have used Antony Flew's excellent edition (La Salle, IL: Open Court, 1988). Flew's helpful introduction and notes make this a most useful edition, particularly for beginning students. The "problem of induction" has been much discussed. For a very helpful overview of the attempted "solutions" to the problem by one of the twentieth century's top philosophers of science, see *The Foundations of Scientific Inference* by Wesley C. Salmon (Pittsburgh: University of Pittsburgh Press, 1966). The online *Stanford Encyclopedia of Philosophy* has the article "The Problem of Induction" by John Vickers, which offers a more advanced presentation of the problem and its proposed solutions.

Hume's famous remark about reason being the slave of the passions is found in his *Treatise of Human Nature*, edited by L. A. Selby-Bigge (Oxford: Oxford University Press, 1975).

Sir Karl Popper's most famous and important statement of scientific method is his *Logik der Forschung*, published in Vienna in 1934 and translated into English (by Popper himself) as *The Logic of Scientific Discovery* (New York: Harper Torchbooks, 1959). The quotes from Popper in this chapter come from his *Objective Knowledge: An Evolutionary Approach* (Oxford: Oxford University Press, 1972). Bryan Magee's very lucid and succinct exposition of Popper's thought is *Popper* (Glasgow, UK: Fontana, 1973). An excellent and somewhat more detailed study is Geoffrey Stokes's *Popper: Philosophy, Politics, and Scientific Method* (Cambridge: Polity Press, 1998). A very useful and interesting examination and critique of Popper's falsificationism is in A. F. Chalmers's *What Is This Thing Called Science?* third edition (Indianapolis and Cambridge: Hackett, 1999). A more general, and, I think, cogent critique of Popper's philosophy of science is in W. H. Newton-Smith's *The Rationality of Science* (Boston: Routledge & Kegan Paul, 1981), pp. 44–76.

On the chaotic nature of solar-system orbits, see Ivars Peterson, *Newton's Clock: Chaos in the Solar System* (New York: W. H. Freeman, 1993). J. C. A. Gaskin's explanation of Hume's "natural belief" is in his chapter "Hume on Religion" in *The Cambridge Companion to Hume* (Cambridge: Cambridge University Press, 1993), pp. 313–44. As mentioned in the "Further Readings" section of the previous chapter, the most complete and authoritative discussion of inference to the best explanation is *Inference to the Best Explanation* by Peter Lipton (London and New York: Routledge, 1991).

CHAPTER SEVEN

IF YOU HAVE SCIENCE, WHO NEEDS PHILOSOPHY?

THE LIMITS OF SCIENCE?

Nothing succeeds like success, as the saying goes, and by any reasonable standard natural science has been stupendously successful. Science is so successful, both with respect to practical achievements (vaccines, smart phones, smart bombs, space telescopes, nanotechnology, etc.) and in the construction of sophisticated and powerful theories (e.g., quantum mechanics, special and general relativity, big bang cosmology, the neo-Darwinian synthesis, plate tectonics, etc.) that we naturally ask: "What, if anything, can science NOT do?" The ongoing expansion of science raises important philosophical questions: Are there limits to scientific knowledge? That is, are there some matters that are important and knowable but not accessible by the empirical methods employed in the natural sciences? Are there questions, some perhaps of central significance to human life, that lie beyond the competence of science? What kinds of questions would they be? Are there aspects or regions of reality in principle impenetrable by scientific methods? Concomitantly, do other fields of human inquiry—philosophy, theology, criticism, or jurisprudence, perhaps—have resources that science lacks that enable them to address questions that are "beyond" science? Are there alternative "ways of knowing," fundamentally distinct from science but just as legitimate in their own domains?

These are deep questions. Any adequate answer would require many more pages than we have here. Therefore, in this chapter I will not address these questions in a general way, but by focusing on two particular cases, which are illustrative of the broader issue of what science can and cannot answer.

In increasing numbers and with increasing aplomb, some philosophers have suggested that many important questions typically designated as philosophical, and traditionally addressed by philosophers using philosophical methods, should in fact be examined with the tools and methods of natural science. Examples included epistemology and ethics, which are core areas of philosophy. Epistemology examines human knowledge in the most general sense. It asks questions like: What is knowledge? Are there criteria of truth? How do we rationally ground, warrant, or justify our beliefs? How do we respond to skeptical challenges that question our basic claims to know? Ethics asks questions like: What is the best way to live a human life? What is the basis of moral obligation, that is, why ought we to do some things and not others? Is there a highest good that is the basis or foundation for all goodness?

Such epistemological and ethical questions are the topics of perennial debate, and this is the rub. For over 2,500 years the Western philosophical tradition has debated these questions endlessly, but it is not clear that any definite results have ever been achieved. For instance, epistemologists still argue over the role that sense perception plays in knowledge and how, or whether, we know things *a priori*. Ethicists still divide along lines that would have been recognizable to Mill or Kant—or Aristotle. Science is progressive in the straightforward sense that issues get settled and scientists move on. Philosophers keep returning to the same basic issues and seemingly never reach consensus. Could progress be made at last if these questions were addressed by science instead of by philosophy? Could emergent scientific disciplines offer solid answers about knowledge and morality instead of endless controversy?

It has happened before. Psychology emerged from philosophy as a scientific discipline in the late nineteenth century with works like William James's *The Principles of Psychology*. Previously, questions now regarded as the domain of psychology were addressed by philosophers. Leading philosophers, such as John Locke in *An Essay concerning Human Understanding*, and David Hume in *A Treatise of Human Nature*, made no distinction between questions now regarded as psychological and those now seen as philosophical. Psychology eventually emerged as an empirical science with its own distinctive methods and established results. Likewise, could a scientific epistemology and ethics supplant, or at least significantly displace, the philosophical study of these topics? Some philosophers answer yes, and others think that it has already largely happened.

While recognizing that there are now independent academic disciplines that used to be encompassed by philosophy (a doctorate in any liberal-arts field is still called a "doctor of philosophy," a PhD) many philosophers think that epistemology and ethics cannot be turned over to science. The reason, as they see it, is that science deals with *fact* whereas philosophy properly deals with *value*. In other words, science is fundamentally descriptive, not normative; it is very good at judging what *is* the case, but not for judging what *ought* to be. Psychology might tell us how we reason or how we make ethical decisions, but it takes philosophy to tell us how we should reason if we want to be rational beings and how we should act if we want to be moral beings. Epistemology discusses epistemic values like truth, justification, or rationality and the appropriate criteria for judging whether and to what extent those values are realized. Ethics conducts analogous debates with respect to moral values and norms. For instance, epistemologists might argue about whether truth or merely justified belief is the appropriate goal of human cognitive activity, and, concomitantly, how such goals are to be realized. Ethicists, on the other hand, might argue about whether moral acts should be judged by their consequences or by the nature of the act itself, for example, whether it was done out of pure respect for the moral law. Addressing such questions, many philosophers say, is the job of philosophy and science cannot settle them.

Opposing these more-traditional philosophers are advocates of naturalized epistemology and ethics, that is, philosophers who hold that questions of epistemic and moral values and norms can and should be addressed by the tools and methods of natural science. These philosophers deny the alleged dichotomies between fact and value and between descriptions and norms. They argue that humans are so constituted that what they find intrinsically valuable is just as much a biological fact as that they are bipeds or have hair—and that it is a fact that certain norms guide them toward the realization of such values. In this case, they argue, psychology, neuroscience, cognitive science, and evolutionary biology might have much of importance to say about questions of epistemology and ethics. Indeed, they hope that a naturalized epistemology or naturalized ethics might break some of the ancient deadlocks.

THE BREAKDOWN OF PHILOSOPHY

The project of naturalizing philosophy is actually a response to the perceived breakdown of traditional ways of doing philosophy. In the late 1970s and early 1980s, two books appeared that claimed that philosophical projects that had long reigned were now dead and that radical new approaches were needed. One book was Richard Rorty's *Philosophy and the Mirror of Nature*. The other was Alasdair MacIntyre's *After Virtue*. Both created quite a sensation at the time, becoming bestsellers (to the extent that books of academic philosophy are ever bestsellers). Each was written by a philosopher regarded as among the best in his field. Rorty and MacIntyre both had distinguished records of publication and held positions at top universities, yet each argued that the field to which he had devoted a lifetime of scholarship had gone seriously awry.

Rorty argued that the project of foundationalism had long been the heart of epistemology (1979, 155–64). The term *foundationalism* is based on a metaphor. A building must rest on a solid foundation, or it will collapse. Philosophers have also often held that the edifice of human knowledge must also have a solid foundation; it must rest on a set of indubitable beliefs, or the whole structure will be unsound. The archetypical foundationalist was René Descartes, whose *Meditations on First Philosophy* attempted to provide a once-and-for-all foundation for knowledge by resting all of his beliefs on two indubitable truths—that he exists and that he is a thinking thing—and a supposedly infallible cognitive aptitude—clear and distinct perception. With these meager tools, Descartes set out, first, to prove the existence of God, since only the existence of a benevolent, all-powerful deity could ensure that we are not in a state of pervasive intellectual darkness and deception. Unfortunately, hardly any other philosophers have accepted Descartes's arguments for the existence of God, and, if those fail, his entire project fails.

Descartes's failure did not discourage other philosophers' search for foundations because the intuition behind foundationalism is a strong one. Knowledge is traditionally defined as justified true belief. Justification, the rational authorization we have for holding a belief, is often regarded as conferred by a grounding relationship between beliefs. Belief *A* is grounded on belief *B* if and only if *B* is itself justified and *A* is legitimately inferred from *B*. The intuition underlying foundationalism is that our beliefs are justified only if they are ultimately grounded on basic beliefs. A basic belief is a belief that is justified even though it derives its justification from no other beliefs. Many beliefs are nonbasic, that is, they are based on (justified by) other beliefs. I have the nonbasic belief that Ulan Bator is the capital of Mongolia. I have heard this from a number of sources. My belief that Ulan Bator is the capital of Mongolia is therefore based on my further belief that these sources are reliable. My belief that these sources are reliable is also nonbasic and is based on other beliefs—such as that when independent sources say that something is so, and they are in a position to know, and have no reason to deceive, then it is probably so. And so on.

Thus, if we let the arrow, "→" mean "is believed on the basis of," and *A*, *B*, *C*, and so on stand for beliefs, we can represent the justifying relationships between nonbasic beliefs like this: $A \to B, B \to C, C \to D,$ $D \to$ . . . and so on. But this chain cannot extend back *ad infinitum*, or none of our beliefs will have any rational basis. As we discussed somewhat in the previous chapter, the reason is that the justificatory relationship is provisional: A can derive its justification from B only if B itself is justified, and B in turn is justified only if C is; and so on. But if this chain goes back forever, without being based ultimately on a belief that just IS justified and derives its justification from no more basic belief, then no member of the chain is finally justified. We must therefore postulate a set of basic beliefs, beliefs that are self-justifying and that serve as the ultimate foundation, the source from which the justification of our entire edifice of beliefs derives. Such basic beliefs stand on their own in a state of absolute, self-sufficient justification; in their absolute rational autonomy, other beliefs can neither support nor undermine them.

Foundationalist epistemologists broadly agreed on candidates for such foundational beliefs. They identified beliefs that are (a) self-evident, (b) incorrigible, or (c) evident to the senses as candidates for basic beliefs, the kind that can ground all human knowledge. A self-evident proposition is one that we recognize as true the moment we understand it. "Anything that is square has a shape" or "1 + 1 = 2" would be examples of propositions that express self-evident beliefs. An incorrigible belief is one that we cannot be wrong about; that is, if we honestly think it is so, then it must be so. If I honestly think that I perceive bright red, then I actually perceive bright red, even if it is a hallucination and there is not really anything red in my visual field. Finally, anything that is an immediate deliverance of the senses is something I can be sure of. If I have the taste of licorice or hear the sound of thunder, I cannot be mistaken that this is what I taste or hear—even if I am having a gustatory or auditory hallucination. The project of foundationalism was to show how all of human knowledge could ultimately be based on such secure, indubitable basic beliefs. If successful, this project would show how human knowledge could rest on a secure, certain base. Descartes's dream would be realized at last.

Foundationalism sounds too good to be true, and it is. For one thing, the distinction between "basic" and "nonbasic" beliefs appears simplistic and naïve. We do treat some of our beliefs as relatively basic in the sense that we take them for granted and see no reason at a given time to inquire about their justification. I see no reason to seriously question that I am seated at a desk in front of my laptop computer as I type these words. But which beliefs really are basic in the sense that they are indubitable and not even in principle subject to rejection or revision in the light of new information? Even some beliefs regarded as "self-evident" have been discarded over time. For many, it used to seem self-evident that Euclidean geometry is *the* true geometry of space. Then non-Euclidean geometry came along and proved, in certain contexts, to be a better model for physical space. Physicists took it as self-evident that light can be a wave or a particle but not both. Quantum physics showed that there are times that light can be described as a wave and other times where it is correctly describable as a particle. Even "raw feels" can be misunderstood and misreported. Trying to get patients to describe their symptoms precisely is a source of frustration for doctors. Is that feeling pain or just discomfort? Is it a stinging or a burning sensation?

Some philosophers say that *none* of our beliefs should be regarded as in principle immune to revision. Be that as it may, it is clear that any residue of absolutely indubitable beliefs would constitute far too meager a basis to constitute the rock-solid foundation of all knowledge. Trying to base the entire structure of human knowledge on a paltry set of indubitable truths truly would be to construct an upside down pyramid!

For Rorty, the demise of the foundationalist project had far-reaching implications for philosophy (357–94). If philosophy could no longer promise an ultimate grounding for knowledge, what real authority could it claim? Rorty pushed this point, questioning the very self-image of philosophy. Philosophy, Rorty charged, had long regarded itself as the Tribunal of Pure Reason, the court of last appeal on all matters of human rationality. Such grandiosity could be justified if philosophers did indeed hold the key to ultimate epistemic grounding. But once this promise was debunked, what is left to justify philosophers' inflated view of themselves and their mission? What, for instance, especially qualifies philosophers to comment on the standards, values, or methods of fields in which, frankly, they are amateurs? Surely, for instance, the physicist is the best judge about what counts as good evidence in physics. If a philosopher disagrees, the attitude of the physicist is appropriately disdainful.

Even in areas like ethics, does expertise in ethical theory make a philosopher the most qualified to make ethical decisions? In real-life situations, where practical, first-hand experience is really needed to make the delicate and difficult decisions, is the approach of the ivory-tower philosopher appropriate? For instance, who should make tough decisions about medical ethics, doctors who have struggled for years, face-to-face with the issues, or a philosopher who has only a sophisticated theory and little practical experience? What finally separates the ivory-tower philosopher from the ideologue who obeys the dictates of theory blithely unhindered by the complexities of practical realities?

Such questions are difficult, and, for a philosopher, disquieting. Rorty recommends a radical deflation of the role of philosophy. According to him, the demise of foundationalism leaves us with no viable epistemological alternative to conceptual relativism. He recommends what he calls "epistemological behaviorism," that is, that we should seek no deeper grounding for our beliefs than the reigning epistemic standards of our given time and place. Say what your culture lets you say. There can be no deeper "grounding" for our beliefs than that. Living in a scientific culture, we Western intellectual-types will naturally justify our beliefs in the light of science. However, that is merely our cultural practice, and is not in any sense inherently or absolutely superior to the practices of other cultures. If, by contrast, a conservative Islamic culture settles questions by appeal to the Qur'an, we Westerners are in no position to criticize.

Given cultural and cognitive relativism, and the resulting impossibility of grounding our discourse in some set of foundational metastandards, what can the role of the philosopher be? Rorty holds that the philosopher should now be a sort of intellectual jack-of-all-trades who specializes in promoting the "conversation of mankind" (389–94). The philosopher, for instance, could promote dialogue between the Western, scientific, secular worldview and the conservative Islamic worldview, thus promoting mutual understanding and tolerance.

However, I think that there are strong reasons for regarding Rorty's recommendation of "epistemological behaviorism" to be wrongheaded and pernicious. The nightmare world of George Orwell's *1984* is precisely a society in which you say just what your culture—that is, Big Brother—lets you say. No, Orwell's Winston Smith was right: The most basic freedom is the freedom to say that 2 + 2 = 4—just because it *is* so, whether or not your society approves. Rorty thinks that cognitive relativism is the default position that we revert to when foundationalism fails. I, on the contrary, think relativism to be just as doctrinaire and artificial as foundationalism. Further, I consider Rorty's deflated view of the role of philosophy to be too deflated. I would let only some of the air out, not all of it. This not the place to develop a critique of Rorty. Rather, I will just point out that the failure of foundationalism need not entail a turn to cognitive relativism. There is another alternative, namely epistemological naturalism. The epistemological naturalist agrees with Rorty's claim that human knowledge neither needs—nor can have—the sort of grounding the foundationalist sought. However, he disagrees that cognitive relativism and epistemological behaviorism are the correct responses to the failure of foundationalism. We will see below how the naturalistic alternative in epistemology has been developed.

Alasdair MacIntyre's *After Virtue* makes as bold a claim as Rorty's book. MacIntyre argues that the "Enlightenment Project" in ethics has failed and was bound to fail. The Enlightenment Project was to ground ethics in reason alone. The prototypical proponent of the Enlightenment approach to ethics was Immanuel Kant (1724–1804). Kant held that ethical obligation—duty—is absolute, not conditional. The commands of duty are categorical, not hypothetical. Nonmoral commands are only hypothetical, that is, they tell us what we must do *if* we want to achieve some end or accomplish something. For instance, it is a merely hypothetical imperative that *if* you want to go to law school, *then* you should prepare to take the LSAT. The imperative to prepare to take the LSAT is imposed only on those who want to go to law school; the rest of us are spared that burden. By contrast, moral imperatives are categorical: do not murder; do not cheat; do not steal. These commands are not conditioned by any "if . . . then" clauses specifying any particular aims or goals. They are pure commands, binding on all rational creatures *qua* rational and independently of particular circumstance.

Because duty is an unconditioned, categorical demand, Kant held that it cannot be based on anything parochial, contingent, or uncertain but must have a universal, necessary, and certain basis. Therefore, it cannot be that the ground of ethics is to secure human happiness. Such an aim is limited, since it only encompasses human happiness, whereas Kant thought that moral duty must bind *all* rational creatures, even any nonhuman ones. Further, though humans may in fact seek happiness, this is a contingent fact about them; it does not *have* to be so but just is so as a matter of fact. Finally, even if it is universally true that humans seek happiness, the nature of happiness is vague and uncertain while the basis for ethics has to be proven. In short, nothing merely empirical or factual can ground ethics; its grounding must be *a priori*, that is, based on the formal nature of pure moral reasoning itself:

> It is the essence of reason that it lays down principles that are universal, categorical, and internally consistent. Hence a rational morality will lay down principles which can and ought to be held by *all* men, independent of circumstances and conditions, and which could consistently be obeyed by all rational creatures on every occasion. (MacIntyre, 1984, 45)

The guiding principle of morality, what Kant calls the categorical imperative, is simply that moral rule that I act upon must be one that I can consistently will that *everyone* act upon.

Such a concept of morality sounds noble and high-minded; the supreme importance of morality means that it must be based on the highest authority, and this is the authority of reason itself. However, as MacIntyre and other critics of Kant have noted, the grandeur of the structure of Kant's ethics is matched by the emptiness of its content. Really, if our only rule of morality comes down to logical consistency (i.e., whether I can, without contradiction, will the same rule for myself and everyone), then we are told very, very little about how we should act. All sorts of absurd or pernicious rules seem to meet that criterion. A religious fanatic could require that *everyone* have the same level of zeal that he has for his god. Indeed, you could consistently require that everyone worship *you* as a god! Kant tries to give more content to his abstract imperative by offering another version that says that you should treat others as ends in themselves and never as means only. In other words, don't just use people; treat them with the dignity and respect that every person deserves. Fine, but as MacIntyre points out, I could with perfect consistency will that everyone be treated as means except me (46)!

Kant's failure is not idiosyncratic, MacIntyre holds, but is typical of all the major Enlightenment and post-Enlightenment ethical thinkers. The basic problem, as MacIntyre sees it, is a loss of any idea of a human *telos*, a concept of the true, completed, fulfilled, human nature—humans as they *could* be rather than as how they are. That concept of a state of self-actualized completion was a crucial element of the pre-Enlightenment ethical tradition that began with Aristotle's *Nicomachean Ethics*. Aristotle's ethic was a fully naturalistic one. As he saw it, the human *telos* is determined by the biological and metaphysical nature of the human organism; our specific and distinctive nature as rational animals determines what sort of life will be ideally desirable for us. For Aristotle, the *telos* provided criteria for determining, from among the many things people do desire, the ones that are actually desirable, that is, the ones that are worthy of desire. The truly desirable is that which is conducive to the realization of the human *telos*.

I will expand below on the nature of this natural *telos*. It was the loss of this concept in Enlightenment ethical thought that doomed the entire project (MacIntyre, 1984, 54–55). If ethical imperatives no longer aim at a natural *telos*, then what grounds them? How do we distinguish the desirable from the merely desired? Without a *telos* to ground them, moral imperatives are reduced to free-floating commands, arbitrarily imposed and devoid of authority. The logical consequence is ultimately the view that our moral prescriptions rest on nothing but feeling, and that ethical "discourse" can be nothing more than emotive assertion. I plump for my values and you plump for your values. This, says MacIntyre, is the actual, terrible situation we are in, living in the wreckage and detritus of what was once a coherent ethic, flinging its shards at one another in fierce polemic, but with no basis in rational argument and no hope for rational agreement (1–2). Vivid confirmation of MacIntyre's analysis is provided by any of the TV programs where public issues are supposedly debated but the actual content is canned talking points, posturing, spin, and *ad hominem* abuse.

MacIntyre's book combines the rigor of philosophical argument and the resources of vast learning with the ringing moral authority of Old Testament prophecy. It concludes by offering us a stark choice: Either Nietzsche or Aristotle (256). That is, either we admit that moral rules are really only the disguised expressions of the human desire to acquire and wield power (Nietzsche), or we return to the naturalistic ethical tradition that began with Aristotle. MacIntyre recommends that we get back to the naturalistic, Aristotelian tradition.

Of course, Rorty's and MacIntyre's sweeping attacks can be and have been opposed. However, I will here take it for granted that the project of epistemological foundationalism is now defunct, as Rorty argues, and, further, that MacIntyre has offered a cogent analysis of the obvious extreme disarray of ethical discourse in our culture. Naturalistic alternatives have been proffered as suitable remedies for these breakdowns. In the next section we will look at a salient attempt to articulate a naturalized epistemology. Then, in the section that follows, we will look at Aristotle's naturalistic ethics and a notable recent attempt to update that position.

NATURALIZING EPISTEMOLOGY

In 1969 the eminent American philosopher W. V. O. Quine wrote an influential essay recommending the naturalization of epistemology. Quine's essay is densely argued and draws on much detail from the history of analytic philosophy, including some of his own best-known arguments (the indeterminacy of translation and the rejection of analyticity). There is no need to recapitulate the entire argument here and I will merely indicate a few of its main points. Quine held that the job of epistemology is to elucidate the foundations of science, including the foundations of mathematics (Quine, in Kornblith, 1985, 15). Such elucidation was to be achieved by reducing the abstruse to the familiar by showing that even most recondite theory can, in principle, be translated into the more intellectually palatable terms of sensory experience, logic, and set theory. An outstanding example of the project of analytic epistemology was Rudolf Carnap's 1928 work *The Logical Construction of the World*. Yet even Carnap's rigorous effort to achieve this failed and Quine holds that the project of reducing science to sense data, logic, and set theory is hopeless (20).

In that case, that is, given the past failure and poor prospects of analytic epistemology, Quine recommends that epistemology simply be replaced by psychology. Psychology, a branch of natural science will take over the job of epistemology by showing how science arises from sensory input:

> [Psychology] studies a natural phenomenon, viz., a physical human subject. This human subject is accorded a certain experimentally controlled input—certain patterns of irradiation in assorted frequencies, for instance—and in the fullness of time the subject delivers as output a description of a three-dimensional external world and its history. The relation between the meager input and the torrential output is a relation that we are prompted to study for somewhat the same reasons that always prompted epistemology; namely, in order to see how some evidence relates to theory, and in what ways one's theory of nature transcends any available evidence. (24)

This is Quine's "replacement thesis"—that natural science, in the form of psychology, can simply replace epistemology. Psychology can do the job of elucidating the relationship between evidence and theory. This was the job epistemology promised but failed to do.

Yet the replacement thesis faces an obvious, glaring problem. As Hilary Kornblith notes, psychology can tell us how we *do* arrive at our beliefs, but we can still ask, "Are the processes by which we do arrive at our beliefs the ones by which we *ought* to arrive at our beliefs?" (1; emphasis added). *Prima facie*, it seems, replacing epistemology with psychology is as implausible as replacing ethics with social science. Could we replace our ethical accounts of how people *should* act with sociological and anthropological accounts of how they *do* act? This seems strongly counterintuitive, to say the least. The obvious fact of the matter is that people do not always (or perhaps even usually) act the way that they should. So, if the goal of ethics is to tell us how people should act, then the idea that we could replace, without loss, the field of ethics with the factual behavioral descriptions of the social sciences seems preposterous.

However, might there be a reason for holding that the way we do arrive at our beliefs is, in fact, the way that we should do so? In another essay, Quine suggests that, for instance, inductive inferences about kinds and their qualities might have a Darwinian validation:

> If people's innate spacing of qualities is a gene-linked trait, then the spacing that has made for the most successful inductions will have tended to predominate through natural selection. Creatures inveterately wrong in their inductions have a pathetic but praiseworthy tendency to die before reproducing their kind. (38–39)

How do we explain the fact that our inductive inferences up to now have been reliable? Because if they had not been, we would never have survived long enough to ponder the question!

Quine's appeal to Darwin is plausible and intriguing. However, as Kornblith notes, to justify the replacement thesis, Quine would have to repose a great deal of confidence in his answer:

> The attempt to defend the replacement thesis by way of the Darwinian argument requires that the conclusion of that argument be given a very strong reading. Someone who concludes on the basis of natural selection that the process by which we acquire beliefs must be roughly like the process by which we ought will not be in a position to defend the replacement thesis. If the psychological investigation is to be able to replace epistemological theorizing, there must be a perfect match between the processes by which we do and those by which we ought to acquire beliefs. Without such a perfect match, the results of psychological theorizing will only give an approximate answer to question 1 ["How ought we to arrive at our beliefs?"], and epistemology will be called upon to make up the slack. (Kornblith, 1985, 5)

Yet Quine admits that Darwinian considerations only give a "partial" explanation of why inductive inference should be considered reliable (Quine, 39).

Actually the situation is much worse than that for a straightforward replacement thesis. Studies in cognitive science show, unequivocally, that the way we very often, indeed usually, arrive at our beliefs is not in any sense a rational process. On the contrary, we seem to be strongly disposed to irrationality, as examples below will illustrate. Therefore, the idea that we can understand how beliefs *should* be formed by looking indiscriminately at how they *are* formed is a hopeless task. *Inevitable Illusions: How Mistakes of Reason Rule Our Minds* (1994), by cognitive scientist Massimo Piattelli-Palmarini sums up much of this evidence.

According to him, we are prone to cognitive illusions that fool our thinking just as optical illusions can fool our seeing. He often draws on the work of psychologists Daniel Kahneman and Amos Tversky, who have done a number of famous experiments showing how people, even the highly educated and intelligent, are prone to bad thinking. Some of the most notorious examples involve our spontaneous judgments about probabilities.

For instance, imagine that you have heard of a highly unpleasant, fatal, and incurable but fortunately rare disease that afflicts only one in ten thousand people. You are a slight hypochondriac and decide to get tested for the disease even though you have no symptoms and do not belong to any high-risk subgroup that would make it more likely that you have the disease. The test for the disease is highly accurate. Ninety-eight percent of the time, if a person has the disease, the test report is positive, and only 2 percent of persons with the disease will test negative. In other words, the rate of false negatives is 2 percent. Also, the test has a very low rate of false positives, so that there is only a 2 percent chance that you test positive if you do not have the disease.

You are appalled when the test comes back positive. What should your reaction be? Should you (A) start writing your last will and testament, or (B) don't get too excited since you very probably don't have it? Amazingly, the answer is (B). Though the test is very accurate, with very low rates of false positives and false negatives, it is, on the basis of just one test, highly unlikely that you have the disease. How can this be?

Do the math: Out of a randomly selected population of 10,000, on average 9,999 people will not have the disease. The test has a false positive rate of 2 percent, so of those 9,999 disease-free people, 9,999 × .02, or on average about 200 people, would get false positives if tested. Out of those 10,000 randomly selected persons, on average only one person would have the disease, and that person would very likely test positive for it. So, if 10,000 randomly selected persons were tested for the disease, on average there would be about 200 disease-free persons who test positive for every diseased person that tests positive. If you tested positive for the disease, you are far more likely to be in the group of those who got a false positive simply because the number that do not have the disease (9,999) is so much higher than the one that does. In other words, you would only have about a 1/200th or one-half of 1 percent of a chance of having the disease even if you tested positive.

Piattelli-Palmarini notes that when presented with a similar example, even most hospital doctors get it wrong and think that there is a much higher chance that the person getting the positive report has the disease (1994, 81). What goes wrong here is that our spontaneous estimates of probability strongly tend to ignore base rates. The base rate is the background incidence of the phenomenon in question. In the above example, the base rate is that the disease afflicts only one in ten thousand. This base rate is so low that, as the math shows, even a highly accurate test will produce far more false positives than true positives, meaning that the probability that you have the disease given that you tested positive will be quite low. Generally, though, we don't do the math. We go with what our "gut" feeling is, and our "gut" tells us that surely the results of such a reliable test must be right.

Such examples seem simple enough once they are explained, but when people follow their spontaneous hunches, they overwhelmingly get them wrong. The mistakes in reasoning presented by Piattelli-Palmarini are just a tip of the iceberg of spontaneous human irrationality. A popular book by skeptical activist Michael Shermer, *The Believing Brain*, reports the really bad news about how we form and hold our beliefs. Early in the book, Shermer summarizes his claims:

> The brain is a belief engine. From sensory data flowing in through the senses the brain naturally begins to look for and find patterns, and then infuses those patterns with meaning. The process I call *patternicity: the tendency to find meaningful patterns in both meaningful and meaningless data*. The second process I call *agenticity: the tendency to infuse patterns with meaning, intention, and agency*. We can't help it. Our brains evolved to connect the dots of our world into meaningful patterns that explain why things happen. These meaningful patterns become beliefs, and these beliefs shape our understanding of reality. Once beliefs are formed, the brain begins to look for and find confirmatory evidence in support of those beliefs, which adds an emotional boost of further confidence in the beliefs and thereby accelerates the process of reinforcing them, and round and round the process goes in a positive feedback loop of belief confirmation. (2011, 5; emphasis in original)

Of course, Shermer can make his case only if *some* human beliefs are formed rationally. If *all* human beliefs are formed irrationally, then so are Shermer's beliefs about human rationality! If we draw on cognitive science to show us that human beliefs are all too often formed irrationally, we must still trust that cognitive science itself has the resources to draw rational conclusions about those instances of irrationality. Still, Shermer's analysis is acute. He has quite a bit to say about the deleterious effect of confirmation bias, one of the idols of the tribe Bacon noted four centuries ago. Shermer calls confirmation bias "the mother of all cognitive biases" (259).

One great advantage that researchers now have over Bacon is that with modern imaging techniques they can actually see how the brain works when subjects are exercising cognitive bias. Shermer reports that during the 2004 election season, thirty subjects, half Republicans and half Democrats, were given brain scans and asked to assess statements by the candidates, Republican George W. Bush and Democrat John Kerry, and judge which had contradicted himself. Naturally, Republicans condemned Kerry and excused Bush, while Democrats did the opposite. Shermer comments:

> What was especially interesting were the neuroimaging results: the part of the brain most associated with reasoning—*the dorsolateral prefrontal cortex*—was quiescent. Most active were *orbital frontal cortex*, which is involved in the processing of emotions, and the *anterior cingulate cortex* . . . which is so active in patternicity processing and conflict resolution. Interestingly, once subjects had arrived at a conclusion that made them emotionally comfortable, their *ventral striatum*—a part of the brain associated with reward—became active. (260; emphasis in original).

"A rational man proportions his belief to the evidence," Hume said. The brain study cited by Shermer indicates that we too often do the opposite. We proportion our beliefs to our emotions and process the evidence as our feelings demand.

The upshot is that a wholesale and uncritical displacement of epistemology by psychology will leave us with no way of distinguishing good practices of belief formation from bad ones. Surely this is an unacceptable outcome. But maybe Quine means that we should not focus on how we generate beliefs *in general*—which is bound to include much silly stuff—but only on how *scientists* arrive at their *scientific* convictions. We would like to know, for instance, just how Einstein arrived at his theories of relativity or how Darwin became convinced of natural selection. However, even a topnotch scientist will all too often leap to a conclusion, become obsessed with an *idée fixe*, appeal to *ad hoc* hypotheses, or ignore, dismiss, or misconstrue countervailing evidence. Further, the rationality and objectivity of science has much less to do with the psychological processes that transmute sensory experience into theory in a particular brain than with the communal norms and collective practices of scientific communities that systematically guard against bias. For instance, a double-blind methodology is an effective guard against confirmation bias and other distortions precisely because it and other effective methods reduce and rigorously constrain the role of personal judgment in belief formation. Science is objective not because scientists are just naturally more objective thinkers—or due to any other quirk of the scientific brain (including intelligence)—but because scientists acknowledge their biases, and construct methods to counteract them.

Any viable naturalized epistemology will have to accommodate epistemic norms (and the methods that we design to implement those norms). If the goal of a naturalized epistemology is to provide a scientific account of human knowledge, and if, as we say, such an account must encompass epistemic norms, then norms must be scientifically justifiable. Norms are imperatives. They are sets of prescriptions and proscriptions. How can an imperative be scientifically justified? Some philosophers argue that they cannot and that certain essential epistemic norms must be justified *a priori*. If this is so, then a naturalized epistemology is impossible in principle since natural science can provide us only with empirical and not *a priori* knowledge. In his stimulating book *Knowledge and Its Place in Nature* (2002), Kornblith addresses the claim that epistemic norms must appeal to *a priori* intuitions.

The standard practice of philosophical epistemology is something like this: Epistemology is concerned with propositional knowledge, "knowing that" rather than "knowing how." Knowing how to juggle or how to ride a bike is not the kind of knowledge of interest to epistemologists; knowing that the sun is bigger than the moon or that the price of gold has just risen or that fish-oil supplements improve memory (if they do) is the kind of knowledge that engages epistemologists. Put another way, epistemologists are interested in knowledge that can be expressed in propositions of the form "*S* knows that *P*," where *S* is a knowing subject and *P* is what he knows.

As summarized by philosopher Scott Sturgeon, the way that epistemologists approach the study of propositional knowledge is by proposing schemas like this:

(K) *S* knows that *P* iff (if and only if) *S* and *P* have features F_1, F_2, ... (1995, 11)

Sturgeon continues:

> By completing (K) we uncover the essence of propositional knowledge. To see how this works, notice that (K) is a *biconditional*. This means (K) is two claims asserted in tandem. The word "and" signals that (K) is a conjunction of two claims. Specifically, (K) is a conjunction of two if-then statements, or conditionals. There is a left-to-right direction of (K) and a right-to-left-direction of (K). (11; emphasis in original)

In one direction, (K) says that *S* knows that *P* *only if* *S* and *P* have features F_1, F_2, . . . , which is equivalent to saying, "If *S* knows that *P*, then *S* and *P* have features F_1, F_2, ..." These propositions tell us that it is a necessary condition for *S* to know that *P* that *S* and *P* have *each* feature F_1, F_2, ... To say that these features are necessary means that if *S* or *P* lacks *any* of the features F_1, F_2, . . . , then *S* does *not* know that *P*. In the other direction, (K) tells us that *if* *S* and *P* have features F_1, F_2, . . . , then *S* knows that *P*. This proposition tells us that it is a sufficient condition for *S* to know that *P* when *S* and *P* have *all* the features F_1, F_2, ... When we say that these features are sufficient, this means that if *S* and *P* have *all* of these features, then *S* *does* know that *P*. So (K) asserts that features F_1, F_2, . . . are individually necessary and jointly sufficient for *S* to know that *P*.

The practice of epistemology is then, in principle, quite straightforward (though often very complex in the details). One philosopher proposes a set of features F_1, F_2, . . . that are alleged to be both necessary and sufficient for *S* to know that *P*. Other philosophers reply with *counterexamples*, i.e. instances intended to show that features F_1, F_2, . . . are not necessary and/or not sufficient for *S* to know that *P*. For instance, suppose I were to say that the necessary and sufficient conditions for *S* to know that *P* are these: (1) *S* believes that *P*, and (2) *P* is true. Someone could then offer a counterexample purporting to show that conditions (1) and (2) could be met but *S* would still not know that *P*.

Suppose, as a counterexample, that *S* believes that *P*, where *P* is the belief that the Vikings were the first Europeans to reach North America, and *S* believes it only because *S*'s name is Ole Olsson and he is proud of his Norse heritage. In that case, *S* would have no rational reason to believe that *P*. Mr. Olsson's belief is not based on any evidence, but in a feeling of pride in the supposed accomplishments of his ancestors. So, *S* (Mr. Olsson) would have a belief *P*, and *P*, the belief that the Vikings were the first Europeans to reach North America, would be true (as archaeologists presently hold). However, Olsson could hardly be said to *know* that *P* because he doesn't believe it for any rational reason. Surely, it seems, we can't be said to *know* something, even if it is true, when we have no rational grounds for believing it. Hence the counterexample seems to indicate that the features of *S* and *P* that I recommended, that *S* believes that *P* and that *P* is true, are not sufficient for *S* to know that *P*.

How do epistemologists supposedly recognize a good counterexample when they see one? That is, what is the basis for their judgment that some feature or set of features might or might not be a necessary or sufficient condition for knowledge? Generally, they appeal to *intuitions*. When philosophers talk about intuitions, they do not mean anything like when people used to talk about "women's intuition." Such talk was intended to indicate that men think logically whereas women leap to conclusions on the basis of emotion. When philosophers appeal to their intuitions, they do not think that they are being illogical and emotional. Rather, they see intuitions as deep, fundamental, immediate awareness of truth. We have ethical intuitions, such as that causing pain for one's own amusement is bad. There are also metaphysical intuitions. For instance, some philosophers have held that it is intuitively obvious that nothing exists unless there is a sufficient reason for its existence (the principle of sufficient reason, as it was called by Leibniz). Other philosophers (such as the present author) do not share that intuition.

There are also epistemological intuitions, such as that knowledge must involve rational belief. In epistemology, if someone proposes necessary and sufficient conditions for knowledge, the game is to devise a counterexample that appeals to our intuitions about knowledge to show that the proposed conditions fail. Note that the counterexample does not have to be anything realistic, or even remotely so. It may be a completely far-fetched scenario such that if it *were* so, then we *would* make a certain kind of judgment. For instance, suppose that we assume (like Descartes) that knowledge is identified with absolute certainty. In that case, if someone were to propose that what we plainly see is certainly true, then we could respond with a far-fetched counterexample like this: Suppose that, instead of having a body as you think you do, you are a brain suspended in a vat of liquid and hooked up to a supercomputer that feeds you sensory stimuli that perfectly mimic ordinary bodily experience. In that case, you would have the experience we call "seeing" and it would be indistinguishable from the experience of ordinary seeing. Yet your experience would deceive you. You aren't actually a human seeing, for example, trees, but a pathetic brain in a vat being made to have the experience equivalent to seeing trees. Since you cannot be *absolutely* sure that you are not at this moment a brain in a vat, you cannot be absolutely sure that what you see is real. Hence, an admittedly very far-fetched scenario serves as a counterexample.

But where do our intuitions come from? What justifies them? Isn't it quite possible for two equally intelligent and informed people to have conflicting intuitions about things? For instance, it may seem to you intuitively clear that if anyone knows something that he should be able to state his reasons for holding it. To me, on the other hand, it seems that somebody might well know something but be unable to articulate reasons. For instance, an experienced pilot might know that something is wrong with an aircraft before being able to articulate just what it is. Further, don't intuitions change over time? For instance, 150 years ago, it seemed even to the best educated a truth too obvious to question that the different races of humans differed in innate intelligence. Now, educated people, at least, generally lack this intuition. Intuitions, indeed, seem to be conditioned and informed by one's overall experience and knowledge. After thirty years of teaching in higher education, I have pretty well-honed intuitions about what an undergraduate student can produce in a written assignment. Therefore my intuitions about whether a piece is plagiarized are probably much more sensitive and accurate than a brand-new instructor's would be.

Such considerations indicate that intuitions are not *a priori* insights, as some philosophers have claimed. If they were, we would expect them to be universal and permanent, and so not variable from time to time or from one person to the next. Are there universal, permanent intuitions about knowledge that are given *a priori* and that are necessary for understanding the essential nature of knowledge? If so, then the project of naturalistic epistemology is doomed. No scientific account of human knowledge could be adequate but would have to be supplemented by nonscientific, *a priori* insights. There is no doubt that naturalistic philosophers do appeal to intuitions as non-naturalistic philosophers do. The onus is therefore on them to give an account of intuitions as *a posteriori* and empirical, and so consistent with a scientific account of knowledge.

Kornblith offers such an analysis (2002, 8–21). Imagine an amateur rock collector. He finds different rocks and groups them by an intuitive set of similarities. Perhaps rocks of similar color, density, and hardness go together. As time goes on and the collection builds, many rocks in numerous groupings are represented. For instance, there may be group of clear crystals and a group of dense green rocks. Now suppose that our hobbyist takes such an interest that he decides to become a professional mineralogist. As his studies progress, he learns that some rocks that look alike are not really that close in chemical composition. Others that look different are really chemically close or identical. For instance, he learns that diamond and graphite, though entirely different in appearance, are really just forms of carbon. On the other hand, fool's gold and real gold, though they look alike, are entirely different substances.

The point is that as our collector's knowledge grows, his standards about which rocks to group with others will change. His informed opinion will often be counter to his original, naïve intuitions. In fact, he will develop new intuitions and start to automatically group things in a way more conformable to his theoretical knowledge. Kornblith holds that intuitions are like the pretheoretical standards whereby the rock collector arranges his collection. As the collector's knowledge grows, the original intuitions are replaced—to some extent—by more-sophisticated scientific information. This does not mean that the original intuitions about rock similarities were epistemologically useless or simply mistaken. Clearly, they had *some* purchase on the underlying reality and were useful in getting things organized before deeper knowledge was acquired.

The same seems to hold for our intuitions about knowledge. There is no reason to think that they are simply mistaken. At a certain stage of inquiry, such intuitions may be all we have to go on. However, if knowledge is a natural phenomenon, as Kornblith and other epistemological naturalists take for granted, then it is appropriately studied with the same sorts of tools that are used to study other natural phenomena, namely, the tools of natural science. As our knowledge about knowledge increases, our original intuitions may have to be modified or abandoned. For instance, if we recognize that nonhuman animals have knowledge that is like human knowledge, then the intuition that you have to be able to articulate reasons for what you know obviously will no longer hold. Nonhuman animals cannot articulate reasons for what they know.

Whether we attribute knowledge to animals—knowledge in a sense compatible with the kind of knowledge that humans have—will depend on the success of our theories about animal knowledge where the success of those theories is judged in the same way that all scientific theories are judged, that is, by accuracy, simplicity, theoretical support, comprehensiveness, and so on. If such well-confirmed theories are opposed by allegedly *a priori* intuitions—then it is too bad for those so-called intuitions. The history of science is littered with the detritus of many confident, allegedly *a priori* pronouncements on what science cannot do.

The upshot is that naturalist philosophers can also appeal to intuitions and, like other philosophers, use them to provide an initial categorization of things and organization of subject matter. However, the naturalist does not regard intuitions as *a priori*, but as fallible and corrigible, that is, subject to revision or rejection as our theoretical knowledge builds. These theories will give us new ways of grouping, characterizing, and conceiving the phenomena that may or may not be consistent with our initial intuitive ideas. Thus, for naturalists intuitions may be a good place to begin an inquiry, but, unlike many other philosophers, they think that they are a bad place to finish up.

If our aim is to have a scientific model of human knowledge, where do we start? Kornblith notes that one promising and already-thriving branch of science might be a good place to start (28). The study of cognition in nonhuman animals, cognitive ethology, is already a well-developed science. In order to survive, animals have to face many challenges that require them to gain information from their environments and respond appropriately. The "four 'F's'" of animal behavior—fleeing, fighting, feeding, and . . . reproducing—all depend on acquiring information and responding appropriately. Should the information not be received or should it be ignored or elicit an inappropriate response, the animal's chances in the Darwinian competition will decline precipitously. An animal that fails to recognize a stalking predator or a receptive mate, or responds by fighting rather than fleeing a much more formidable adversary, or overlooks a source of nutritious food, will display what Quine in an earlier quote called "a pathetic but praiseworthy" tendency to die before passing on its genes.

In fact, animals often display an astonishing aptitude for getting and using information. Some of their abilities for getting and using information far surpass those of humans. For instance, a salmon spawned in a tiny stream in British Columbia will spend most of its adult life in the deep ocean. When it comes time to create a new generation, however, it can return upriver by detecting faint chemical signatures in the water until it comes to the precise stream where its life began. Some migrating birds can accurately navigate thousands of miles, much of it over open ocean. A shark can detect quite minute traces of blood in the water. Bees can see ultraviolet light, and a rattlesnake can detect the body heat of warm-blooded prey. But, for all these amazing information-gathering feats, can a nonhuman animal be meaningfully said to have *beliefs* or *knowledge* in anything like the way that humans have beliefs or knowledge?

Of course, some philosophers would immediately dismiss this possibility since beliefs seem to be required for knowledge and a belief is generally classified as a propositional attitude, that is, an attitude toward a proposition. A proposition is defined as the content of a declarative sentence; it is what we are expressing when we utter such a sentence. Different sentences of the same language or of different languages can express the same proposition. Thus, "My name is Keith," "Keith is my name," and "*Mein Name ist Keith*" all express the same proposition. To believe that my name is Keith is to entertain that proposition and to adopt an attitude of acceptance toward it. Further, to entertain a proposition one seemingly must understand the sentence expressing that proposition. Now, my cat has many amazing abilities, but, so far as I can tell, she cannot understand declarative sentences and so cannot entertain propositions. So, she would seem to have no beliefs and so, seemingly, no knowledge. Therefore, if it is necessary to entertain propositions to have beliefs, we are forced to the (highly counterintuitive) conclusion that animals and small children have no beliefs. Surely, though, my cat can believe that there is food on the supper dish and a toddler can believe that mommy is nearby.

One way to avoid the counterintuitive consequence would be to say that cats, dogs, and toddlers do understand propositional contents in terms of their own private mental language. Thus, Fido can believe that the squirrel is up the tree, though he cannot understand the English sentence "The squirrel is up the tree" or any synonymous sentence in any human language. Better, I think, would be to reject the definition of *belief* as a propositional attitude. Rejection of that definition is motivated by the observation that not all—or even most—human beliefs involve entertaining propositions. I believe that the garlic powder is where it usually is—on the first row of the spice rack, right next to the cayenne—because that is where I put it last night. I have believed it since I put it there last night. Yet, I never articulated, even to myself, the sentence "The garlic powder is on the first row of the spice rack, right next to the cayenne" or any synonymous sentence until just now when I wrote out the sentence. Someone might say that I did not know it until I articulated the sentence just now. However, in that case it would be hard to explain how I can do things like this: When I want the garlic powder, I can go right to it and get it without first articulating its location to myself. I don't even have to be thinking about garlic powder at the time; I just get it "automatically." Indeed, the vast majority of our everyday beliefs seem to be subliminal. How could you do anything if you had to take the time to articulate every belief relevant to your activity?

Further, it seems to me that we all have perfectly good beliefs that bear specific information that is not expressible in propositions. For instance, it certainly seems to me that I plainly know (and hence believe) that wintergreen mints have a particular strong, distinctive taste, but I cannot articulate what that taste is—other than to state the truism that they taste like wintergreen. So, perhaps not all information-bearing mental states are expressible in propositional form. Maybe, then, we need a better definition of *belief*, one that does not entail propositional content.

Perhaps a better understanding of belief would be to view it holistically, as a *system* of interconnected beliefs rather than as individual beliefs expressed as propositional contents. Also, belief clearly is fundamentally connected to behavior of all sorts, including but not limited to linguistic behavior. Someone who strongly believes something and who is strongly motivated to act on that belief can be expected to behave in certain ways. For instance, if someone holds as a deep and sincere belief that dancing is immoral, we can expect that person to speak disapprovingly of dancing in certain contexts, avoid dance clubs, not take dance lessons, and not watch *Dancing with the Stars*.

Philosophers David Braddon-Mitchell and Frank Jackson propose that an individual's system of beliefs not be regarded as a set of propositions in an internal mental language (1996). Information can be represented in many ways other than linguistically. For instance, maps can provide detailed but nonlinguistic information about physical spaces. Braddon-Mitchell and Jackson therefore suggest that belief systems are more like internal maps, vastly complex and comprehensive nonlinguistic representations of reality that serve to direct behavior, including human linguistic behavior. The physical embodiment of such an internal map is a hugely intricate set of neuronal connections and interactions in a brain. The informational content of an inner map will be (roughly speaking) a global representation of reality, that is, how one takes *everything* to be. Kornblith does not explicitly endorse the Braddon-Mitchell/Jackson view of beliefs, but it certainly seems congenial to his epistemological naturalism. The "map" view is certainly not anthropocentric, since other animals with brains could have internal maps constituted by neuronal organization, just as people do.

Kornblith does argue that there are circumstances in which we rightly attribute beliefs to nonhuman animals. He notes that there are many occasions where animal behavior is explicable only in higher-level informational terms rather than as lower-level physiological terms (2002, 41). Many, perhaps most, behaviors of animals (and humans) are explicable in low-level terms. Scratching, sneezing, coughing, yawning, sighing, wincing, smiling, frowning, stretching, fidgeting, blinking, and making yummy noises when you taste something good are probably all explicable as behavioral responses to physiological prompts. On the other hand, some behaviors—even behaviors of quite humble organisms—are explicable only in terms of the accurate internal representation and processing of information.

Kornblith mentions a desert ant that will emerge from its mound and wander apparently randomly in search of food. Once the food item is located, the ant takes it, and—however circuitous its meanderings have been—it returns in a straight line to the mound. Clearly, the ant's tiny brain has kept an accurate record of every twist and turn of the ant's peregrinations and correctly computes the straight-line direction home. Insects can record and act on quite complex information, though it is implausible to attribute beliefs to them. In fact, the ant's behavior shows none of the adaptability characteristic of beliefs. Beliefs can change to track changes in the data. However, if an experimenter moves the ant, the ant cannot adjust, but will rigidly follow the course that would have returned it to the mound before it was moved. The ant's brain simply executes an algorithm without the ability to process additional information about the ant's position.

Then there are the much more complex behaviors, ones that fit no rigid pattern like the dead-reckoning abilities of the ant but that are explicable only in terms of what an animal thinks, believes, or knows. Anyone watching the ant can predict with assurance just what the ant will do with the information it has. As soon as it finds food, it heads straight home. Beliefs are different. Beliefs underdetermine behavior. That is, from the mere fact that some creature has a belief, no reliable prediction is possible. Just because Fred believes that ice cream is present does not mean that he will eat the ice cream. He also has to desire the ice cream. Further, he might believe the ice cream is present and want the ice cream but not eat it because he also believes it belongs to someone else.

Still, we do very often explain behavior in terms of beliefs and desires. The behaviors we explain in terms of beliefs and desires are quite heterogeneous and do not fit a stereotyped pattern. What they often have in common is that they are explicable only in desiderative and doxastic terms, that is, that the animal acts those ways because it *desires* a certain end and *believes* that certain kinds of behavior will achieve that end. Kornblith cites a number of remarkable examples.

For instance, ravens will work together in pairs to steal a meal from a hunting bird. The most plausible way to describe this behavior is that while one bird distracts the hawk, the other snatches the hawk's prey. The ravens' behavior is naturally interpreted by saying that the ravens acted this way in the belief that certain types of behavior (and the behaviors are quite heterogeneous and follow no set pattern) would distract the hawk and let the other make off with the prey. Hawks are considerably larger, stronger, and better armed than ravens, and if a raven tried to steal prey from an undistracted hawk, it would probably only succeed in adding its protein to the hawk's dinner.

Another avian example is the behavior of piping plovers in leading predators away from their nests. Piping plovers live and breed in exposed sites on the sand and gravel of beaches. Sometimes a predator, like a fox, will be heading toward the nest site. The plover will then perform an elaborate series of behaviors, which again have no set pattern, including dragging her wing, uttering cries of apparent distress, and running along the ground just ahead of the fox. The fox follows the plover until it is far away from the nest, at which point the plover takes to wing, leaving the fox hungry. Again, it seems natural to interpret these behaviors in terms of a belief/desire psychology: The plover wants the fox to go away from the nest and believes that by pretending to have a broken wing, the fox will be led to follow her on a path away from the nest. The fox wants to eat and believes that the plover has a broken wing and will be an easy meal.

Someone might object that natural selection has the power to produce complex behaviors in animals, like insects, that do not seem to be sentient. For instance, honeybees do a complex dance to indicate sources of pollen to other members of the hive. These dances convey detailed information about locations and distances, but it seems far-fetched to attribute beliefs or desires to bees. Could we not explain the behavior of birds in similar ways without having to postulate beliefs or desires? Ockham's razor, as well as the nagging fear of anthropomorphism that always haunts studies of animal behavior, could be invoked by skeptics of belief/desire attributions. Why go out on a limb and make such claims, when lower-level explanations might do?

First, birds are far more neurologically sophisticated than bees, and therefore would be much better candidates for attributions of sentience. Second, the described behaviors of birds are much more like human behaviors than the stereotyped, rigid patterns of the honeybee dance. When humans are trying to distract someone, what do they do? There is no set pattern of behavior; someone determined to distract you could do a silly dance, make faces, tell a joke, rave and rant, disrobe, brandish a weapon, or all sorts of things. Really, the ravens act very much like human pickpockets. Like ravens, pickpockets generally work in tandem. I heard of one case where a woman walked up to a man and spat in his face. While the startled gentleman was reacting to that, the woman's child was lifting his wallet. However, spitting is only one of many different distracting behaviors practiced by pickpockets. Similarly, ravens exhibit a variety of behaviors that have little in common except that they are apparently aimed at distracting hawks.

Likewise, the plover does not do the same "act" each time a predator draws near her nest, but has to monitor the intruder and adapt her behavior depending on what the predator does. As Kornblith notes, unlike lower-level information-bearing states (like the ant's), beliefs provide information that can be used in any number of different ways, permitting a wide variety of behaviors, not a stereotyped set. When we walk from a dark area, like a movie theater, into bright sunlight, we squint and blink. This behavior is an automatic response to the dazzling light. It requires no beliefs or desires. On the other hand, knowledge of the sun's location can account for many different behaviors, depending on our desires. If we are cold and want to be warmer, we can move into the sunlight. If we are afraid of melanoma, we can put on sun block. If we want to observe the sun with a solar telescope, we move into the sun and point the telescope toward it. The only thing that these diverse behaviors have in common is that they can be explained in terms of our beliefs about the location of the sun, plus our different desires and other beliefs at the time. So, if nonhuman animals exhibit a significant diversity of responses to information, and if different responses appear to be different means of achieving the same useful end, then explanation of the behavior in terms of a belief/desire psychology is strongly motivated. Some wits have noted that their aptitudes for talking and thieving have already marked ravens as similar to human beings. Maybe then it is not so far-fetched to explain some of their complex behaviors in ways that we are used to describing human behavior.

Let's assume that Kornblith has made a good *prima facie* case for using animal cognition as a model for studying human cognition. That is, if we are convinced that some animals do have cognitive states like ours—beliefs and even knowledge—then an examination of just how animals acquire those cognitive states could elucidate the processes of human cognition. We are therefore on the way to a naturalistic account of human knowledge, one that employs the tools of science—cognitive ethology in this case—to investigate that remarkable natural phenomenon of human knowing. But we still face the glaring problem of normativity. Animals' beliefs are no more uniformly correct than humans'. Dogs famously often bark up the wrong tree. The fox thinks the plover is injured. When I want to keep my cat from running out the door, I can (sometimes) fool her into thinking I am entering one door when I enter another. So, an uncritical substitution of cognitive ethology for epistemology would seem to gain nothing over Quine's replacement thesis.

How do we preserve epistemic norms yet study knowledge scientifically? Put another way, how does a naturalized epistemology address the traditional epistemological task of articulating the justification or warrant for our beliefs? Terms such as *justification* and *warrant*, when used in an epistemological context, refer to the rational reasons or grounds that we have for our beliefs. Justified or warranted beliefs are ones that are rational and reasonable to hold; beliefs that are not justified or warranted are not reasonable beliefs. We secure the rational grounding of our beliefs by respecting epistemic norms in the way that we form, hold, or support our beliefs.

What, then, is a norm? A norm is an imperative; it tells us "do this" or "don't do that." But what kind of norms could *science* disclose to us? Well, clearly, they would have to be what Kant called hypothetical, not categorical, imperatives. Categorical imperatives, if any, will fall within the realm of conclusions derived by *a priori* reasoning. A categorical imperative can depend on no contingent conditions of the sort that natural science can disclose. Categorical imperatives tell us what *has* to be; science can only tell us what, in fact, *is*.

A hypothetical imperative has the form "If you want *X*, then do *Y*." If there is some goal you want to attain, then you must do certain things. An epistemic norm would therefore be a hypothetical imperative telling us that if we want to attain a certain epistemic goal, then we should do certain things to achieve it. But what epistemic goal *should* we target? Is there any epistemic condition or quality that is that is supremely good or preeminently valuable among epistemic goods?

Kornblith argues that truth must be our preeminent epistemic goal since, whatever other, nonepistemic goals we have, we will need to know truly how to achieve those goals (2002, 158). In other words, for Kornblith, truth, in the sense of having the ability to make accurate cost-benefit analyses of the proposed means to our nonepistemic ends (whatever they may be), is rationally required in order to pursue *any* such ends at all. In short, having goals requires having the means to reach those goals, which requires knowing the true means to the goals:

> It is thus of the first importance that our cognitive systems remain suitable for the purpose of performing the relevant cost-benefit calculations. And this requires that our cognitive systems be accurate, that is, that they reliably get at the truth. (158)

One apparent problem with Kornblith's suggestion is that, as Bas van Fraassen argued (recall chapter 5), to make true predictions (e.g., about the means for achieving an end), a hypothesis only needs to be empirically adequate, not true. Empirical adequacy, not truth, would seem to be the epistemic goal required for having any other goals at all.

A further complicating factor is that it might sometimes be the case that cognitive systems that systematically generate false beliefs are *more* efficacious in helping us reach our goals than systems that reliably lead to truth. Consider two prehuman ancestors who hear a sudden rustle in the grass. One, Conclusion-Jumper, immediately thinks "lion!" and scampers up a tree. The other, Truth-Seeker, sticks around to investigate. If the rustling was just the wind, then Truth-Seeker is spared the minor inconvenience of an unnecessary climb. But if it is a lion, then Truth-Seeker enters the food chain while Conclusion-Jumper is safe. Better to be wrong and safe than right and dead. Daniel Dennett has suggested that possession of a HADD—a "hyperactive agent-detection device"—that would often generate false positives with respect to the presence of hostile agents would have conferred a selective advantage on our ancestors. Frequently, being wrong would promote the goal of survival.

Kornblith might reply that he admits (158) that truth need not be our *only* epistemic value, but that truth must nevertheless be our preeminent epistemic value, that is, the salient criterion for judging cognitive systems. I agree for various reasons. For one thing, even for an antirealist like van Fraassen, the conditions in which we have to settle for empirical adequacy instead of truth are quite circumscribed. Van Fraassen was concerned with situations where we posit theoretical entities that are, in principle, unobservable. When we are dealing with in-principle observables, there seems to be no problem with making truth our aim rather than mere empirical adequacy.

But what about nonhuman animals? Humans can reflect on their cognitive systems and change them in accordance with chosen goals, like truth. For instance, we can recognize the limitations of our "gut" estimates of probability, as when we ignore base rates, and can refuse to trust our "gut" and instead scrupulously follow the probability calculus. Animals, though, have to work with the cognitive aptitudes they have. Can we expect that natural selection will favor the evolution of truth-tending cognitive systems over those that generate merely empirically adequate representations? After all, it is a meaningful tautology to say with respect to natural selections that whatever works, works; that is, what matters for natural selection is whether an internal representation promotes survival, not whether it is true.

The straightforward answer is that, in innumerable circumstances, nothing is more useful to have than truth. To see truly means that you see something as an *x* if and only if it really is an *x*. As noted, there may be circumstances where experiencing harmless phenomena as harmful agents, and so perceiving falsely, has survival value. On the other hand, there are many circumstances when animals have to get it right. Consider the piping plover again. When a predator approaches her nest, she needs to recognize the animal as a predator and monitor its motions very carefully and accurately as she leads it away. She has to know when to change her behavior in response to the predator's behavior, and when it is safe to end the performance. It seems that what the plover needs is perception that is highly accurate and reliable. Nothing would work any better, and it is hard to imagine any biologically feasible alternative that would work as well.

Even if something more effective than truth could be imagined, natural selection results in functionality but not necessarily optimality. Natural selection is an undirected trial-and-error process that can only work on the real, not the ideal. If an ancestor of the piping plover stumbled on the representation of reality as an effective survival tool, natural selection would likely work to preserve and enhance that capacity to meet the ongoing and pressing exigencies of survival. In the real world, organisms do not have the time to wait around on the off chance that an ideal system will develop.

It seems that for both humans and animals, truth is very often a most useful thing to have. I would add a further consideration in favor of truth as the preeminent epistemic goal for human beings. I would assert—against the claims of some philosophers—that for human beings truth has value *per se*, whether or not it is efficacious in serving specific desires. Why? Because truth is the only epistemic good genuinely commensurate with the human capacity for wonder. So far as we can tell, humans, and only humans, can not only ask "What use is it?" but also "Why *is* it?" If we are looking for something distinctly and uniquely human, it seems to be that we are capable of valuing knowledge *for its own sake*. That is, humans are capable of valuing the truth *just because it is the truth*, and not for what it can get us. We value truth because discovering it fulfills our essential human nature, as Aristotle recognized long ago when he uttered the first line of his *Metaphysics*: "All humans, by nature, desire to know." It is this human quality—simple curiosity—that is the saving grace in the midst of all of our irrational propensities. Insofar as humans are capable of disinterested wonder, truth must be our prime epistemic desideratum.

A picture is worth a thousand words, they say, and sometimes it is worth a thousand arguments also. I once saw a photograph that persuades more cogently and eloquently than any philosophical argument. In this photograph, two intelligent beings are contemplating a fish in a bowl. One being is a large tomcat; the other is a human toddler. Both are fascinated by the fish, but in entirely different ways. The cat's expression is plain: "I want it." The toddler's expression is one of awe-struck wonder. The curiosity of the cat can be expressed by the question "How do I get it?" The toddler's curiosity is expressed by "How can it *be*?"

If, then, truth is our goal, our epistemic norms will have the form "If you want truth, do *X*" or "If you want truth, do not do *Y*." In other words, efficacy at arriving at truth—what Kornblith calls "reliability"—will be our norm. Epistemic practices and cognitive systems that are reliably truth-tropic will meet our approval, and those that are unreliable will not. Furthermore, with reliability as our norm, the traditional distinction between description and prescription dissolves. It will be a matter of *fact* that some epistemic practices and cognitive systems are conducive to truth while others are not. In other words, for a naturalized epistemology, whether or not our beliefs are warranted will be an objective matter, one determined by the facts about how our beliefs are formed. Those formed by reliable processes will be warranted; those formed by unreliable processes will not be. Further, empirical inquiry can disclose the causal connections that make some processes causally efficacious in generating truth and others not. Human knowledge—how humans reliably represent the world—is therefore a natural phenomenon and, as such, is a proper object of scientific inquiry. The project of naturalized epistemology is therefore validated.

NATURALIZING ETHICS

The project of naturalizing epistemology is a recent one. However, Aristotle proposed a fully naturalized ethical theory over 2,300 years ago. In fact, as MacIntyre noted, Aristotle's ethical theory, adapted to the framework of Christian theology by Thomas Aquinas and other medieval thinkers, was the mainstream ethical view prior to the Enlightenment. Aristotle's naturalistic ethical theory was developed in his *Nicomachean Ethics*, still universally regarded as one of the classic texts on ethics. In fact, it is safe to say that the Aristotelian view of ethics has undergone a major revival and is now more popular than it had been in hundreds of years.

Aristotle was "The Master of Those Who Know," as Dante called him. He made first-rate contributions in many fields, and, in fact, was the founder of several intellectual disciplines. For instance, he was the first formal logician. His theory of the syllogism was the focus of Western logic for over two thousand years. Aristotle was probably the greatest all-around genius the human race has produced.

Most of all, though, Aristotle was a biologist. He was probably the foremost theoretical biologist prior to Darwin. Yet he was not only a theorist. He would wade out into tidal flats and inspect the catches of fishermen to see if he could find new specimens to dissect. Some of his anatomical discoveries were not replicated until centuries later. The most distinctive thing about Aristotle's ethical theory is that it is rooted in biology. Ethics is the study of the good, specifically, the human good. For Aristotle, the central question of ethics is this: What is the best way to live a human life, that is, to achieve the human good? Aristotle held that you cannot answer this question until you understand the essential nature of the human organism. For Aristotle to understand an organism is to grasp its *telos*, the essential set of functions that nature has set for each type of organism. Nature has endowed each type of organism with a distinctive set of potentialities, which, when fully actualized constitute the full functioning of that organism. The good for any creature consists in the complete actualization of its distinctive biological potentialities, that is, in doing what it is designed to do and doing it as well as possible.

In the classic Steven Spielberg shocker *Jaws*, the marine biologist Matt Hooper, played by Richard Dreyfus, is trying to convince the idiot mayor of the resort community Amity to close the beaches in the wake of several horrendous shark attacks. He says that the attacks were due to a great white shark. At one point, Hooper comments that the great white shark is a "miracle of evolution" because its whole life is spent, swimming, eating, and reproducing. Through evolution, nature has adapted the great white shark to be a fearsome predator. To fulfill that role, the shark has to swim, eat, and reproduce, and it has to do those things very well. The great white shark is thriving when it is highly successful in doing those things that nature designed it to do. A shark that is a strong swimmer, is an effective hunter, and passes on its genes to a new generation of sharks is living as well as a shark can live.

What about humans? What is our natural function? Obviously, we are far more complex than sharks, with many more and much more diverse needs and desires. Still, we are just as much a part of the natural world as the sharks are. What, then, is the human function, that is, with what characteristic, human-specific potentials has nature endowed us, so that in fully actualizing these potentials we achieve the best life for a human being? To answer this question, Aristotle begins with a description of that ideal human condition, the state of fully self-actualized thriving that constitutes the *telos* of human beings.

As to the name of that state, says Aristotle, everybody agrees that it is called "happiness." The word that is always translated as "happiness" here is *eudaimonia*. Really, better translations of that beautiful Greek word would be "well-being," "thriving," or "flourishing." In our culture, unfortunately, the meaning of "happiness" has been corrupted so that it means something like a "state of pleasurable feeling." Aristotle certainly thought that *eudaimonia* was a state that involved many pleasures, and that those pleasures were the ones most naturally satisfying. But *eudaimonia* is more than a feeling. It may be best to regard *eudaimonia* as a state of mental and moral excellence, combined with the enjoyment of good health and a modicum of material prosperity.

As the definition of *eudaimonia* implies, it is an essential part of the human function to be rational, that is, to achieve excellence in the performance of one's intellectual functions. A rational person is one who thinks logically and clearly and who is willing to be guided by reason. Aristotle also makes it clear that humans are designed by nature not to live solitary lives. In his *Politics* Aristotle says that a human is a "political animal." This (fortunately) does not mean that we are each cut out to be politicians, but rather that humans are designed to live in a polity, a politically organized society with other human beings. Human well-being or flourishing can only occur in such a setting. To thrive entirely on one's own, Aristotle also observed, you would have to be a beast or a god. Humans need each other. Another essential aspect of the human function is therefore the ability to live in society harmoniously (or as harmoniously as possible) with other human beings.

Another beautiful Greek word used by Aristotle is *arête*, which is generally translated as "virtue." The virtues for Aristotle are states of excellence in performing a distinctive function. The virtue of a racehorse is to run swiftly; the virtue of a draft horse is great strength in pulling loads. For human beings there are two types of virtue, intellectual and moral. The intellectual virtues are states of excellence in fulfilling our natural function as rational beings. The moral virtues are states of excellence in fulfilling our natural function as social beings.

Probably the current term that best captures at least part of the idea of intellectual virtue is *critical thinking*. The critical thinker is one who possesses the skills to analyze arguments and information and tell what is trustworthy and credible from what is not. As I have heard it put, a critical thinker is good at distinguishing gold from bullshit. This is much harder than it sounds since politicians, lobbyists, corporate spokespersons, advertisers, and ideologues are so good at disguising bullshit and making it look like gold. The critical thinker also can make decisions based on the best information available rather than the passions of the moment. When you think clearly and precisely and decide wisely, you have achieved intellectual virtue and you are functioning rationally.

The *Nicomachean Ethics* is mostly about moral virtue. Moral virtues are those states of character that, when exercised, constitute the condition of excellence in interacting with our fellow humans. The moral virtues include such states as courage, temperance (self-control), generosity, and justice. In general, those who act courageously, temperately, generously, and justly will get along far better with others than those who act in a cowardly, intemperate, greedy, or unjust way. When, as a fixed state of character, you are disposed to act courageously, temperately, generously, and justly, you have achieved moral virtue and are functioning morally.

The upshot is that the best life for a human, the life of *eudaimonia*, is to live virtuously, that is, to achieve excellence in the performance of the human function. The happy person is one who has attained intellectual and moral virtue.

The intellectual virtues can be acquired by learning; education should mainly be about developing the intellectual virtues. What about moral virtues? What exactly are they and how to we achieve them?

A virtue, as we say, is a fixed state of character. It is a permanent disposition to act in certain ways. A virtue is not an emotion, but it deeply concerns the emotions. Consider anger. Anger, by itself, is neither a vice nor a virtue. There are times and circumstances when it is appropriate to show anger and times when it is not. To have the virtue of temperance is to exercise control over our anger. Anybody can get angry, Aristotle observes. That is easy. What is hard is to have the discipline to get angry only at the right time, toward the right person, to the right degree, and for the right reason. When we learn to do that, we have achieved the virtue of temperance. Similarly, courage relates to the feeling of fear. A coward is one who too quickly gives in to his fear. On the other hand, a foolhardy person is one who acts incautiously; one should be afraid to face dangers foolishly. A courageous person knows what dangers to face and which ones to avoid.

As the example of courage indicates, a virtue is a mean state, that is, it is a state of character that is in the middle between two other states, each of which is a vice. Put differently, a virtue is a mean between two extremes of which one is an excess and the other a defect. One who is too tight with his money is a miser; one who throws his money away is a spendthrift. Stinginess, the vice of misers, and prodigality, the vice of spendthrifts, are two opposite vices with the virtue of generosity between them. A generous person doesn't throw his money away, but he doesn't hoard it, either. He gives his money at the right time, in the right amount, to the right person, and for the right reason.

But what is the "right" time, place, amount, and so on? What determines that? Aristotle says that finding the mean in ethics is not a mathematical operation like finding the mean between two numbers. What we have to do is trust the judgment of the practically wise person. The person with practical wisdom is the one who has lived long enough and built up enough experience to see what works and what does not. This is why wisdom does (or should) come with age—as you live, you see what works and what does not. Someone who is stingy, like Dickens's Ebenezer Scrooge, lives a miserable life and inflicts misery on others. Likewise, the spendthrift, like the prodigal son of the parable of Jesus, also lives miserably. The generous person, the one who has achieved the mean state between stinginess and prodigality, has achieved well-being in that regard.

We may sum up the Aristotelian view of ethics as follows: To be happy is to be good, and to be good is to be virtuous. To be virtuous is to display, as a fixed trait of character, the tendency to act generously, bravely, moderately, justly, and so forth. Virtues are qualities of human excellence, excellence in fulfilling the human function. What is the human function? Nature has adapted human beings to live a life of rationality in society with other human beings. All creatures flourish best when they possess and exercise the qualities of excellence that enable them to do well what they are adapted to do. Thus, a successful hunting dog is successful because it possesses the qualities of a keen sense of smell, eagerness for the chase, speed, strength, endurance, and obedience to its master. Likewise, human beings flourish when they consistently do well what nature has designed them to do, that is, to think rationally and to live in peaceful and mutually beneficial society with other humans. The intellectual virtues are the excellences whereby people learn to think rationally and the moral virtues are the excellences whereby people live successfully in society with other human beings. For instance, a person who is generous, courageous, temperate, and just will, in general, live more successfully among his fellow human beings than one who is selfish, cowardly, ill-tempered, or unjust. The former, even in difficult and humble circumstances, will live better than the latter, even if the latter is rich and powerful.

The Aristotelian view of ethics has many strong points. For one thing, it offers a straightforward and simple answer to the most basic question about ethics: Why be good? Why is it more reasonable to be good than to live a life of total amorality? Aristotle's answer is that the only way to be happy is to be good. That is, virtue is a necessary condition for happiness. Is this so? Cannot the wicked be happy? Many wicked people live to a ripe old age, fat, sassy, and unrepentant. I think that Aristotle would reply that the wicked can, of course, experience pleasure, but that the happiness he has in mind will elude them. The reason is that the pleasures of the wicked are not natural pleasures, and so are inherently unfulfilling. A natural pleasure is one that is comes from activities that are conducive to the human *telos*. An unnatural pleasure is one that impedes the fulfillment of the human function. For instance, vengeance gives pleasure, but it is not a natural pleasure because those who give themselves over to vindictiveness are consumed by it and are harmed more than the victims of their vengeance. Aristotle could point to Clytaemnestra in Aeschylus's *Oresteia*. She is so obsessed with achieving vengeance against her husband Agamemnon that vindictiveness eats her, devouring her from the inside like a parasite until her humanity is consumed. In the end, she is transformed, literally, into a spirit of vindictiveness, more implacable than the Furies. The pleasures of exercising the intellectual and moral virtues are authentic pleasures, however, that are genuinely satisfying and enriching.

Aristotle's ethic is also objective, that is, it is grounded in external biological fact, not anything subjective. As MacIntyre observed, the tendency of post-Enlightenment ethics is toward ethical subjectivism, the idea that, ultimately, our judgments of right and wrong come down to feeling. Yet, if MacIntyre's analysis is correct, it is this prevailing subjectivism that accounts for the shattered, balkanized, dysfunctional state of what passes for moral discourse in our society. A biologically based ethic, one founded on the nature of the human organism, provides factual basis from which to begin ethical discussion and a final *telos* to guide ethical deliberation.

Finally, a fully naturalized ethic is one that is scientifically knowable. The human good, the proper end of our moral striving, is not a matter of speculation, feeling, or allegedly *a priori* insight. It is subject to empirical investigation by the standard methods of natural science. It can be scientifically demonstrated that there are some conditions of the human organism that constitute maximal functionality. There can be a science of well-being. It can be shown that humans who habitually behave in certain ways are healthier; accomplish more; think more clearly; evince less stress, anxiety, anger, or depression; report higher levels of personal satisfaction; and live more harmoniously with themselves and others. Further, it can be shown that certain conditions or behaviors are conducive to those desirable states and that others impede their attainment. Thus, there can be a reliable basis for judging moral precepts, laws, habits, lifestyles, customs, social conditions, religious dictates, or anything else that has a bearing on the attainment of human well-being.

The main drawback of Aristotelian ethical theory is that it is based on outdated biology. The Scientific Revolution of the seventeenth century encompassed a sweeping rejection of the Aristotelian notion of final causes and teleology. That is, the idea that there could be in the natural world anything akin to human purpose or aims was swept away in favor of a mechanical model of causation. Physicists could no longer think that bodies had a natural place or natural motion toward such a place. For the Newtonian cosmology that supplanted the Aristotelian worldview, the natural state of a body not subject to a net force is to remain either at rest or in uniform rectilinear motion. There is no particular locale where a body "should" be.

The loss of an idea of natural place or *telos* was also carried into biology. The modern biologist does not understand an organism's anatomy, behavior, or abilities as explicable in terms of an inherent set of potentialities for the realization of a fixed, natural *telos*. Robins do not have wings *because* it is their essential nature to be flying creatures. A post-Darwinian biologist would explain *both* the robin's wing and its flying aptitudes in terms of a complex historical account in which selective pressure over countless generations drove wing development, and wing development in turn enhanced flying abilities. In short, we now understand an organism's characteristic aptitudes and anatomical adaptations in terms of a lengthy evolutionary process that proceeded toward no particular goal but depended at every stage on the vagaries of environment and variation. Natural selection, as Richard Dawkins puts it, is a blind watchmaker; it achieves marvelous organic adaptations without having an end in view (1987).

The fact that humans have certain distinctive capabilities can no longer be explained in terms of an intrinsic function, but as the end result of an evolutionary process that did not have us in mind or serve any aim or purpose at all. If we can no longer speak of a human function, a notion key to Aristotle's whole ethical theory, can there any longer be a naturalistic basis for ethics? Apparently, talk of a "human function" is now passé, and so is Aristotelian ethical theory.

The conclusion of the above paragraph is far too hasty, noted author on politics and philosophy Larry Arnhart would argue. Arnhart's *Darwinian Natural Right: The Biological Ethics of Human Nature* (1998), develops a naturalistic ethical theory that is both Darwinian and Aristotelian, arguing that, in fact, Darwinian biology supports the Aristotelian project of basing ethics on human nature.

Arnhart notes that, in addition to phylogenetic explanations of animal behavior, explanations in terms of the sort of Darwinian history mentioned above, modern biology also encompasses functional explanations (1998, 102). When we want to explain why an organism has some anatomical, physiological, or behavioral trait, one kind of explanation goes like this: Organism *O* has trait *T* because *T* has the function *F*. For instance, we can say that humans have lungs because lungs have the function of providing the oxygen necessary for metabolism. But how does the fact that a trait has a given function explain the existence of that trait? Aren't we right back to the notion of an Aristotelian *telos*? No, functional explanations fit right in with a Darwinian framework. We say that organism *O* has trait *T* because *T* has the function *F* and the performance of function *F* has survival value for organisms like *O*. Essentially, a functional explanation for the existence of an organic trait identifies a selective pressure on a population of organisms and explains the existence (and retention and improvement) of that trait by noting that the function of the trait is to cope with the selective pressure.

Thus, antelopes are cursorial creatures, built to run fast. We explain their cursorial features (e.g., long legs, lean bodies) by noting that such anatomical traits have the function of allowing them to run with supreme swiftness and so cope with the selective pressure applied by swift predators. Over untold generations, natural selection has created, retained, and improved those cursorial features in antelope populations because those features have the function of permitting antelopes to run rapidly and so escape predators. Darwinian biology does not eliminate teleology but puts it in an entirely different conceptual framework. Can that new framework still do the job *vis-à-vis* ethics that Aristotelian teleology did?

Arnhart argues that it can. For instance, why do we find that a particular bond of parents with biological children is found across cultures and throughout history? Why do attempts to disrupt that bond—like utopian communities and Israeli kibbutzim that attempted to make child-rearing a communal responsibility and downplay the role of biological parents—have such a poor record of success, as Arnhart documents? Why do such communities, despite stringent efforts to socialize communal parenting, inevitably produce frustration, unhappiness, and failure? Arnhart argues that humans are extreme "K-strategists":

> In nature there is a trade-off between parental nurture and fecundity, so that some organisms produce many offspring but invest little energy in each (r-strategy), while others produce fewer offspring but invest more energy in each (K-strategy), a difference in reproductive strategies noticed by Aristotle. . . . Compared with most animals, human beings have moved far in the second direction by producing a small number of offspring and then nurturing each one intensively for a long period. Typically, human parents remain attached to their children for their entire lives. (103)

That is, a high degree of parental (especially maternal) bonding with children is a characteristic human behavior because such behavior has the function of serving a particular reproductive strategy, a strategy that enhances reproductive fitness. A functional explanation in an evolutionary context therefore tells us why strong parental bonding with biological children is a transcultural, transhistorical human universal.

When, among human beings, certain ends are universally regarded as desirable, even across societies—otherwise very different in customs and outlook—and when achievement of those ends is plausibly connected with the enhancement of reproductive fitness, inference to the best explanation leads to the conclusion that those ends are rooted in human nature. Of course, as we noted in chapter 3, many will stringently deny that there is a biologically determined, universal human nature and will insist on some version of Locke's "blank slate" theory. Here we cannot enter into the many intricacies of that debate and will simply take for granted that Arnhart and others have identified a number of ends regarded as desirable by all human societies and plausibly based in biological human nature.

Arnhart identifies twenty of them. They include parental care, sexual identity, sexual mating, social ranking, justice as reciprocity, political rule, religiosity, intellectual understanding, health, beauty, and wealth. Some of these, such as religiosity, may at first seem to have little to do with conferring a selective advantage. However, there now exists an extensive literature, by authors such as Pascal Boyer, Daniel Dennett, David Sloan Wilson, and Scott Atran, arguing that religiosity either confers a direct selective advantage to groups that evince it, or that religiosity is a by-product of psychological factors that were selectively advantageous. In short, many authors now agree that there are certain ends universally regarded as desirable among human beings and that the desirability of those ends has a biological, evolutionary explanation.

If certain ends are inherently desirable, then the achievement of those ends will be an intrinsic good for human beings. The good for human beings will be the desirable. Humans flourish when they achieve those ends. The achievement of the inherently desirable is enriching and fulfilling and constitutes the "natural" pleasures that Aristotle mentioned. We are happy, in a state we can call after Aristotle's *eudaimonia*, when we have achieved the ends that are naturally and intrinsically desirable. Of course, different people will find such ends desirable to different degrees. Some people are intensely competitive in seeking social status; others show little interest in the game. Some people dote on children; others take the attitude of W. C. Fields. Still, it is hard indeed to imagine anyone happy who had achieved none or only a few of those natural goods. Someone without health or wealth or family or social standing or friendship or a sense of purpose would seem to be in a very bad way, perhaps suicidally depressed.

Natural desiderata provide the basis for moral norms. Actions, desires, laws, customs, social arrangements, and religious dictates can all be judged with respect to their tendency to promote or frustrate human enjoyment of natural goods. For instance, we can say that the laws, customs, and social arrangements of a society like Pakistan or North Korea appear less conducive to human well-being—the achievement of naturally desirable ends—than a society like Denmark. We can say that certain laws or customs, like Jim Crow and segregation, prevalent in the southern United States until recent decades, were bad because they unfairly frustrated the attainment of natural goods by a large segment of the population. Individual actions, like student plagiarism, can be seen as wrong because those acts disrupt the relationships of trust that are vitally necessary for educational processes.

Ethics, then, can have a basis in human nature, even in a Darwinian context. Modern biology does not undermine the project of a naturalized ethic but gives it deeper support. Further, by identifying the good with the naturally desirable, we have a solid, objective basis for ethics that can be explicated and clarified by the human sciences. Attempts to base ethics on a vacuous categorical imperative or feeling are unnecessary and, indeed, wrongheaded. The good for humans is a topic of empirical inquiry and rooted in the facts of human biology. Philosophy since the Enlightenment has led us into the moral morass noted by MacIntyre. Science, apparently, can restore sanity.

PHILOSOPHY IN AN AGE OF SCIENCE

We have seen that there are two-quite plausible programs for putting science at the heart of enterprises traditionally put within the bailiwick of philosophy. Naturalized epistemology proposes that philosophers employ scientific fields like cognitive science or cognitive ethology to supplement—or replace—standard philosophical methods. The scientific study of how beliefs are formed by processes that reliably lead to truth promises to elucidate the ancient problems of epistemology, specifically how beliefs are to be warranted in such a way that they can constitute knowledge. If warrant is achieved by objective, observable psychological processes of belief formation that are certifiably reliable, that is, truth-tropic, then the question of how humans acquire knowledge is one that science is in the process of answering. Philosophy's empty promises to provide accounts of justification in strictly philosophical terms—for example, the foundationalist project—are redeemed by straightforwardly empirical investigation. Likewise, it seems, naturalistic ethics can provide an empirically based, objective, and scientifically explicable basis for morality. In this case—if we admit that these naturalizing projects are at least *prima facie* plausible and promising—what are we to say about the future of philosophy? Will it simply be displaced? Is its 2,500-year career ending?

In his excellent introduction to philosophy, *Connections to the World*, Arthur C. Danto defends a more-traditional view of the nature and purpose of philosophy:

> What philosophy offers and nothing other than it does is a view of the whole of things—of consciousness, objects, existence, and truth—of the world in its entirety and our relationship to that entirety from within. Somewhere in his or her studies, a student ought for once to gain the perspective that only so global a vision allows, in which the basic orders of things and the relationship between them mapped. Nothing basic can be left out, everything must be accounted for. (1989, xv)

Surprisingly, perhaps, Kornblith says much the same thing. He quotes and endorses the view of philosopher Wilfrid Sellars:

> Along with Wilfrid Sellars, I believe that, "The aim of philosophy, abstractly formulated, is to understand how things in the broadest possible sense of the term hang together in the broadest possible sense of the term." (171)

These are impressive pronouncements, but what exactly do they mean? It seems to me that to say that philosophy deals with the whole of things, or things in the "broadest possible sense of the term," means that philosophy is ultimately concerned with the articulation and examination of worldviews.

Heavy-duty academic terms always sound more impressive in the German. In English, we call it a *worldview*; in German, it is a *Weltanschauung* (*Weltanschauungen*, pl.) Your *Weltanschauung* is the set of basic ideas and guiding concepts that frame your view of everything. Unlike computers, humans are not programmed to generate automatic output in response to input. Our interactions with the world are complexly mediated by concepts, that is, by axioms, theories, assumptions, and conjectures that tell us what to see, what to ignore, how to react, what to expect, what matters, and how we should feel about it. Without some set of ideas to organize our thoughts and guide our cognitive endeavors, we would be as innocent as infants. Your *Weltanschauung*, then, is your basic take on reality—the overarching perspective that gives you the fundamental categories you use to organize your mental world and even shapes your basic intuitions that underlie your spontaneous, unreflective judgments about what is or what should be. Clearly, your *Weltanschauung* largely defines who and what you are.

Such a description of philosophy seems to be at odds with what you actually find in the philosophy journals. There you find dense argument that consists of detailed, often highly technical analyses of fine points and distinctions. There are no sweeping, comprehensive pronouncements on "the whole of things." However, even the most abstruse and seemingly *recherché* philosophical analyses ultimately contact philosophers' broadest commitments. For instance, a fine point about the interpretation of the semantics of possible worlds might bear on commitments as broad as theism versus atheism.

When *Weltanschauungen* clash as, for example, when a conservative, religious perspective clashes with a liberal, secular one, generally more heat than light is generated. Science cannot effectively address such conflicts because the standing of science and the authority of scientific modes of rationality will differ widely between worldviews. If your worldview says that a literal interpretation of the Bible or the Qur'an is the final word on everything, then this conviction will require you to deny or distort science when its conclusions conflict with your purported revelation. Yet even when we cannot turn to science to settle our disagreements, because our disagreements are too deep, we would still like for our views to be articulated and defended as rationally as possible, and this is where philosophy comes in. Philosophical argument can at least aim to show that our deepest intellectual commitments are clear, coherent, consistent, plausible, and free from paradox or self-refutation. Philosophy, then, can be seen as an effort to achieve rationality in the broadest contexts and in the broadest sense of rationality.

Talk of *Weltanschauungen* takes us back to the discussion of the arguments of Thomas Kuhn that we examined in earlier chapters. I argued in chapter 2 that Kuhn's various notions of incommensurability do not entail that scientific theory change is in any sense irrational or that it involves sudden "conversions" or leaps of faith. Rather, I argued that, at every stage the parties engaged in reasoned debate that appealed to evidence and methods recognized as rational considerations by all sides of the controversy. Admittedly, at the end of scientific revolutions things look very different than before, but these big changes in perspective were brought about step-by-step and at no point did the parties to the debate simply talk past each other or have to rely on intimidation, emotional appeals, or bandwagon effects. The reasons were *there* and they mattered.

Yet Kuhn correctly noted that the roots of scientific disputes often go deep, deeper than they appear on the surface. The very term *paradigm*, though lacking a precise meaning, does indicate that scientific disagreements are often embedded in deeper disagreements that can only be called philosophical. Scientific disputes, implicitly or explicitly, often involve disagreements that are epistemological or metaphysical. Scientists sometimes have to become *de facto* philosophers in these disputes. In fact, if you look carefully at scientific disputes, you will see that they often turn on points that are properly regarded as philosophical and can only be addressed by philosophical argument. Science is a "first order" kind of inquiry; it asks questions about nature. Philosophy of science is a "second order" kind of inquiry; it asks questions about science. Which of two explanations is better supported by the evidence is a scientific question. The question of what *constitutes* a legitimate scientific explanation is a question of philosophy, not science.

In *On the Origin of Species*, Darwin frequently addresses philosophical as well as scientific questions. For instance, as we noted in chapter 2, he devotes considerable space to examining the hypothesis of special creation as a theory competing with natural selection. He realizes that special creation is, of course, more than just a purportedly scientific claim and impinges on deep philosophical and religious commitments. In addressing special creation, therefore, Darwin often speaks the language of philosophy and even theology. Darwin's remarkably comprehensive intellect, his background that encompassed both theology and natural science, and his sensitivity to cultural context, made him a particularly qualified commentator on the relation between science and religion.

Darwin addresses the nature of design hypotheses, showing why they are perennially problematic within science. By noting numerous examples, Darwin shows that design arguments raise particular problems for scientific explanation. It is not simply that natural selection provides a physical *modus operandi* and design hypotheses do not. Darwin does not beg the question in favor of materialism. Rather, as Darwin repeatedly shows, design hypotheses are so exiguous in content that they fail in the most basic task of scientific explanation, namely, providing some new piece of information that tells us why the phenomenon was to be expected.

On the other hand, if design hypotheses are given content, as by invoking a benevolent creator, then they raise the perennially thorny problem of how natural evil can be reconciled with the existence of a provident creator. Darwin would sometimes play the role of "Devil's chaplain" by noting some natural facts that seem to us particularly horrifying or repugnant, like wasps that lay their eggs in the paralyzed, but living, bodies of caterpillars so that the wasp larvae can slowly devour them from inside. Darwin notes that such survival strategies are expected if living things evolved by an unthinking, unplanned, amoral process like natural selection. However, they are deeply problematic if we see all of nature as the product of an intelligent and benevolent creator. Another problem with intelligent-design hypotheses is that many alleged designs look unintelligent. Organisms often solve problems with anatomical or behavioral features that do the job, but hardly with ideal efficiency. This is understandable if such features are the result of a blind watchmaker—natural selection—but is hard to explain as a product of a creator with superhuman intelligence and power. When some replied that God might have a good reason for permitting such apparent instances of bad design, Darwin replied that if we hypothesize that God planned things, but then the "plan" we allege looks no different from no plan at all, then the hypothesis is vacuous. This is an astute *philosophical* observation.

Philosophical differences also divided Darwin from some of his scientific critics, such as Richard Owen. As noted in chapter 2, Owen coined the term *homology* to designate deep structural similarities in anatomical features that are superficially very different. A bat's wing, a human hand, a bear's paw, and the flipper of a porpoise look very different and have very different uses. However, when we look at the underlying skeletal structure, we see that these structures match very closely. Darwin adduced homologies as strong evidence for evolution. Shared structure is evidence of descent with modification from a common ancestor. Owen, however, was a biological Platonist who held that individual organisms reflect an ideal archetype. Hence, for Owen, homologies are not evidence of a common ancestry but show that many different species have been created according to the same plan.

The difference between Darwin and Owen on homologies therefore *could not* be settled by a straightforward appeal to evidence. Each recognized the evidence but claimed it for his theory. The fundamental difference between Owen and Darwin was not scientific but philosophical and chiefly concerned the kind of explanations that should be invoked by scientific theories. Like practically all current scientists, Darwin was committed to the sort of methodological naturalism discussed in chapter 3. Only physical causes are admissible in scientific explanations, and appeal to transcendent objects, like archetypes, was ruled out. Again, what constitutes an acceptable scientific explanation is not a scientific but a philosophical question.

In cases like this, where the evidence will not settle the dispute, scientists *must* employ philosophical arguments. And they do. Therefore, the suggestion that science can simply replace philosophy is wrong for the reason that, as Kuhn observed, scientific debates often embed—or are embedded within—philosophical debates. These philosophical differences often cannot be settled by straightforward empirical means, but must be addressed with philosophical argument. *Science cannot replace philosophy because philosophy is an essential part of the scientific enterprise.* Kuhn was wrong about many things, but on this point he was absolutely right.

In conclusion, perhaps philosophy should be viewed as an intellectual enterprise that is continuous with science in the sense that it does not claim any special intellectual faculties or tools unavailable to science but is distinguished by the kinds of questions it asks. Philosophers should not claim any special *a priori* insight or intuition into the nature of things. Further, as Kant cogently argued, when philosophy attempts to appeal to "pure reason" in contexts removed from empirical constraint, it inevitably flies off to cloud-cuckoo-land. Amateur armchair theorizing, which philosophers have committed too often before, should definitely be a thing of the past. Philosophy should be done only by those who *really know* the relevant science. Those who specialize in the philosophy of mind, for instance, should really know their neuroscience. Those who do philosophy of science should have additional degrees in a particular science or in the history of science. Without a solid grasp of the relevant science, philosophers consign themselves to irrelevance by asking the wrong questions and giving answers that go nowhere.

Yet the legitimate questions that philosophers do ask will still be those that cannot be answered in a straightforwardly empirical way. Conflicts between worldviews cannot be settled by an appeal to empirical evidence, since what counts as evidence and how it is to be weighed will differ from one worldview to another. The second-order types of questions that philosophy asks also cannot be settled by a direct appeal to evidence. In fact, as Kuhn showed, the difference between two theories will often not involve a simple disagreement over the evidence but a deeper disparity between concepts of what should *count* as a good theory. Likewise, we cannot decide which of two scientific explanations is better if we have radically divergent views about what a scientific explanation should be or should do. Insofar, then, as philosophical questions arise in science—and they do—science will need philosophy as much as philosophy needs science.

FURTHER READINGS FOR CHAPTER SEVEN

A balanced, clear, and comprehensive introduction to epistemology is Louis Pojman's *What Can We Know? An Introduction to the Theory of Knowledge*, second edition (Belmont, CA: Wadsworth, 2001). Pojman covers both the traditional approaches to epistemology and the newer, naturalized treatments. An equally excellent introduction to ethics is James Rachels's *The Elements of Moral Philosophy*, fourth edition (Boston: McGraw-Hill, 2003). Rachels's book is a model of readability.

Richard Rorty's *Philosophy and the Mirror of Nature* (Princeton: Princeton University Press, 1979), though it is written for professional philosophers, is one of those philosophy books that had an impact on the broader intellectual community. When I began doctoral study in philosophy in 1982 at Queen's University in Kingston, Ontario, Canada, I was appalled to find that Rorty's book was all the rage. It struck me then, as it does now, as a farrago of *non sequiturs*. While Rorty certainly is right that classical foundationalism is dead, it simply does not follow that the project of correctly and objectively representing the world is now defunct. In fact, as I claim above, it seems to me that a turn to epistemological naturalism is the most reasonable response to the demise of the foundationalist project. A succinct and trenchant critique of Rorty is in Susan Haack's *Evidence and Inquiry: A Pragmatist Reconstruction of Epistemology*, expanded edition (Amherst, NY: Prometheus Books, 2009).

Alasdair MacIntyre's *After Virtue*, second edition (Notre Dame: University of Notre Dame Press, 1984) impacted me as strongly as Rorty's book, but in a positive way rather than a negative way. To call the book "magisterial" is an understatement. The deep learning evident on every page does not inhibit the clarity of its exposition or the cogency of its argument. Clearly *something* terrible has happened to ethical discourse in our society. An excruciating hour watching supposed "debate" between opposed parties on any important issue shows that rational disagreement has been pervasively replaced with rhetorical posturing, emotional appeals, *ad hominem* abuse, and invincible stupidity. To my mind, MacIntyre's diagnosis of our situation is compelling.

Quine's seminal essay, "Epistemology Naturalized," is available in many places. I got it from *Naturalizing Epistemology*, edited by Hilary Kornblith (Cambridge: MIT Press, 1985). The essay is quite difficult for nonspecialist readers. Beginners should consult the Pojman book mentioned above for a lucid account and evaluation. As Kornblith notes in the introduction to the above book, the "replacement thesis" obviously raises the problem of norms. How do we tell good reasoning from bad if we ditch epistemology? Due to the work of Daniel Kahneman and Amos Tversky, it is now clear that spontaneous ways of thinking are often grossly erroneous. The book by Massimo Piattelli-Palmarini, *Inevitable Illusions: How Mistakes of Reason Rule our Minds* (New York: John Wiley & Sons, 1994) is a highly readable and enjoyable introduction to this research. Kahneman himself has recently published *Thinking Fast and Slow* (New York: Farrar, Straus, and Giroux, 2011), his own introduction to his research with Tversky. This book has rightly received many plaudits but is considerably longer and more detailed than the Piattelli-Palmarini book, which is still recommended for a quick review.

All of Michael Shermer's books are entertaining and instructive. *The Believing Brain* (New York: Henry Holt, 2011) is a real eye-opener. Some of its claims may be exaggerated, but the basic point can hardly be doubted. Belief almost always comes first and then a process of rationalization, and Shermer's message is that we have to face up to the fact that this is true for us as much as anyone else. Nobody is exempt from the effects of confirmation bias.

Scott Sturgeon's clear and instructive presentation of the methodology of standard epistemology is found in the outstanding anthology *Philosophy: A Guide through the Subject* (Oxford: Oxford University Press, 1995), pp. 10–26. Though nearly twenty years old, this anthology remains an excellent introduction to many fields of philosophy.

Hilary Kornblith's *Knowledge and Its Place in Nature* (Oxford: Clarendon Press, 2002) is an important book. It offers a strong statement of epistemological naturalism and a clear and cogent defense of its main tenets. Its criticisms of standard coherentist and foundationalist epistemologies are especially powerful. Kornblith also provides strong support for a reliabilist theory of justification by supporting a comprehensive view of knowledge that encompasses both human and animal cognition. The accounts of the cognitive abilities of nonhuman animals are riveting and persuasive. As I indicate in this chapter, he offers an analysis of intuition that is quite compatible with a naturalistic perspective and far more plausible than viewing it as a faculty of *a priori* insight.

David Braddon-Mitchell and Frank Jackson develop their "internal map" theory of belief in *Philosophy of Mind and Cognition* (Cambridge, MA: Blackwell Publishers, 1996), pp. 179–85. This book is an excellent overview of the philosophy of mind and cognition at a rather-advanced level. For those wanting more beginner-friendly books on a subject often dauntingly technical, see *What Is a Mind: An Integrative Introduction to the Philosophy of Mind*, by Suzanne Cunningham (Indianapolis: Hackett Publishing, 2000). And *Matters of the Mind* by William Lyons (New York: Routledge, 2001). Both are quite readable and informative; the Lyons book is especially enjoyable because of its detailed background information.

There are many translations of Aristotle's *Nicomachean Ethics*. This is a book that every educated person should read. You can probably pick up a very inexpensive edition at your local used bookstore, or even find it for free online. A recent translation that has gotten much praise is by Robert C. Bartlett and Susan D. Collins (Chicago: University of Chicago Press, 2011). Most of Aristotle's works are very dense and tedious, but not this one. The neo-Aristotelian tradition in ethics is alive and well. Larry Arnhart's *Darwinian Natural Right: The Biological Ethics of Human Nature* (Albany: State University of New York Press, 1998) shows how to adapt Aristotelian ethics to a Darwinian context. My opinion is that the book is brilliant and a very important contribution to current ethical theory. It is also quite readable, with many interesting examples. Of course, the eternal nature versus nurture debate continues, and many will think that Arnhart and other neo-Aristotelians come down too far on the side of nature. Arnhart's book and others, such as Stephen Pinker's *The Blank Slate* (New York: Viking

Penguin, 2002), provide powerful evidence that the continued opposition to the idea of a biological basis for human behavioral proclivities is more rooted in ideology than in biology. Richard Dawkins's luminous *The Blind Watchmaker: Why the Evidence for Evolution Reveals a Universe without Design* (New York: W. W. Norton, 1987) is both entertaining and instructive, as Dawkins's books always are.

Books that argue for a biological basis of religiosity include *Darwin's Cathedral: Evolution, Religion, and the Nature of Society*, by David Sloan Wilson (Chicago: University of Chicago Press, 2002); *Religion Explained: The Evolutionary Origins of Religious Thought*, by Pascal Boyer (New York: Basic Books, 2001); *In Gods We Trust*, by Scott Atran (Oxford: Oxford University Press, 2002); and *Breaking the Spell: Religion as a Natural Phenomenon*, by Daniel Dennett (New York: Viking Penguin, 2006). Though these books disagree with each other on a number of points, they support the quite-fascinating thesis that religiosity is a natural, biologically based human feature. Ironically, though some of these authors are atheists, if they are right, their conclusion implies that atheist arguments will never succeed! If they are right, arguing against religion would be like arguing against sex.

Arthur C. Danto's *Connections to the World: The Basic Concepts of Philosophy* (Cambridge: Cambridge University Press, 1989) is an ideal introduction to philosophy. It is accessible to beginners but can be read with profit and enjoyment by professional philosophers. Anyone tempted to take a dismissive or disparaging attitude toward philosophy should read this book. It convincingly shows that philosophy is far from defunct and still offers much richness to our intellectual lives. Another book that shows how instructive and enjoyable philosophy can be is Robert Fogelin's *Walking the Tightrope of Reason: The Precarious Life of a Rational Animal* (Oxford: Oxford University Press, 2003). Fogelin's third chapter, "Pure Reason and Its Delusions," is a brilliantly clear defense of Kant's attack on "pure reason."

CHAPTER EIGHT

SCIENCE, SCIENTISM, AND BEING HUMAN

Since its beginnings in the early seventeenth century, modern science has inspired and intimidated. It is inspiring to recognize that we humans, for all our manifold and grievous flaws, are capable of learning something about the universe, including that part of the universe of personal concern to us—ourselves. It is intimidating to think that science might be able to explain *everything* when that "everything" includes us. Why? Because science explains in terms of impersonal (unthinking, unfeeling) entities, processes, and laws. We, however, are clearly personal beings with thoughts and feelings and self-awareness, and we resent it when science "reduces" us to something impersonal. Who wants to be told that the rapturous, delectable yearning of a first love is really just glandular secretions? Can we really imagine that the thunder of Beethoven's *Eroica* symphony or the sublimity of his *Moonlight* sonata arose from the massed firing of neurons?

Another problem with scientific explanations is that they are generally deterministic. At the quantum level, there is no determinism, only probability. However, for bigger things—and even the nerve cells in our brains are much larger than quantum objects—determinism is generally presumed. We may use statistics and probability to describe the behavior of big things, but this is seen as a concession to the complexity and inscrutability of deterministic causes, which are nevertheless presumed to hold. For instance, we use probability to describe the outcome of multiple tosses of fair coins, but we presume that the actual trajectory of each toss is determined by a complex set of initial conditions, applied forces, and ambient circumstances like air currents. However, we like to think of our own behavior as free of deterministic causation. WE make up our own minds as free agents, right? It is insulting to imply that our ostensibly free choices might have been rigidly constrained and that we are really no freer than the tossed coin.

But *does* science have such alarming implications? If we understand ourselves scientifically, does this preclude seeing ourselves in those comfortable, familiar terms as lovers, actors, dreamers, and creators, free to make our own choices and morally responsible for our acts? Does the methodological reductionism of science imply a reduced self-concept? Are we just complex machines afflicted with the tragic illusion that we are more?

Many philosophers and other thoughtful people have feared that science really does have these disturbing implications, and their reaction has been to seek a way to limit the range of scientific explanation. One of the first and most famous of these persons was René Descartes. We met Descartes in previous chapters, and he clearly was no crackpot or ignoramus but one of the leading mathematicians and physicists of his day. Nevertheless, Descartes feared that the advancement of science, already dramatically obvious in his day, threatened to reduce humans to automata, complex mechanisms capable of complex action, but lacking free will and certainly without hope of immortality.

Descartes addressed the problem by developing his famous theory of Cartesian dualism, also often referred to as "substance dualism" (I will just call it "dualism" for short). Descartes distinguished between physical, extended (three-dimensional) objects, which he called "*res extensa*" (extended stuff), and incorporeal, spiritual, nonphysical souls or minds, which he called "*res cogitans*" (thinking stuff). Physical objects, including human bodies, are material and subject to the laws of physics. The human mind, however, is not a physical thing and is not subject to the laws of physics; so the deterministic laws that govern physical things do not rule the mind. Further, while physical things decay and are destroyed, the mind can outlive the destruction of the body.

Physics reigns in its domain, but the human essence, the immortal soul, the seat of intellect, consciousness, and all "mental" occurrences, is an exception to otherwise-omnipotent physical law. The essential nature of mind is to think, where "thinking" is construed broadly to include consciousness and all mental states.

Other philosophers dealt with the apparent threat of science in their own ways. Some, like Irish philosopher George Berkeley, denied the existence of matter altogether and held that the laws of nature are really just the regular and consistent actions of God. Throughout the nineteenth century, a century of unprecedented scientific progress, the leading philosophical schools advocated some form of idealism, the view that reality is ultimately mental. If reality, including "physical" reality is actually mental, then we do not have to worry about being "reduced" to configurations of atoms. Even many of the leading scientists of the nineteenth century, like Louis Pasteur, held that life could not be explained purely in material, mechanical terms, and they posited an *élan vital*, a spiritual "life force" to account for the apparently fundamental difference between living tissue and inanimate stuff.

The development of science and philosophy has not been kind to these efforts to resist the reductive impetus of physical explanation. For instance, the enormous advances in neuroscience of the last few decades make it harder to deny that everything that makes us *who* we are is a matter of brain function, that is, that brain activity is both necessary and sufficient for all mental activity. We think, feel, imagine, perceive, wish, hope, choose, decide, intuit, believe, and doubt with our brains. Our nervous system comprises our organs of thought just as the digestive tract comprises the organs of digestion.

As just one of many lines of evidence for the above assertion, consider the popular books of noted neurologist Oliver Sacks. In books like *Awakenings*, *The Man Who Mistook His Wife for a Hat*, and *An Anthropologist on Mars*, Sacks chronicles many cases, some sweet, some poignant, and some tragic, of people who had undergone major changes to their brains, often damage due to disease or trauma. Each of these persons experienced major changes in their mental lives, sometimes changes so deep that they seemingly became different persons. For instance, he describes the case of "the last hippie," a man who had suffered a brain tumor, which surgeons removed, but which left his brain damaged, particularly portions of the temporal lobe that affect memory. This patient was perpetually stuck in 1967 and had no memories from after the late 1960s. For him, Lyndon Johnson was always president, the Vietnam War was ongoing, hippies were turning on, the flower children were having love-ins, and the Grateful Dead, Janis Joplin, and Jimi Hendrix were still performing. I guess the Age of Aquarius is not the worst era to be stuck in. In case after case, Sacks shows how thought, feeling, and perception—the very constituents of personhood—are profoundly and fundamentally tied to brain function.

Philosopher Owen Flanagan, while admitting that neuroscience has not yet offered a complete account of how mind arises from the brain, argues that there has been so much progress in the neurosciences that at least we should accept as a regulative idea, that is, as a heuristic principle to guide further inquiry, that our mental lives are wholly explicable in terms of brain function (2002, 74–79). Let's call this regulative idea the "mind to brain" (MTB) thesis. The MTB thesis is now accepted by nearly all brain scientists and most philosophers of mind, but there are notable exceptions. We begin this chapter by looking at a statement of such dissent by two Christian philosophers, Charles Taliafero and Stewart Goetz. The Secular Web hosted a debate between Taliafero and Goetz on one side and philosopher Andrew Melnyk on the other. Melnyk defends the thesis that the mental is fully realized in the functioning of the brain. Taliafero and Goetz support Cartesian dualism, the view that the mental is not reducible to the physical but entails the existence of a nonphysical, substantial soul or mind.

I hold that Melnyk wins this debate and will say why. To some extent I go off on my own, beyond what Melnyk claims. If we accept the MTB thesis, we are left with some very deep and troubling questions. Must we conclude that our traditional picture of ourselves as rational, free, and autonomous agents is wrong?

Philosopher Alex Rosenberg bites the bullet and defends what he calls "scientism." He thinks that, given the current state of scientific knowledge, we are compelled to alter radically our traditional ideas about ourselves. We must recognize that free will is an illusion and that there is no rational basis for morality. He frankly asserts that the results of science entail ethical nihilism, though he does qualify it as "nice" nihilism. Rosenberg naturally opposes the sort of supernaturalism defended by Taliafero and Goetz, but he equally opposes the claims of many secular philosophers. Rosenberg's conclusions are very different from Flanagan's. Flanagan, though he is a strong defender of the MTB thesis and lifts the banner of science as high as Rosenberg, holds that morality and all the freedom worth having are compatible with the implications of science.

I will present these contrasting views and come down on the side of Flanagan. I finish with a meditation on Benjamin Disraeli's famous query: Is man an angel or an ape? I conclude that we must firmly reject both horns of Disraeli's false dilemma—repudiate both dualism and scientism—and return to a more Aristotelian conception of what it is to be a rational animal.

MIND: PHYSICAL OR SPIRITUAL?

Philosopher Andrew Melnyk argues that the mind is "fully realized" in the physical functioning of the brain. Melnyk holds that we must distinguish between a mind and its particular realization, that is, the specific form, embodiment, or configuration that it takes. He holds that *mind* should have a functional definition. Many things are defined by their function. Just a few years ago, we would have defined *book* in terms of its physical constitution, print on paper. Now we have handheld electronic devices, like Kindle, and Nook, that do the job that printed books do. *Book*, then should now be given a functional definition. To have a function is to have a specified job to do, some particular set of tasks that the object exists to perform, whether the specifying is done by human design or "the blind watchmaker," natural selection (recall the discussion of functional explanations in the last chapter). The job we want a book to do is to present an extended text in a manner convenient for reading. A book, then, is something—anything—that has the function of presenting an extended text in a manner convenient for reading. A book can still be realized in the form of print on paper, or it can be realized digitally and displayed on the screen of a handheld electronic device. The physical embodiment of the book is secondary; what matters is what a book *does*.

What holds for a book also holds for organs of the body. A kidney, for instance, is something that functions to filter the blood. If a natural kidney fails, and surgeons replaced it with an internal artificial device, one that served to filter the blood in the same way that a healthy natural kidney does, then we would call that device a "kidney" even though it is made of plastic and metal. Melnyk suggests that *mind* also be given a functional definition. What we call "mind" is something—anything—that functions to perform mental activities—thinking, feeling, perceiving, imagining, and so on. A mind is whatever does mental things. We then say that the physical realization, the specific embodiment, of the human mind is the brain. For humans, mental objects, like thoughts, are internal representations created by brains. Similarly with all other "mental" events; they are one and all also physical events, that is, acts done by brains.

Melnyk puts it like this:

> Humans have minds but it does not follow that minds are genuine persisting objects. Just as humans have minds, they also have personalities, but no one imagines that human personalities are persisting objects, not even nonphysical ones. In fact, however, physicalism about the human mind can easily treat minds as persisting objects—by treating an organism's mind as its *mental system*, on a par with its digestive system or immune system. As a rough-and-ready first approximation, we might suggest that an organism has a mental system just in case it houses some complex object or other that can use sensors to form representations about the organism's environment and internal states, that can store some of these representations, and that can

> undergo internal processes in which these representations interact with one another and with representations of goal-states so as to produce organismic behavior, which behavior, often enough, achieves the organism's goals. ("Case," 3)

Whether a being has a human gray-matter brain or something entirely different, like the brain of an android or extraterrestrial, what matters is what that brain-stuff *does*, not what it is made of.

What, then, will be a mental property on this account? Melnyk explains:

> Such mental properties as thinking, hoping, wishing, and doubting can be understood as different kinds of mental representing. For example, thinking that one's keys are in the ignition might be a mental property that a human possesses just in case he or she internally represents that his or her keys are in the ignition, where this internal representation plays one kind of causal role with regard to effecting behavior (the belief role) in his or her overall mental system. By contrast, hoping that the gas gauge is broken might be a mental property that a human possesses just in case he or she internally represents that the gas gauge is broken, where this representation plays a different kind of causal role with regard to effecting behavior (the hope role) in his or her overall mental system. ("Case," 3–4)

Brains create representations that function in different causal roles. A belief is an inner representation that plays a certain causal role, and a hope is another representation that plays a different role.

This is Melnyk's version of the MTB thesis: mind is a set of functions fully realized in the internal representations created by brains and the causal relations of those functions. Let's suppose that this is a coherent picture. What reason do we have to think that it is true? Melnyk gives two reasons: an induction from the history of science and evidence from the neurosciences of the dependence of the mental on the neural ("Case," 6–9). Actually, the arguments Melnyk gives here, and two others I will mention, support not just *his* version of the MTB thesis but also that thesis broadly conceived, that is, *any* MTB thesis. Therefore, I will consider these arguments successful if they support *any* MTB thesis, and not just Melnyk's.

Melnyk's particular claim is, of course, not the only kind of MTB thesis that has been or could be defended. I will take the MTB thesis broadly as any claim that, for all creatures with brains, all mental predicates, processes, or phenomena, reduce to, are realized by, are identical with, supervene on, are epiphenomena of, are predicates of, or are exhaustively caused by types or tokens of brain states or occurrences. In short, the MTB thesis will encompass any of these or any other theses that the mental depends entirely on the physical occurrences in the brain.

If we take a broad view of the history of science, we can see that one of the dominant themes has been that phenomena once thought too wonderful or mysterious to have a physical explanation have been found to be explicable in physical terms. Progress in explaining nonmental phenomena as physically realized should led us to expect the same sorts of explanations for mental happenings:

> Of course, that all nonmental phenomena are physical or physically realized doesn't deductively entail that mental phenomena are too. But it can still be some evidence that they are; and it is. Consider an analogous case: the discovery that all *non*carnivorous plants are biochemically realized is clearly some evidence that *all* plants, including therefore all carnivorous ones, are biochemically realized. . . . Likewise, the discovery that all *non*mental phenomena are physical or physically realized is some evidence that *all* phenomena, including therefore all mental phenomena, are physical or physically realized. ("Case," 6)

Though Melnyk does not mention this specific example, one good instance is that, as mentioned above, well into the twentieth century some leading biologists thought that life was such a distinctive phenomenon that it could not be explained exclusively in physical terms. Hence, a supernatural "vital force" was posited. As time went on, though, detailed experimental evidence showed that even the most distinctive functions of living things were explicable purely in physical terms. For instance, the discovery of the Krebs cycle showed how animals are powered and energized by the metabolic processing of starches and sugars into glucose, and how the cellular powerhouses, the mitochondria, function to provide that energy. Further, the invocation of occult forces simply did not provide the sorts of explanations that scientists seek. Postulating such an inscrutable force serves only to deepen our mystery. In fact we are left with two mysteries instead of one—the mystery we were trying to explain and the mystery invoked to explain the first mystery. Likewise, advancing science certainly seems to indicate that all of the distinctive mental functions are physically explicable, and postulating a soul seems increasingly to be as obscurantist as postulating vital forces.

Melnyk's second argument deals with the burgeoning evidence of the detailed dependence of mental events on neural occurrences. Melnyk notes some of this research:

> For any (human) person you like, and for any mental state or mental process that person might be in our might undergo, in order for that person to be in that mental state or to undergo that mental process, there is a neural activity of some distinctive kind that has to be going on—simultaneously—in that person's brain. In the numerous studies that support this finding, a brain-imaging technique is applied to the brain of a human experimental subject who has been subjected to some stimulus (e.g., spoken words, objects placed in the hand) or instructed to perform some mental task (e.g., to read silently, to clench a fist). What is found in the studies is that in each subject distinctive regions of the subject's brain are especially active when a particular stimulus is presented or a mental task performed. ("Case," 7)

In short, *all* of the experimental evidence we so far have indicates that neural activity of distinctive sorts is *always* present when (and only when) certain mental actions are being performed. These results indicate that neural activity is both necessary and sufficient for mental activity.

Put simply, if we do not think with our brains, we would expect to see some instances of mental activity with no brain activity. We would, for instance, expect to see someone carrying on an intelligent conversation when his EEG was flat, indicating no brain activity. Why would not such occurrences be common?

Some people have claimed the occurrence NDEs (near-death experiences) in which people supposedly felt, saw, and heard remarkable things when clinically dead. However, the claim that these events occurred when there was no brain activity is controversial in the extreme. It is safe to say that no such claims have been accepted by the relevant scientific communities. Even if there is a "residue" of unexplained reports of NDEs, it is hard to see how this is any different from the "residue" of unexplained UFO sightings. About 1 percent of reported UFO sightings remain unexplained, but, given the fact that the other 99 percent have mundane explanations, do we conclude that the "residue" must be extraterrestrials in the mother ships? Likewise, if innumerable cases indicate a consistent connection between mental and neural activity, what are we to make of a few highly disputed claimed exceptions?

Melnyk could have offered several other arguments for a physical mind had space allowed. I will mention two standard ones. First, what about animal minds? Do nonhuman animals have souls? For the sake of argument, let's suppose that everyone, even the hard-bitten dualist, agrees that the mental states of animals are entirely due to brain activity. Let's assume that no one questions the MTB thesis as applied to nonhuman animals. However, as we noted in the last chapter, animals seem capable of a number of quite-sophisticated mental representations. It even seems reasonable to attribute true beliefs and, given a reliability account of warrant, even knowledge to them. Animals give every indication of being able to feel a great variety of emotions, including some of the "moral" emotions like empathy and courage. Some animals clearly grieve when one close to them is killed. Others sometimes display great tenderness in caring for the young of other species.

Given, then, (1) that animals do have such mental capacities, and

(2) that animal and human mental functions are similar, and (3) that nonhuman animals perform these functions with their brains, surely this creates a strong presumption in favor of explaining human mentality in similar terms. Anyone disputing this presumption, and offering a radically different explanation—like the dualist one—would have to bear a heavy burden of proof. Such a person would have to indicate which characteristics of human intellect or feeling make for an in-principle distinction between merely animal capacities and those that demand a completely different kind of explanation. How do we draw a line between animal and human mentality that is not arbitrary and *ad hoc*? Of course, animals cannot do calculus or understand *The Critique of Pure Reason* or compose in iambic pentameter, but neither can most humans. What is it that all humans and no animals can do that demands a wholly different kind of explanation?

The problem is made much worse when we consider that human beings descended by an evolutionary process from nonhuman ancestors. At what point in the evolutionary development of human mentality did physical explanations become inadequate? When did humans get souls? Did *Australopithecus afarensis* have a soul? What about *Homo habilis*? *Homo erectus*? Will there be Neanderthals in heaven, or were they soulless? It is hard to see how any nonarbitrary answer could be given to these questions. If a dualist objects that this argument begs the question against him by assuming the truth of evolutionary theory, then we may admit to begging the question. However, nothing would discredit dualism more than if its advocates were to ally it with creationist pseudoscience.

If, on the other hand, dualists are willing to assert that some nonhuman animals have nonphysical minds, the above problem of arbitrariness and *ad hoc*ness is just relocated. Surely oysters have no souls. If bonobos do, then we have to have some principled answer to the question of which animal capacities demand explanations in terms of a soul and which do not. Do frogs need souls, but fish not? Given the spectrum of animal abilities from the desert ant mentioned in the last chapter to chimps, where do you drive the golden spike and say that here, just here, is where physical explanation becomes impossible? In the absence of such a specification, the presumption in favor of the MTB thesis stands.

Another way of supporting a hypothesis is to discredit its competitors. If the main competitor to the MTB thesis is dualism, then discrediting dualism will *ipso facto* support the MTB thesis. The oldest and most persistent difficulty with dualism is the interaction problem. In a letter from one of his patrons to Descartes, Princess Elizabeth of Bohemia, clearly a woman of considerable perspicacity, makes the following objection to the Cartesian dualism of mental substance and material substance (I quote from William Lyons, including his parenthetical comments):

> How can the soul of man, being only a thinking substance [i.e., pure consciousness], determine [i.e., causally interact with] his bodily spirits [in modern terms, the neurotransmitters of his nervous system] to perform voluntary actions? (2001, 25)

In other words, if, as Descartes claimed, the whole nature of mind is to consist of pure consciousness and it lacks any physical parts or predicates, how is it conceivable that there could be interaction between mind and body? After all, physical entities interact by the exertion of physical forces such as gravity and electromagnetism. All chemical phenomena, for instance, are ultimately explicable in terms of electromagnetic force. How, then, do we account for the apparent interaction between mind and body, as when the desire for something leads to bodily motions to obtain it, or when bodily trauma causes a sensation of pain?

Despite heroic efforts, Descartes could offer no explanation, and neither can modern dualists. Mind/body interaction on the dualist theory is just as mysterious now as it was in Descartes's day. Dualists often respond by conceding that the interaction of mind and matter is a mystery, but they reply with the *tu quoque* that physical actions are equally mysterious. If physical causation of physical effects is just as mysterious as mental causation of physical effects, then it is unfair and arbitrary to stigmatize dualists as mystery-mongers. Dualists often invoke David Hume here. Hume argued that causation is really nothing more than the consistent conjunction of types of events; *A*s are consistently followed by *B*s and that is the whole story. Thus, when it comes to causation, physicalists can say no more than dualists: *B*s do follow *A*s.

However, physicalists should not be frightened by Hume's ghost. Science has made enormous progress since Hume's day in providing deeper and richer and *detailed* causal accounts. I will quote myself from a review published a few years ago:

> We now possess many well-confirmed and copiously detailed explanations of *how* physical effects are brought about. Indeed, one of the major challenges facing a student in a field such as molecular biology is the sheer weight of detail that has to be mastered to comprehend how molecular processes accomplish their effects. Perhaps [dualists] would reply that such accounts, however detailed, merely scratch the surface and do not tell us what is really, fundamentally going on when physical causation occurs. At the most basic level, they might claim, at the level of our theories of fundamental forces and their interactions, all we can say is that things *do* happen in a given way.
>
> But the point is precisely that with many scientific causal accounts, there is a great wealth of explanatory detail *before* we reach causal bedrock. Even at the presumably rock-bottom level of quarks, electrons, and photons we have well-confirmed, mathematically precise theories, like quantum electrodynamics, that often make astonishingly accurate predictions. These theories do not just tell us *that* fundamental particles interact, but give us much information about *the way that they do*. With supernatural causal explanations, on the other hand, our inquiry simply hits a wall. An advocate of the Cartesian theory of mind can only say that mind does move things by a power which . . . is occult. (Parsons, 2008)

Owen Flanagan sums up these points by noting that the defender of Cartesian dualism must say that the mind performs psychokinesis every time we commit a voluntary action (2002, 58). By referring to psychokinesis, Flanagan is not resorting to the cheap tactic of lampooning dualism by associating it with flaky paranormalism (cf. Stephen King's psychokinetic heroine, Carrie). Rather, the point is that we understand *how* physical objects interact; we have identified the basic forces that mediate those interactions and know a great deal about those forces. With respect to the purported interaction between spiritual minds and physical objects, there are speculations but no confirmed theory, and no reasonable prospect of getting one. The occult nature of the mind/body interaction appears to be a permanent and ineradicable feature of dualism, and this will always be the Achilles' heel of that theory. Supporters of dualism must argue that, nonetheless and despite this limitation, dualism is the best theory for accounting for *all* the phenomena, and it is to arguments we now turn.

Stewart Goetz holds that dualism uniquely accounts for one very large and essential class of data—our first-person experiences. He quotes Descartes from *Meditation VI* to the effect that when he considers himself as a thinking thing he recognizes that his mind (which was identical with his soul for Descartes) is a simple thing without parts. That is, the different mental faculties are all powers and capacities of a single, unified subject. Goetz comments:

> Descartes notes that when he introspectively considers himself, he apprehends himself to be one and entire, by which he seems to mean that he is aware that he has no substantive parts. . . . Moreover, because Descartes is aware that he is a substantively simple soul, when, for example, he thinks about what he is, he is aware that *all* of him thinks. He thinks as a *whole* or in his *entirety* because as an entity that lacks substantive parts, it is not possible for one part of him to think while another does not. And when he *simultaneously* thinks about his being a soul, hopes that he will survive death, and feels a pain in his foot, and is aware that it is one and the same soul in its entirety that is the subject of each event. (2013, 263–64)

Goetz argues that our awareness of ourselves as simple, unified wholes provides one argument for dualism:

1. I (my soul) am (is) essentially a simple entity (I have no substantive parts).
2. My body essentially is a complex entity (my body has substantive parts).
3. If "two" entities are identical, then whatever is a property of the one is a property of the other.
4. Therefore, because I have an essential property that my body lacks, I am not identical with my body. (2013, 264)

Goetz and Taliafero, in their reply to Melnyk on the Secular Web, claim that we also have an immediate, first-person awareness that when we choose things for reasons, the choice is an uncaused mental event and the reason that explains the event is irreducibly teleological, that is, purposive, in nature (Goetz and Taliafero, 1–2). What they seem to mean is that when you choose to do something for a reason, say to keep a promise, you are immediately aware of two things (a) that your choice is not caused, and (b) that the (noncausal) explanation of your choice is your reason for making it, that is, that you did so for the purpose of keeping a promise. They comment:

> If physicalism is true, important elements of this first-person point of view are mistaken. If M [Melnyk] made a choice to write his paper, then that choice was not ultimately and irreducibly explained teleologically by a purpose. Instead, M's choice was ultimately and irreducibly explained causally and it or something else it caused in turn causally produced events in his brain, which in turn causally produced the movements of his fingers and the production of his paper, where no event in this sequence ultimately occurred for an irreducible purpose. . . . Given these implications of physicalism, why think it is true? (2)

In other words, Melnyk's theory of the physical realization of the mind (PRM) entails that our seemingly immediate first-person experience of making choices for reasons must be wrong. We seem, for instance, to do things like this: We choose to go to the zoo this Sunday because we promised little Susie that we would take her. Our experience is not that we are *caused* to take little Susie to the zoo by our promise; rather we seem to make a free choice to honor that promise and that is our reason for making the choice. If PRM is true, however, that appearance of free choice is an illusion. Given PRM, what must really happen is that a sequence of neural events in your brain was the proximate cause of your "choice" to take little Susie to the zoo. In fact, you had no choice at all. On PRM, all mental events are realized as physical events, and physical events have physical causes. Your "choice" was no freer than the kick you give when the doctor taps your knee. Yet, surely, it seems, any claim that goes so blatantly against plain experience must be false.

An even deeper problem concerns the causal determination of belief. It appears that if PRM is true, it cannot be rationally believed!

> What this [PRM] seems to imply is that everything that occurs in our mental lives, including our beliefs, is ultimately explicable in terms of physical causation alone without any explanation of that which is irreducibly mental by something else which is irreducibly mental. If physicalism is true, it seems that the ultimate explanation of M's belief that physicalism is true will mention only neural phenomena and/or more fundamental physical constituents. There will be no mention in that explanation of anything that is ultimately and irreducibly mental. So, even if there are irreducible contents of apprehensions and beliefs which are physically realized . . . physicalism entails that M believes that physicalism is true, not because his apprehensions of irreducible contents ultimately cause his belief with it irreducible contents but because physical (neural) realizers of those apprehensions causally produce physical realizers of that belief. In short, physicalism seems to entail that apprehensions of contents are explanatorily impotent. Given the truth of epiphenomenalism and the explanatory impotence of apprehensions of contents, why should any of us try to reason with others about anything with the hope of changing their beliefs by having them apprehend contents of conceptual entities? (Goetz and Taliafero, 4)

This important and subtle "argument from reason" (AFR) must be explained carefully: A common view of rationality holds that what makes many of our beliefs rational is that we base them on reasons. That is, for many *P*s, you rationally believe that *P* when, and only when, your belief that *P* results from having fairly considered the reasons for (and against) *P* and it seems to you that the reasons, on balance, are sufficient for the truth of *P*. For instance, if I want to know the capital of Mongolia, I pull down my atlas and see that the capital of Mongolia is Ulan Bator. My recognition that a standard reference book, which would be very unlikely to be wrong about such a fact, tells me that the capital of Mongolia is Ulan Bator seems to me sufficient reason to accept that claim. Hence, my belief is rational. The essence of this account of rational belief is that one irreducibly mental event, belief that *P* is explained by another irreducibly mental event, the apprehension of the reasons for *P* and their sufficiency for the truth of *P*.

If you ask Melnyk why he accepts PRM, then, like anyone else, he will give reasons that he holds are sufficient for the truth of his claim. Let's call those reasons *R*1, *R*2, *R*3 . . . *Rn*. Goetz and Taliafero argue that if PRM is true, Melnyk cannot hold PRM *because* of reasons *R*1, *R*2, *R*3 . . . *Rn*. Suppose that Melnyk apprehends reasons *R*1, *R*2, *R*3 . . . *Rn* (where we will assume that those reasons include the recognition that the other reasons are jointly sufficient for the truth of PRM). It follows that, since PRM is assumed true, the cause of Melnyk's belief is not his *mental* apprehension of those reasons but the *physical* realization of those mental apprehensions in his neural states. This is because, if PRM is true, the mental state "belief that PRM is true" is fully realized in a physical (neural) state. Further, that neural state is caused by other neural states, namely, the physical realizations of the apprehension of reasons *R*1, *R*2, *R*3 . . . *Rn*.

The upshot is apparently that, on PRM, the mental act of understanding a reason is an epiphenomenon of the physical cause. To say that *A* is an epiphenomenon of *B* means that *B* causally determines *A*, but *A* does no causal work, either back on *B* or on anything else. For instance, the noise that a car engine makes is caused by the running of the engine but plays no part in making it go, so engine noise is an epiphenomenon. So, on PRM, it cannot be the mental act of understanding *per se* that explains the belief; rather, it is the physical realization of that act that does *all* the causal work. The mental act therefore just seems a useless appendage, an epiphenomenon. Hence, if PRM is true, then Melnyk *cannot* believe the truth of PRM because of the reasons he cites. Melnyk does not believe because of those reasons. Instead, his belief was caused by a sequence of neural, that is, purely physical, events in his brain. But how can such a belief be rational if it is not based on reasons? Melnyk's belief that PRM is true seems to be like the breaking of a glass bowl when it is dropped onto a hard floor. Both belief and breakage seem to be caused by brute, blind force (to indulge in a bit of rhetoric), so neither the belief nor the broken bowl can be explained in terms of reasons.

It seems to follow from this argument that the truth of PRM precludes any rational basis for believing that PRM is true. This is because the truth of PRM precludes the possibility of having any reasons for believing it because it relegates all causes of belief to a sequence of physical causes. Surely, it seems, there must be something profoundly and irremediably wrong with a theory that, if true, ensures that it cannot be rationally believed to be true!

I think that the AFR articulated in the above few paragraphs is Goetz and Taliafero's strongest argument. However, given the limitations of space, Melnyk's reply is rather truncated, so I will say in somewhat more detail what I think is wrong with Goetz and Taliafero's AFR.

First, note that Goetz and Taliafero's arguments, and practically all arguments against the MTB thesis, are *a priori* in nature, whereas the arguments for MTB are mostly empirical. Historically, *a priori* arguments have fared very poorly when opposed to empirical arguments. Philosophers will draw an *a priori* line in the sand and scientists will gleefully jump over it. Zeno "proved" that motion is impossible. Galileo paid no attention when he did his experiments with balls and inclined planes. Leibniz "proved" that there could be no physical atom. Atomic theory progressed nonetheless. Spinoza "demonstrated" that all that happens *must* occur. Quantum physics ignores such determinism. Descartes said that the mind *must* be spiritual, but neuroscience rapidly progresses on the opposite assumption. The dismal track record of *a priori* claims against empirical ones provides some reason to doubt the cogency of arguments like those of Goetz and Taliafero.

Further, human beings would be most unfortunate if a theory as important as PRM were true and could not be rationally believed. Goetz and Taliafero appear to concede that PRM *could* be true, but they hold that the truth of PRM would preclude rationally believing it by our standards of rational belief. However, if our standards of rational belief can preclude us from rationally believing an important theory that (we are assuming) *is* in fact true, then perhaps our standards of rational belief are deficient. Standards of rational belief are supposed to permit, not preclude, rational belief in true theories. If PRM is true—and, again, Goetz and Taliafero apparently concede that it *could* be—then this is a very important truth and there needs to be some way that we can rationally believe that it is true.

A reliability account of warrant would seem to fill the bill. According to reliabilism, all that matters is whether a belief-forming process reliably generates true beliefs. A belief that is formed by a reliable process is warranted and therefore a rational belief. The nature of the process is irrelevant. If I can reliably form true beliefs by rolling a pair of magic dice, then rolling the dice confers warrant on my beliefs. If we accept reliabilism, then it does not matter if Melnyk is caused to believe that PRM is true by a physical process. It only matters that the process was reliable in generating true beliefs, and the reliability of physical belief-forming processes can certainly be accommodated by PRM and would seem to generate no self-referential problem. That is, there could be reliable processes causing the belief that PRM is true.

Some might object that precisely the problem with a reliability theory is that its account of warrant runs plainly against some of our deepest intuitions about rationality, such as that many beliefs are rational only if they held *because of* the reasons for them. But what is the worth of such intuitions when they are opposed by successful theory?

Recall how in the last chapter we described the behavior of the piping plover in leading a fox away from her nest. The plover has to monitor the fox's actions reliably and alter her behavior accordingly to keep him on the path away from the nest. If, as we argued, the best theory to explain many instances of the plover's—and many other animals'—behavior is in terms of true belief, and (given a reliability account of warrant) even knowledge, then the worth of countervailing intuitions is diminished. Kornblith proposed a comprehensive theory of knowledge that encompasses both animal and human cognition. Such a theory entails a reliability account of warrant, and insofar as that theory is supported, so are reliability accounts of warrant.

Besides, everyone admits that very many—I would say the vast majority—of our rational beliefs have nothing to do with apprehending reasons but seem to be a result of reliable processes. I see a cup on my desk and immediately believe that there is a cup on my desk. Apprehending reasons has nothing to do with it. The same thing holds for many beliefs based on testimony. As soon as my wife tells me that it is raining again, I immediately believe that it is raining again. I don't weigh her credibility against my background beliefs or anything like that. I immediately, and rationally, accept it. Even much inference is immediate and, though we *could* spell out reasons for our conclusions, we don't actually do it by apprehending such reasons. We just *see* what follows, like Crusoe immediately knowing that another human had been on his island when he saw the footprint. In each of these cases, what seems to make the beliefs rational is that each was caused by a reliable process.

Goetz and Taliafero do not deny cases like these but only say (private communication) that *some* knowledge must be gained by apprehending reasons and seeing that these grounds provide adequate support for a claim. So, let's cut to the chase and ask, Would reasons be epiphenomena on PRM, useless danglers that have no explanatory role in accounting for our beliefs? Not at all. In fact, it seems to me that to think that reasons would be epiphenomenal given PRM is to fail to take PRM seriously and to continue to think in outmoded and obscurantist dualistic terms.

Let's begin by reviewing what it means to say that the mental is physically realized. PRM first defines the mind in functional terms; a mind is anything that does mental stuff. That is, a mind is *anything* that thinks, feels, imagines, perceives, wishes, doubts, and so on. PRM then identifies the human brain (or, technically, certain physical subsystems of the brain) as the object that, for human beings, performs the function of doing mental stuff (including rational thought). At bottom, then, PRM is a theory about *how* we think; we do it with our brains. We have a mental system fully realized in our mental organ, the brain, just as we have a digestive system fully realized in the organs of digestion. The functioning of our mental system—those orderly, causally linked patterns of neuronal firings—is how we think, just as the functioning of our digestive system—peristalsis, the secretion of digestive juices, and so on—is how we digest. Those neural processes in our brains are not a more fundamental reality to which thought can be reduced. They are thought itself—the genuine article. Further, the fact that we think with our brains does not make thought any less real, significant, or explanatory than it is on dualism.

Let's imagine a case of simple inference by *modus ponens*: Upon arising in the morning, I look out the window and see that the streets are wet. I infer that it rained overnight. Generally, that inference would be immediate, but I have not had my coffee yet, so my thinking process is slow. Let's imagine that I think it out step by step: "I see that the streets are wet. I know that if the streets are wet, then it rained last night. Therefore, it rained last night." According to PRM, what just happened here? What caused my belief that it rained last night? That belief, "it rained last night," is neurally realized and, of course, its proximate cause was other neural events. However, those other neural events were events of a very special sort; they were also the thoughts "the streets are wet" and "if the streets are wet, then it rained last night." So those physical events were also mental events—because the physical events are the *doing* of the mental events—and so an equally good explanation is that my belief was based on my reasoning validly in accordance with *modus ponens*.

Given PRM, there are two ways of individuating a thought-token (note: "token" is philosophical language for speaking of a particular instance of a general "type." Thinking "it is raining" on a particular occasion is a thought-token of a thought-type that many can have on many occasions). We could individuate a thought-token by indicating that a particular thought-content was entertained by a particular person at a given time, for example, "At time *T1* Parsons thought 'the streets are wet.'" Or we could individuate it by a specifying a particular set of neural events in a particular brain, for example, "At time *T1* this particular pattern of neuronal firing occurred in Parsons's brain." But just because we can individuate something in two different ways does not mean that we are individuating two different things. When the soprano hits high C, we can individuate that event in terms of music or of physics, but it is the same event. According to PRM, then, one and the same thought that is individuated in two ways. The neural event *is* the mental event. Thinking "If the streets are wet, then it rained last night; the streets are wet; therefore, it rained last night" just *is* a causally connected series of neural events.

Let me be explicit: It seems to me that the "is" in the above sentence should be taken as the "is" of identity. The mental event and the physical event are identical in exactly the same way that hitting a particular musical note and moving one's vocal cords in a particular way are the same event. Here I go farther than Melnyk, who tells me by private communication that he remains agnostic as to the identity of mental events and their physical realizers. For me, "realization" is best construed as an identity relation between mental-act tokens and physical-act tokens. I think that we need to see that the essentialist Cartesian concepts of "mental" and "physical" as mutually exclusive categories is an obscurantist, religiously based holdover from the seventeenth century, one that should no longer have *any* place in our discussions of mind. The relation of mind to body will always seem unnecessarily mysterious until we thoroughly rid ourselves of that illegitimate bifurcation, often tacitly accepted by dualism's critics as much as by its defenders.

Physical realization therefore does not render the mental an epiphenomenon. On the contrary, it is the physical realization of the mental (i.e., mental/physical identity) that "empowers" one mental event to cause another mental event. Since the only causes are physical causes, it is *in virtue of* their physical realization that mental events can cause other, physically realized, mental events. But doesn't this mean that, as Goetz and Taliafero insist, it is the physical that does all the causing and the reasons are impotent? No, because, again, the physical causing of my belief and the being convinced by reasons are *one and the same thing*. Intellectually grasping that a certain conclusion follows from certain premises *is* a patterned sequence of neural firings. The physical causing *is* the mental causing; that is what PRM means, as I understand it. In short, PRM holds that being convinced by reasons is something we accomplish with our brains.

I think that Goetz and Taliafero read PRM as a thesis of the one-way causation of the mental by the physical, rendering the mental into an epiphenomenon. But this is wrong. PRM, as I understand it, is not saying that the physical produces the epiphenomenal mental like a car engine produces useless noise. It seems that Goetz and Taliafero misconstrue PRM because they continue to think in dualist terms. Dualism must regard the mental and the physical as two opposite and irreconcilable properties, whereas PRM erases that distinction and asserts that the mental is something that is done when certain special physical things are done. If the assumption that "mental" and "physical" are mutually exclusive categories is deeply entrenched for you, the claim of PRM will be hard to understand, much less accept.

Melnyk offers cogent responses to the other arguments by Goetz and Taliafero. Concerning the argument that PRM is counterintuitive because it denies our first-person experience of choice, Melnyk notes that this can be taken in a stronger or a weaker sense. He calls the weaker sense TG1 and the stronger sense TG2:

> (TG1) By introspecting one's own . . . choices, one acquires *some reason* to think that they are uncaused mental events.

> (TG2) By introspecting one's own . . . choices, one acquires an *indefeasible reason* to think that they are uncaused mental events. ("Reply," 2; emphasis in original)

Melnyk begins with the stronger claim, TG2, and shows that it is far too strong. To say that a reason that is "indefeasible" is to say that it provides a reason for a claim that is so strong that the claim is proven and no further evidence can overturn it.

> TG2 . . . is open to serious doubt. The doubt arises because, on the face of it, introspection is a perceptual faculty like vision or touch, different from more familiar perceptual faculties by being directed upon one's mental states rather than upon one's surroundings. But perceptual faculties like vision and touch can ever only yield defeasible reasons to believe their deliverances—partly because their deliverances are influenced by fallible background theory, partly because they are prone to malfunction, and partly because they are perfectly reliable even when they're functioning properly. Therefore, barring any reason to think that introspection is special, we should suppose that introspection too can only ever yield defeasible reasons to believe its deliverances. ("Reply," 2–3)

That is, unless Goetz and Taliafero can offer some reason for thinking that introspection is infallible and its deliverances indefeasible, TG2 must claim far too much. We might add that, on the contrary, it is clear that introspection is often wrong. Hilary Kornblith mentions an experiment in which four very similar items of the same price were laid out for display in a store, arranged from left to right (2002, 111). The experimenters found that, far more than would be expected on the basis of chance, people selected the rightmost item, whichever one it was. When asked to judge by introspection whether they might just be selecting the item just because it was placed to the right, the customers strongly denied the suggestion and insisted that their choice had been based on the quality of the item. Numerous experiments have shown that introspection is highly fallible and that we are highly prone to rationalizing when we introspect. In introspecting, as in ordinary perception, we often see what we want to see or expect to see.

With respect to the weaker claim, TG1, that introspection gives us some reason to think that our choices are uncaused, Melnyk notes that it is one thing not to perceive a cause and something else to perceive that there is no cause:

> We can agree that introspection does not represent our choices as having causes; but that doesn't entail what [Goetz and Taliafero] are claiming, i.e., that introspection reveals our choices as *not* having causes. ("Reply," 4)

It is one thing to fail to see a cause; it is something else entirely to see that there *is no* cause. Suppose Bob wears his argyle socks today. When asked why, he introspects and can find no reason prompting his fashion choice. However, the fact that Bob did not discover a reason does not mean that he saw by introspection that there was no reason. Perhaps subconsciously, Bob was prompted to wear them to irritate his office mate Sam, who hates argyle socks. Since Goetz and Taliafero offer no reason why the causes of our choices should be visible to introspection, then their weaker argument fails also.

What about the argument that we know from introspection that we are simple (i.e., lacking parts), substantial, unified selves, as dualism claims? Melnyk replies:

> Are [Goetz and Taliafero] right that introspection represents our minds (or ourselves) as entities that lack parts. Apparently not, because introspection has a very limited representational repertoire. It represents us as having a variety of mental properties—as undergoing perceptual and bodily experiences; as thinking various thoughts and as feeling various emotions, but it doesn't represent us as having any *other* properties than mental ones. Therefore it no more represents us as lacking parts than it represents us as having parts, or as spatially located, or as electrically charged, or as divisible by two prime numbers. And, as usual, we mustn't conclude that introspec-

> tion represents us as lacking parts just because it doesn't represent us as having parts. ("Reply," 6)

Once again, Goetz and Taliafero ascribe more to introspection than can be justified.

In sum, Melnyk clearly seems to get the better of the argument and to offer a cogent defense of PRM. Most of his arguments and rebuttals to Goetz and Taliafero also serve to support any MTB thesis. I will therefore take it for granted that the scientific evidence as well as philosophical argument support the claim that we are physical creatures in a physical world endowed, by evolution, with a marvelous organ that enables us to do those amazing things we call "mental." I opened this chapter with the rhetorical question: "Can we really imagine that the thunder of Beethoven's *Eroica* symphony or the sublimity of his *Moonlight* sonata arose from the massed firing of neurons?" The answer is yes. Yes, we can.

SCIENCE AND THE HUMAN IMAGE

The problem facing us now is that, given the truth of the MTB thesis, and everything else science tells us about ourselves, how does this impact our understanding of the human situation? What, if anything, are we going to have to change about our self-concept and self-image?

Lots and lots, says philosopher Alex Rosenberg. We have to give up free will and morality, for starters. How should we then live? Rosenberg's answer, seemingly paradoxically, is summarized by the words of the old reggae song: "Don't worry. Be happy." We will see below just how Rosenberg defends these claims, but first we have to ask just what aspects of our self-image seem to be endangered by the findings of science. Owen Flanagan spells out the details of what he calls "the manifest image," that is, the seemingly original, naïve, uncritical view of what it is to be a human being that people just naturally had before the rise of modern science. The question we address in the remainder of this chapter is how much of the manifest image must be replaced by the scientific image, the understanding of human beings that our science appears to endorse.

Is there a spontaneous, pre-scientific human self-image, a way that everyone just naturally thinks about themselves and others? This is a very hard question to answer, but it seems to me that the answer is no. One of the problems I have with Goetz and Taliafero's arguments is that they seem to assume that the self apparently revealed by their own introspection is universal, that is, that all humans throughout history would have seen the same thing when they looked inside. I'm not sure. I am not sure that an introspective hero of the homeric age or Chinese gentleman of the time of Confucius would have spontaneously seen the same thing that Goetz and Taliafero see when they look inside. I am not even sure that everyone today would affirm that introspection reveals to them that they are simple, substantial, unified selves. With Melnyk, I suspect that what we see when we look inside, like what we see by visual perception, is largely influenced by our beliefs, desires, and emotions.

Perhaps all we need to say about the alleged manifest image is that for very many people both educated and uneducated, both in the past and at present, it has seemed that humans are distinguished by having a faculty of free will and an awareness of moral concerns. For these persons (the vast majority of us, I assume), freedom and morality are essential components for a meaningful life. Nonhuman animals, on the other hand, seem to have no ability to choose freely, but act on instinct or impulse. Also, though they may have a strict "pecking order" that defines social roles, and certainly display love and empathy, they seem to lack the capacity for genuine moral reflection. Only humans can display virtue or vice. Animals cannot sin; even "bad" dogs don't go to hell. Further, only humans can *reason* about morality and seek a rational basis for our ethical judgments. So, let's limit ourselves to this question: Given that we are physical creatures and that our whole mental life is attributable to brain activity, and given everything else science tells us about ourselves, in what sense, if any, do we have freedom of the will, and is there still a rational basis for morality?

What does it mean to have free will? What are the specific attributes of such freedom? Flanagan lists ten attributes of free will that are important to us:

self-control
self-expression
individuality
reasons-sensitivity
rational deliberation
rational accountability
moral accountability
the capacity to do otherwise
unpredictability
political freedom (Flanagan, 2002, 104–105)

Each of these points should be briefly explained and their value to us indicated. Self-control does not mean declaring in ringing tones, "I am the master of my fate. I am the captain of my soul," like in William Ernest Henley's "Invictus." It is the less grandiose claim that *we*, our decisions, not external circumstance or internal compulsion, make the big difference in what we do. Nobody wants to think that they are helplessly blown about by forces outside of their control, like debris in a tornado. *We* want to be in charge of ourselves.

Self-expression and individuality mean that we want to be recognized as unique and not reducible to a type. We want the freedom to talk, act, and adopt a personal appearance and style that expresses who we really are and what makes us different. This is why teenagers, who are just developing these traits, are so resentful if told that they cannot dye their hair blue or get a nose piercing. As adults we are equally resentful if the homeowners association tells us that we must plant roses and that zinnias are unacceptable. At a deeper level, this is why we value the freedom to speak our minds on things that concern us.

Reasons-sensitivity, rational deliberation, and rational accountability are essential aspects of rationality. We consider ourselves rational people and we want others to see us the same way. We like to think that we are capable of making decisions about beliefs or actions on the basis of reasons and not as knee-jerk reflexes or conditioned responses. We prize the ability to reflect on our beliefs, desires, and values and to make decisions on the basis of those.

Moral accountability means that we want to be held responsible for our moral choices and to hold others responsible for theirs. Our whole legal system is based on the idea that each adult has the capacity to do or refrain from illegal activity and that, except for the hopelessly insane, nobody is compelled to break laws or act antisocially. Therefore, punishment is appropriate for lawbreakers.

We all value a measure of personal unpredictability and dislike it when people come to assume that we will always act in exactly the same way. We like to be able to surprise, if only in minor ways, say, by wearing a Hawaiian shirt every now and then or listening to music of a kind we don't usually favor. Also, we like think that our choices are not secretly compelled and that we could just as easily have done things other than we did. Yes, I voted for candidate A, but I like to think that I could just as easily have voted for candidate B; so, if elected, A had better work to keep my loyalty.

Finally, having all of the above would be worthless if we lived in a society where everything that is not forbidden is mandatory. No matter how much freedom and dignity you inherently possess, you cannot enjoy them if you are born in North Korea. Totalitarian societies seek to control all aspects of human behavior. Though they never succeed in imposing total control, they can create widespread misery.

So do we enjoy free will in a sense that encompasses the above desiderata? First, a bit of background: A hundred years ago, with tongue in cheek, Bertrand Russell called his *Introduction to Philosophy* a "shilling shocker." It costs a bit more that a shilling, but Alex Rosenberg's *The Atheist's Guide to Reality* is intended to be an unabashed, for-real shocker. Rosenberg espouses scientism. *Scientism* is normally a word with pejorative connotations. The accusation of scientism is made against those who appear to be saying that the vocabulary and concepts of science are the only ones that matter and that all nonscientific ways of thinking are bogus.

For instance, advocates of eliminative materialism (EM) have often been scorned as scientistic. EM is the claim that all purely mental vocabulary, such as talk about beliefs, feelings, ideas, pains, and the like belongs to the discredited and outworn concepts of "folk psychology" and should simply be scrapped and replaced by the accurate terms of neuroscience. Thus, we would systematically replace any locution expressing a "folk psychological" judgment or claim with one that makes reference only to neurological occurrences. Example (two eliminative materialists in a tender moment): "Dear, that flimsy negligee you are wearing is stimulating a strong response in my hypothalamus!" Needless to say, many people still hold that we need the "folk" vocabulary of love (or lust) and that adoption of EM would impoverish our lives immeasurably.

Rosenberg freely embraces the label "scientism" and holds that a very great deal of what Flanagan called the manifest image will have to go. His tone is uncompromising and even flippant. Here is how he thinks science requires us to answer the "big questions" about the meaning of life:

> *Is there a God?* No.
> *What is the nature of reality?* What physics says it is.
> *What is the purpose of the universe?* There is none.
> *What is the meaning of life?* Ditto.
> *Why am I here?* Just dumb luck.
> *Does prayer work?* Of course not.
> *Is there a soul? Is it immortal?* Are you kidding?
> *Is there free will?* Not a chance!
> *What happens when we die?* Everything pretty much goes on as before, except us.
> *What is the difference between right and wrong, good and bad?* There is no moral difference between them.
> *Why should I be moral?* Because it makes you feel better than being immoral.

Is abortion, euthanasia, suicide, paying taxes, foreign aid, or anything else you don't like forbidden, permissible, or sometimes obligatory? Anything goes.
What is love and how can I find it? Don't look for it; it will find you when you need it.
Does history have any meaning or purpose? It's full of sound and fury but signifies nothing.
Does the human past have any lessons for our future? Fewer and fewer, if it ever had any to begin with. (Rosenberg, 2011, 2–3)

Wow. Atheist debaters often face religious apologists who make precisely some of the charges against atheism that Rosenberg enthusiastically affirms. Anyone trying to get a hearing for atheism might think that with "friends" like Rosenberg, we don't need enemies! In fact, though, he rejects theism and he equally repudiates secular humanism (277–82). Clearly, Rosenberg has taken on a big task and has a heavy burden of proof. Since we have introduced the topic, let's consider his arguments against free will.

Rosenberg says that the scientific case against free will is simple and direct:

> The mind is the brain, and the brain is a physical system, fantastically complex, but still operating in accordance with all the laws of physics—quantum or otherwise. Every state of my brain is fixed by physical fact. In fact, it is a physical state. Previous states of my brain and the physical input from the world together brought about its current state. They were themselves the result of even earlier brain states and inputs from outside the brain. All these states were determined by the laws of physics and chemistry. These laws operated on previous states of my brain and states of the world going back to before my brain was formed in embryogenesis. . . . When I make choices—trivial or momentous—it's just another event in my brain locked into this network of processes going back to the beginning of the universe, long before I had the slightest "choice." Nothing was up to me. Everything—including my choice and my feeling that I can choose freely—was fixed by earlier states of the universe and the laws of physics. End of story. (236)

Goetz and Taliafero could hardly have put it any better in making their case against the MTB thesis!

Someone might object that Rosenberg is assuming strict determinism, when we know that indeterminism rules at the quantum level. At the quantum level the old bumper sticker slogan applies (cleaning it up): Stuff happens. Might not quantum indeterminacy make room for free choice? However, Rosenberg notes that if a random, uncaused quantum event in the brain were to initiate a chain of causes resulting in an act of choosing, that act would be no freer than the effect of a deterministic causal chain. The "choice" would still just be something that happened to me, not something *I* did. Personal agency is a necessary aspect of free choice (237).

But what about introspection? Goetz and Taliafero based much of their argument on the data seemingly given by self-reflection. Rosenberg debunks such appeals:

> Introspectively it just feels like you choose; it feels like it is completely up to you whether you raise your hand or stick out your tongue. That feeling is so compelling that for most people it tips the scale against determinism. They just know "from inside" that their will is free. (238)

However, for Rosenberg, introspection is no more reliable than it was for Melnyk.

Rosenberg cites famous experiments done by Benjamin Libet and replicated often since (152–54). In these experiments, subjects were asked to perform a simple task, like pushing a button whenever they wished. They also recorded when they first consciously decided to push the button. It takes about 200 milliseconds from the time of one's decision to push the button to the actual flexing of the wrist. Yet 500 milliseconds before the wrist flexed, Libet detected activity in the subjects' motor cortex that initiated the wrist flexing. In other words, the "choice" came *after* the process was already initiated by the brain! Apparently, the brain initiates both the "choice" and the movement! Hence, the subjects' subjective perception of freely choosing was an illusion. We think that "we" make conscious choices but our brain makes them for us—and then makes us think that our conscious choices had something to do with it!

I think that many philosophers, myself included, would reply to Rosenberg as follows: "Quite unaccountably for a professional philosopher you seem to overlook the compatibilist position on free will, namely, that causal determinism is quite compatible with free will in the everyday sense, and, in fact, that this mundane sense encompasses all the freedom we need. In the everyday sense, being free to choose means that I am not compelled, either externally or internally, but get to decide a course of action based on *my* beliefs, *my* values, and *my* desires. This is quite compatible with determinism. Freedom consists not in being exempt from causation, but in the ability to deliberate, either with ourselves or with others on the proper courses of action to realize our ends. Philosophers going back to Aristotle have identified human autonomy—not with indeterminism—but with our ability to make and execute rational plans for the realization of our purposes."

Rosenberg, however, will have none of this idea of freedom as deliberation either:

> Since the brain cannot have thoughts about stuff, it cannot make, have, or act on plans, projects, or purposes it gives itself. Nor, for that matter, can it act on plans that anyone else favors it with. There are no plans. That's just more of the illusion Mother Nature exploited for our survival. (238)

Is he serious? We never think about things? We never make plans? We never have purposes? This seems absurd. However, the history of philosophy is replete with ideas that appear quite absurd on their face (e.g., Berkeley's claim that matter does not exist; Wittgenstein's assertion that you cannot know that you are in pain; Quine's claim that we can never really know what speakers of other languages are talking about), but these assertions were not gratuitous absurdities but offered on the basis of rigorous argument. Hence, even if, in the end, they are actually absurd, they cannot be dismissed but have to be argued out. So we have to examine Rosenberg's arguments for these extraordinary claims. As the saying goes, though, extraordinary claims require extraordinary evidence, and we are justified in putting a heavy burden of proof on Rosenberg's arguments.

Thoughts appear to be about things. For instance, I seem to be thinking about my cat and the endearing but sometimes-irritating way that she will climb onto my chest, even when I am trying to drink my morning coffee and read the newspaper. "Intentionality" is what philosophers call this "aboutness" of thought. Our thoughts seem to have this intentional quality even when we are considering things that do not exist. For instance, it certainly seems to me that I can think about mermaids, centaurs, Santa Claus, honest politicians, and other nonexistent beings. Intentionality seems to just be a datum, a given of our conscious experience. Indeed, intentionality and qualia—the felt qualities of sensations, like the smoothness of silk or the richness of cream—seem to constitute the *essence* of consciousness. Conscious states just *are* those qualitative or intentional states. How could we be wrong about that?

I think that many—perhaps most—philosophers would at this point just dismiss the claim that intentionality is an illusion as a pathological aberration. I think that they would justify peremptory dismissal like this: That all thinking is thinking about is just a datum of consciousness. The idea that there could be a contentless thought seems just ridiculous, in fact, contradictory, like saying that there could be a colorless green nightgown. Therefore, in denying intentionality, Rosenberg is denying thought and asserting that we do not think. Such an assertion is no more worthy of philosophical rebuttal than the assertion that one is a poached egg.

Rosenberg's response would be that the process of introspection that supposedly reveals these alleged data of consciousness is demonstrably highly unreliable. We saw earlier in our criticism of Goetz and Taliafero that philosophers often claim to know far more by introspection than is justifiable. But the fact that introspection does not tell us some things does not mean that it tells us nothing. That thought has objects is not an intuition. It is far deeper than that. Trying to deny that I am thinking about *something* whenever I think is as self-defeating as trying to deny Descartes's *cogito*. The very act of attempted denial just succeeds in introducing a new object of thought, namely the thought that I am not thinking about anything.

Just dismissing Rosenberg's claim is therefore an eminently understandable response. Still, since Parmenides, one purpose of philosophy has always been to push the limits of reason, to see whether reason can justify claims that seem false, even outrageously so. I think that this is an important, if, no doubt, exasperating and fatiguing, enterprise that is worth doing only if, in the end, we return to our familiar beliefs with a deeper understanding of why we hold them.

Rosenberg's basis for rejecting intentionality is that thought is a physical thing—a configuration of neurons and their states in the brain—and no physical thing can be about another physical thing. Let's consider the kinds of thoughts we call memories. My wife and I visited Paris in 2004, and we have many wonderful memories from that trip. Such memories must be encoded in highly complex neural connections, but neural connections are just physical states, like the wires of an old-fashioned telephone exchange. Physical states and things may be related to each other in a number of ways, but none seems to be capable of being *about* another. We may use a key to open a lock, but the key is not *about* the lock. How, then, can my memories, a physical state, be about some other physical entity—the city of Paris?

But are not some physical things in fact about other physical things? Isn't an octagonal red sign with letters spelling out "S-T-O-P" about the physical act of stopping your car? However, a stop sign, considered as a physical entity, is nothing but metal shaped and colored in certain ways. It is no more intrinsically about stopping a car than a pair of green trousers hung in the intersection. Rosenberg puts it this way:

> There is nothing that is intrinsically "Stop"-ish about red octagons. Downward pointing yellow triangles—yield signs—could have been

> chosen as stop signs as well. Red octagons are about stopping because we interpret them that way. We treat them as the imperative . . . expressed in English as "Stop!" (176)

Well, then, might not the brain serve as its own interpreter? Might not the brain interpret some of its own neural states as being about Paris? Yet the interpreter in the brain could be nothing other than *another* neural state, and we face the same problem all over again, namely how one physical state can be about another one. If neural state #2 is the interpreter of neural state #1, interpreting it as being about Paris, then there would have to be a *third* neural state to interpret #2 as being about #1! And then the same problem arises for #3! Clearly, rather than solving the problem, we are on the road to an infinite regress. Our proposed solution just reiterates the same problem, that is, how one physical state can be about another physical state.

The conclusion seems to be that we cannot have any memories about Paris, since memories are physical things and Paris is a physical thing, and physical things do not bear any intrinsic aboutness relations to each other. One would have to be interpreted as being about the other, but because the interpreter in the brain can only be another physical thing, we are just making the problem worse. Further, what holds for memories would hold for any other thoughts, including those that would be about plans or purposes.

This argument seems to rest on an equivocation on the word *memory*. By "memory" we could mean either the physical traces in the brain that encode information and are passively awaiting retrieval, or we could mean the active, conscious memories that we create by accessing that stored information through a process we call "remembering." Paul Thagard describes the process of recalling a concert you had once attended:

> Retrieval of a memory works by reactivating a pattern of firing in a population of neurons. Suppose someone starts telling you about another concert that is similar to the one you went to, perhaps because the bands played the same kind of music. Hearing about

> the new concert may produce a pattern of firing in roughly the same population of neurons that encoded the various aspects of the old concert. The newly generated pattern of firing will then generate additional neural activity by virtue of synaptic connections, possibly producing a pattern of firing that is roughly similar to your original experience. That activation of a firing pattern of neurons constitutes your recalling the memory. (2010, 49).

By reactivating a particular pattern of neuronal firings, the brain draws on stored, encoded information to create an active, conscious memory. Memory is a *creative* process, by the way; remembering is not playing back a recording but an active (and error-prone) reconstruction.

My memories of Paris in the sense of patterns of neuronal connections caused by a trip to Paris are there whether or not I am consciously remembering Paris. Memory in that sense—patterns encoded in the brain—are not intrinsically about Paris any more than the travel guide to Paris is when nobody is looking at it. When nobody is looking at the travel guide, it is just marks on paper, just another physical thing, and likewise for the stored information encoded in the neuronal connections in a brain. However, when we actively remember and that stored information is drawn on to create a conscious recollection, then we have more than passive storage. We have a conscious (self-aware) process of remembering, actively reconstructing memories by drawing on stored information—a process physically realized in the reactivated patterns of neuronal firing noted by Thagard. Memory in this latter sense can most definitely be about Paris or anything else.

At a more fundamental level, Rosenberg seems to conflate the doer with what is done. The singer is the doer, and singing the aria is what she does. She accomplishes this remarkable vocal feat entirely with her physical apparatus for creating and projecting sound, and, of course, physical sound is nothing but a vibration in the air. However, what makes her performance the singing of an aria is not any property of her anatomy or of the air. It is the fact that the sounds she creates realize an abstract pattern, one created by the composer and the librettist. The aria—that abstract pattern—is not a physical thing; it is a particular way of organizing sounds that can be physically realized in innumerable ways. The doer—the singer—is a physical being and she performs her feat entirely by physical processes. Yet what she accomplishes cannot be explained entirely in terms of physical properties or processes but must also mention the abstract, nonphysical, pattern realized in her performance. What is true for arias also seems true of the contents of thoughts like memories. Thought-contents are abstractions that can be realized by indefinitely many thinkers (note: abstract thought-contents that can be realized by many different thinkers are what philosophers call "propositions"). I can think the proposition "The Louvre displays Daumier's *Liberty Leading the People*," and so can my wife; and maybe so can a Martian or a machine.

The conclusion is that speaking of a brain's relation to Paris is not like talking about the relation of one lump of clay to another. Lumps of clay are passive; they just sit there and one definitely cannot have any sort of "about" relation to the other. The living brain, however, is not a lump but is constantly, fantastically active. It is doing things all the time, and Rosenberg has given us no reason whatsoever to think that it is incapable of generating thoughts about things. One of the many amazing things a brain can do is to think about Paris. If you can think about Paris, then you can think about plans, purposes, and the whole intentional shebang.

Rosenberg would not be impressed with the above line of argument. What the brain *does*, he indicates, is no different from what the brains of sea slugs or rats do, and no different, in principle, from what computers do. What sea slugs and rats do when they learn is to rewire their neuronal connections to create new input/output circuits that create a new habit. Thus, sea slugs and rats can be conditioned to acquire new habits, and the conditioning works by rearranging neuronal connections. Rats, for instance, can learn how to locate a life raft in a water tank. There is no reason to think that sea slugs or rats acquire new habits by thinking about them. Their brains simply change to correlate input with different output. The same thing seems to have happened with you when, in early infancy, you learned to recognize your mother's face:

> When the rat acquires and stores information "about" the location of the life raft in the tank, that's just the neurons in its hippocampus being reorganized into new input/output circuits. They have changed in the same way the neurons in the sea slug have changed. Similarly, knowing what your mother looks like or that Paris is the capital of France is just having a set of neurons wired up to an input/output circuit. (185)

Neither the sea slug nor the rat nor you must think about these things to get them right. It is all input and output.

Computers can do very complex tasks that previously only humans could accomplish, like playing chess well enough to beat the human champion or even to excel at the TV game show *Jeopardy!* But computers, like Watson, the computer that plays *Jeopardy!*, do not think about things at all. Once again, it is merely input and output. Watson is cleverly programmed so that when the input is a *Jeopardy!* answer, the output is something we interpret as the appropriate *Jeopardy!* question (in *Jeopardy!* you are given the answer and required to give the appropriate question). Rosenberg says that the brain is just a computer:

> [The brain is] composed of an unimaginably large number of input/output circuits, each one a set of neurons electrically connected with other through their synapses. The circuits transmit electrical output in different ways, depending on their electrical inputs and on how their parts—individual neurons—are wired up together. That's how the brain works. (188)

So, in answer to my suspicion that he is conflating the doing with what is done, Rosenberg could reply that we know precisely what the brain does, and the brain does just what a computer does. If a computer is incapable of thinking about anything, then the brain is incapable of thinking about anything.

Unquestionably many of the things our brains do can be explained in terms of input/output, and intentionality does not enter into it at all. Recognizing your mom's face is certainly one of these; you have been doing it automatically since you were a small infant, and doing it without any thought at all. The same goes for many other operations of our brains. For instance, as soon as I see certain politicians' faces on TV, I immediately begin to mutter expletives. There must be an input/output circuit in my brain such that when the input is an image of politician P, the output is a string of monosyllabic words of Anglo-Saxon derivation.

On the other hand, some of our mental operations, *prima facie*, certainly are not easily explicable *merely* in terms of input/output, that is, stimulus/response reactions. Consider doing philosophy. Can we believe that Rosenberg's production of *The Atheist's Guide to Reality* was *nothing but* a complex concatenation of automatic inputs and outputs? Didn't he have to *think about it*? This is not an *ad hominem* argument or a too-quick "hoist with your own petard" kind of maneuver. Rather, I am just pointing out that reflective thought, and innumerable other things we do with our brains, certainly are not obviously *merely* input/output events. Let's put it this way: If reflective and creative thought (recall Beethoven's *Eroica* symphony again), for instance, are explicable in terms of inputs and outputs, we can say that such an explanation currently exists only as a promissory note. That puts it far too weakly. It is more like a third-party, postdated check drawn on a bank in Burundi.

I think that Rosenberg's basic fallacy is not to mistake the doer with what is done but to think that what the part is doing must be what the whole is doing. Invoking a musical analogy once again, no musician performs a symphony; symphonies are performed by orchestras. Each individual in the orchestra plays his or her part, and the result of a hundred people doing that correctly is a symphony. An individual neuron *is* just an input/output device. However, thinking involves the very complex, hierarchical, multiply patterned, parallel-processed, and causally looped interactions of ensembles of millions or billions of neurons: the cerebral symphony.

With respect to *any* activity involving the complex, coordinated interaction of numerous units, any question about what is being done *can only* be answered by specifying the level that we are talking about. Otherwise the question is meaningless. What is your car doing as you drive down the road? What the car is "doing" depends on what level of organization you are talking about. The fuel-injection system is doing one thing, and the engine-cooling system is doing something else. Locomotion is an emergent property that comes in only when we are talking about the vehicle as a whole. By analogy, asking "What is your brain doing?" is meaningless unless the level of organization is specified. The organization of the brain begins with individual neurons, which are organized into systems, which are organized into systems of systems, and then systems of systems of systems . . . all interacting in astonishingly complex ways with multiple feedback loops and cross-connections. Reflective thought seems to be one of the higher-level activities of the brain. Individual neurons cannot consider philosophical propositions, but millions or billions complexly interacting in the right way maybe can. Perhaps "Philosophy" is one of the tunes the cerebral orchestra can play.

But how can individual events that are about nothing add up to an event that is about something? Suppose that we arrange ten thousand people in a stadium in a 100 × 100 square. On each seat is a specific monochrome card for the occupant to hold up on cue. On cue, each holds up his or her card, and when seen from across the stadium, the cards create a ten-thousand-pixel portrait of, say, Barack Obama. Each individual card is a portrait of nothing; it is just a solid color. However, when displayed all at once and in the correct order, they create a detailed and highly accurate portrait. Of course, this is just an analogy, but it does indicate how a picture of something can be constituted entirely by bits that are pictures of nothing.

It is not even clear that Rosenberg is right about computers and their capacities. Is it certain that no computer, or system of computers, could ever be programmed to think about things? Why not? Perhaps Rosenberg has an argument demonstrating the impossibility of intentional machines, but it is not to be found in *The Atheist's Guide to Reality*.

Before continuing, a nagging question that has been waiting in the wings during this whole discussion must finally be addressed: How can I think that I am thinking about *P* without being able to think that *P*? Conscious experiences have the peculiar property that *appearing* to be in a conscious state *is* to be in that conscious state. If it *seems* to me that I have a splitting headache, then I *do* have a splitting headache. If I *seem* to be hearing the opening notes of *Eine Kleine Nachtmusik*, then I *am* having the experience of hearing the opening notes of *Eine Kleine Nachtmusik*, even if I am having an auditory hallucination. How, then, can I *seem* to be thinking about, say, my cat, unless I am *actually* thinking about my cat? Well, maybe I have made a dreadful mistake and the cat that I am thinking about as mine is actually my neighbor's cat and not mine. But even in this case I am still thinking about *something*, namely, my neighbor's cat.

I think that Rosenberg must at least admit that when I think that I am thinking about my cat, that is, I seem to be thinking about her coloring, her temperament, and so on, then my subjective experience—the way it seems to me—will be the same as if I were *actually* thinking about her. How could it be any different? What content would the state of *really* thinking about her have that only *seeming* to think about her would lack? But if a brain can achieve the one state—thinking that I am thinking about my cat—how can it not be able to achieve the subjectively identical state—actually thinking about my cat? How can one mental act be possible and a second phenomenologically identical one be impossible? Rosenberg may have an answer, but I have not found it.

I think that the above points offer some considerations for doubting Rosenberg's conclusions, but we still need to evaluate his specific evidence. First, let's remind ourselves of just what Rosenberg is claiming and just how heavy a burden of proof we are putting on him. Rosenberg is admirably clear on just what he is claiming; he bites every bullet:

> Having thoughts about the future is necessary for having plans, purposes, designs. In fact all you need to have a plan is to think about something you want in the future and about how to go about getting

> it. That much is obvious. But if your brain cannot think about anything, it can't have thoughts about the future. . . . Since there are no thoughts about things, notions of purpose, plan, or design in the mind are illusory. Farewell to the purpose driven life. Whatever is in our brain, driving our lives from cradle to grave, it is not purposes. (205)

Actually, for Rosenberg *all* the deliverances of introspection and "folk psychology" (i.e., our commonsense ways of understanding human behavior) are illusory. We cannot interpret human action—ours or anyone else's—in terms of plans, purposes, desires, aims, goals, wishes, hopes, fears, or what we normally regard as motives of any sort. We have a stark choice: We can take the illusions of introspection or we can take what neuroscience allegedly tells us:

> Once you recognize that there is no way to take seriously both what neuroscience tells us about the springs of human action in the brain and what introspection tells us about it, you have to choose. Take one fork and seek interpretation of human affairs in the plans, purposes, designs, ideologies, myths, or meanings that consciousness claims actually move us. Take the other fork, the one that scientism signposts, and you must treat all the humanities as the endlessly entertaining elaborations of an illusion. They are all enterprises with no right answers not even coming closer to approximating our understanding of anything. You cannot treat the interpretation of human behavior in terms of purposes and meaning as conveying real understanding. (Rosenberg, 2011, 213)

Shakespeare? Entertaining bunk. History? Hokum. Poetry? Mellifluous malarkey. *Pace* Shakespeare, people are never motivated by greed, lust, malice, jealousy, vindictiveness, or ambition. All such accounts are pseudoexplanations derived from introspection and justified by the cracker-barrel platitudes of folk psychology. What, then, makes us tick? The brain unconsciously makes decisions for us. Consciousness only comes in to fool us into believing that we do things for reasons.

Clearly, we are abundantly justified in placing an onerous burden of proof on such claims. How good, then, has been Rosenberg's evidence? The short answer is: woefully inadequate. Despite his claim to be speaking in the name of science, he has provided little relevant scientific evidence—certainly nothing sufficient to justify such momentous conclusions. He cites the experiments of Libet on choice and Kandel on the neurology of sea slugs and rats, but such results are far too limited to support the sort of sea change Rosenberg wants. The neuroscience he does cite refers mostly to lower-level events such as the input/output functions of neurons. As we noted, however, such information tells us nothing about what brains are doing at higher levels of organization. At those higher levels, at the level of consciousness, Rosenberg admits that the brain does many remarkable things. For instance, it creates massive, obsessive, species-wide illusions of intentionality and generates all of our manifold folk-psychological pseudoexplanations, the whole first-person viewpoint that lies at the base of our laws, literature, and religions. The brain is capable of doing all that, but it is not capable of thinking about a cat.

There perhaps *could* be scientific evidence that apparent deliberation about ends and the decisions we ostensibly base on them are illusory. Consider a decision it seems to me that I recently made: I had known for months that I needed to renew my driver's license by my birthday, but, of course, I procrastinated until my birthday week. I then had to decide which two-to-three-hour (with luck) block of time I could spend at the DMV. Waits at the DMV are shorter in the morning, and I hate standing in long lines or sitting in uncomfortable chairs next to people with screaming babies, so I opted to go some morning. I couldn't do it Monday morning because I had papers to grade. Tuesday I had to be teaching off campus all day. Wednesday there were faculty meetings and committee meetings. It turned out that Friday morning was the only time I had the whole morning free to wait in the line and fill out the forms at the DMV, so I decided to go Friday morning. Or so it seemed to me.

Now I suppose that, in principle, a neuroscientist could have been remotely monitoring my brain all during this time and could observe that my brain had already decided to go on Friday morning before I sat down and consulted my schedule. My brain then created the elaborate illusion that I had made up my mind about this by deliberating. However, so far there is no such evidence and it does not seem that such evidence will be forthcoming soon.

Besides, there is something decidedly odd about Rosenberg's appeal to science. Why trust science? What gives it more authority than our folk notions? Rosenberg says that science is "just common sense continually improving itself, rebuilding itself, correcting itself . . ." (167). So science begins as common sense improving itself. What does this mean? I think it means that our folk epistemology is gradually replaced by a scientific epistemology as our methods get more objective, our techniques more efficient, our observations keener, our tests more rigorous, and so on. But why think that folk epistemology is any more trustworthy than folk psychology? If it is not, then science, by building on common sense would be building on quicksand. In fact, as we noted in the last chapter, epistemology of any sort is expressed in the form of norms, norms that we should follow in our epistemic behavior. Yet Rosenberg eloquently and emphatically denies that we can base decisions on norms. For him, the brain does what it does, and when we invoke norms these can only be *post hoc* rationalizations that serve the illusion of purposefulness.

A deeper worry is whether Rosenberg thinks that science leads to truth. What would "truth" be for Rosenberg? Truth is normally conceived as correct representation—saying of what is that it is and of what is not that it is not. But if there is no aboutness, then there is no representation, and if there is no representation, then there can be no such thing as representing truly. I think that if he is to be consistent, Rosenberg would have to advocate a thoroughly pragmatist conception of truth. That is, if we say that the claims of science are "true," we can only mean that science delivers the goods, that is, it gives us what we want. But if it all comes down to what we want, then it seems a good bet that people will end up wanting the goods of the first-person viewpoint and folk psychology (e.g., Shakespeare, poetry, religion) over the purported benefits of what Rosenberg regards as a scientific view of human nature.

At this point I will merely reassert that Rosenberg has not met the burden of proof. Further, we can affirm most of the desiderata that Flanagan listed as essential for our "manifest image," such as self-control, self-expression, individuality, reasons-sensitivity, and rational deliberation. There is no reason, or at least none that Rosenberg mentions, to see such values as incompatible with a scientific worldview that holds human actions to be caused as much as anything else. For instance, what is self-control other than making decisions based on *my* values, *my* desires, and *my* beliefs? Of course, what I value, desire, and believe may not be matters of free choice. I hope not. I hope that they are determined, determined by my accurate perceptions of what *is* valuable, *is* desirable, and *is* rationally believable.

We still need to consider Rosenberg's rejection of ethics, but we can be much briefer with this. Rosenberg says that science entails nihilism (95). In claiming this, he differs greatly from the neo-Aristotelian ethical naturalists (like Larry Arnhart, whom we considered in the last chapter) who claim to base objective moral values in biology. Rosenberg says that, on the contrary, if we accept what science tell us about ourselves, we must conclude that there are no objective moral truths or values. To the question of whether something is right or wrong there can be no answer. No side in any ethical dispute is correct since there is no moral "right" or "wrong":

> There are questions about the morality of stem cell research or abortion or affirmative action or gay marriage or our obligations to future generations. Many enlightened people, including many scientists, think that reasonable people can eventually find the right answers to such questions. Alas, it will turn out that all anyone can find are the answers that they like. The same goes for those who disagree with them. Real moral disputes can be ended in lots of ways: by voting, by decree, by fatigue of the disputants, by the force of example that changes social mores. But it can never be resolved by finding the correct answers. There are none. (96)

Rosenberg defines *nihilism* as follows:

> Nihilism denies that there really is such [a] thing as intrinsic moral value. People think that there are things that are intrinsically morally valuable, not just as a means to something else: human life or the ecology of the planet or the master race or elevated states of consciousness. But nothing can have that sort of intrinsic value—the very kind of value morality requires. Nihilism denies that there is anything at all that is good in itself or, for that matter, bad in itself. (98)

Rosenberg gives two premises that, he claims, entail nihilism:

> First Premise: All cultures, and almost everyone in them, endorse most of the same moral principles as binding on everyone.
> Second Premise: The core moral principles have significant consequences for humans' biological fitness—for our survival and reproduction. (102)

Rosenberg admits that the first premise may seem highly implausible, given the protracted, bitter, and seemingly irresolvable ethical disputes that roil within and between human societies. However, he reasonably holds that the intractable nature of moral disputation is not due to any fundamental incompatibility or incommensurability of basic standards or intuitions but is due to factual disagreement (105). Even Nazis did not have a *fundamentally* different view of core morality; their brutality was due to wildly false factual beliefs about Jews, Roma, gays, and other of their victims (105–106).

I would put it this way: Nearly all humans, even many of those who are tyrants, fanatics, or murderers, share basic ethical intuitions, but emotion, selfishness, weakness of character, and ideology pervert these intuitions, turning them into justifications for mayhem. For instance, though many want to dismiss terrorists like the 9/11 hijackers as mad dogs, almost certainly each saw himself as a righteous warrior fighting for a holy cause. Surely each would have justified himself much as American Air Force General Curtis LeMay, who ordered the firebombing of Japanese cities in World War II, would have: War is war, and terrible things must be done to serve the higher good. The problem with the 9/11 terrorists was not that their moral sense was deactivated; on the contrary, it was hyperactive. Their problem was that that they had been imbued with an insanely extremist Islamist ideology that identified all Americans as enemies of God who must be killed in the cause of holy jihad.

For the sake of argument, let's accept Rosenberg's two premises; I think that most neo-Aristotelian ethical naturalists like Arnhart would. What follows? Rosenberg says that nihilism follows from these premises. How? If we accept these two premises, we have to accept that humans would universally endorse the same core ethical values and would regard them as correct. If *these* values are the ones that were selectively advantageous for early humans, it is hardly surprising that they are the ones accepted by members of the human species. Neither is it surprising that we would regard them as the *right* ones since natural selection would do its job thoroughly and make sure that these were deeply psychologically ingrained. Therefore, we have an evolutionary explanation of *what* basic moral values humans will share and *why* these will universally seem to be the *right* ones (104–108). But how exactly is nihilism supposed to follow?

Rosenberg presents the moral realist, one who holds that there are objective moral values (like the neo-Aristotelian ethical naturalist), with a dilemma: If you insist that the basic, universal human moral intuitions are *right*, you must either (a) hold that the evolutionary fitness of our core morality makes it right, or (b) that the core morality is independently correct and it just so happened that natural selection coincidentally landed on the correct morality (109). The second alternative is obviously unreasonable. Natural selection is blind to any considerations other than reproductive fitness. It preserves what works to enhance reproductive fitness and destroys what does not. End of story. That it would blindly land on just the morality that had, say, an *a priori* justification would seem to be beyond rational belief.

What, then, is wrong with the first horn of the dilemma, that fitness makes rightness? Rosenberg dismisses this possibility:

> Natural selection filtered out all the other variant core moralities leaving only ours. It won the race, and that is what made the last surviving core morality the right, correct, true one. This makes the rightness, correctness, truth of our core morality a result of evolutionary fitness. But how could this possibly be the answer to the question of what makes our core morality right? There doesn't seem to be anything morally right about having lots of kids, or grandchildren or great grandchildren, or even doing things that make having kids more likely. But this is all the evolutionary fitness of anything comes to. (110)

Being the morality that survives the Darwinian process no more makes it the right one than being the last one standing in a barroom brawl makes someone a good man. Since neither horn of the moral realist's dilemma is acceptable, nihilism must follow. Rosenberg does proceed in his next chapter to try to take the sting out of this conclusion by saying that his nihilism is "nice" nihilism and does not require that we change our usual moral behavior. Following the precepts of ordinary morality will be the easiest and most convenient thing for us. what are the practical consequences of the elimination of morality? None, really. Don't panic and just carry on as before (115–45).

What, then, is Rosenberg's argument against making reproductive fitness, or things conducive to reproductive fitness, the basis of moral value? It looks like an appeal to intuition:

> There doesn't seem to be anything morally right about having lots of kids, or grandchildren or great grandchildren, or even doing things that make having kids more likely. (110)

The apparent appeal to intuition here—how things seem—is a bit odd in a book otherwise devoted entirely to disputing ordinary intuitions about nearly everything. What lies behind such an intuition? Perhaps it is just the old assumption, traceable at least back to Hume, that denies that moral value can derive from fact, that is, that an "is" can imply an "ought." Or perhaps it is the feeling that *this* fact, reproductive fitness, is not the kind of thing that can ground value. How can moral value derive from mere pullulation?

However, ethical naturalists deny the purported is/ought dichotomy and argue that the only, or most reasonable, grounding of morality is in human biology. Human beings are so constituted that they naturally find certain states, conditions, or activities to be intrinsically rewarding, fulfilling, or meaningful. Recall the transcultural, transhistorical desiderata identified by Arnhart in the previous chapter—such things as parental care, sexual identity, sexual mating, social ranking, justice as reciprocity, political rule, religiosity, intellectual understanding, health, beauty, and wealth. Human flourishing consists in the enjoyment of these naturally desirable ends along with the possession of the traits of intellect and character that are conducive to the attainment of these natural goods. A person in such a state would seem to be in the condition that Aristotle called *eudaimonia*.

Why do humans find certain states, conditions, or activities inherently rewarding, fulfilling, or meaningful? Almost certainly because such things are conducive to the reproductive fitness of either the individual or the group and natural selection has made us like what is reproductively useful. Religiosity, for instance, may at first seem unlikely to be conducive to fitness. However, as we noted in the last chapter, some biologists, like David Sloan Wilson, have theorized that religiosity enhances group fitness, giving groups that share religious practices and beliefs an enhanced cohesiveness and cooperativeness over competing groups. Daniel Dennett offers another evolutionary explanation, namely that religiosity piggybacked on the "hyperactive agent detection device" that conferred greater fitness on its possessors. If some such evolutionary accounts are correct, the consequence is that creatures with the biological nature of humans are bound to find some things intrinsically valuable, that is, worthwhile for their own sakes, even though there is an evolutionary explanation of why these things are found naturally desirable. That is how value is grounded in fact.

At this point, Rosenberg and others are likely to object like this: Supposing then that human evolutionary history has made certain states, conditions, or activities inherently rewarding, fulfilling, or meaningful for human beings, what gives these things *moral* significance? Given that we can identify some human state as *eudaimonia*, where does the *moral* imperative come in that says that it *ought* to be promoted?

To the ethical naturalist this is an odd question. What could moral duty be except the imperative to actualize value? If certain natural states are bound to be found to be valuable by human beings, morality would have to consist in following those norms that are conducive to the realization of such value. In other words, for ethical naturalists moral value supervenes on those norms that *in fact* are conducive to naturally valuable things. Moral imperatives are hypothetical imperatives that tell us what we should do if our end is human flourishing.

But such an account will still not *feel* right to many people, perhaps including Rosenberg. I think the reason for this feeling is that, as Alasdair MacIntyre claimed in *After Virtue* (see the previous chapter), since the Enlightenment we have tended to identify moral imperatives as categorical imperatives—pure, *a priori* "oughts" that cannot have any factual grounding. However, as we argued in the last chapter, all that a categorical imperative can amount to is a demand for consistency, and such a demand is far too weak to constitute a basis for ethics. So the idea that morality can be grounded in a categorical imperative is illusory.

It seems, then, that the neo-Aristotelian ethical naturalist can look at the same set of facts that Rosenberg does and draw a very different conclusion. Such a naturalist will see evolutionary theory as not only laying a basis for values but also as explaining their objectivity. Human biology guarantees that certain states, conditions, and activities will be experienced by us as deeply and intrinsically rewarding, fulfilling, and meaningful. Though these things serve the further end of enhancing group or individual fitness, our *experience* of them is that they are *intrinsically* good—satisfying for their own sake and for no higher good. We will therefore endorse as the correct principles of morality those norms that *in fact* are conducive to the realization of those goods for human beings and condemn those behaviors that tend to prevent their realization. Rosenberg's claim that science entails nihilism, even "nice" nihilism, is false. You can have your science and your ethics too.

APE, ANGEL, OR NEITHER?

On the 25th of November, 1864, British prime minister and master orator Benjamin Disraeli, addressed the assembled churchmen of the Oxford Diocesan Conference, and he referred to the controversies that still roiled subsequent to the publication of Darwin's *On the Origin of Species*. The exact wording is not known, but it went something like this:

> What is the question now placed before society with the glib assurance which to me is most astonishing? The question is this: is man an ape or an angel. I, my lord, I am on the side of the angels! (*Wikiquote*)

The "ape or angel" debate has played out again in this chapter. Like Disraeli, Goetz and Taliafero take the side of the angels. In fact, they attribute to humans a truly God-like ability, the ability, as Owen Flanagan puts it, to be an uncaused cause, to make choices that are a kind of creation *ex nihilo* (ix). In their view, we know by introspection that we have a freedom to choose that is unconstrained by empirical circumstance; we are free from the nexus of causality that rules the physical cosmos. As Goetz and Taliafero see it, to understand human persons you must see them as not merely physical, but as possessing—being—a spiritual entity, a soul.

Rosenberg, on the other hand, takes the side of the apes; rather, he goes even farther and takes the side of the machines. You are a computer and what you do is what a computer does. A computer does it all by input and output, and that is how you do everything too. For obscure reasons, evolution has conferred consciousness on us, and the function of consciousness is to pull off the immense and intricate hoax, worthy of a Cartesian evil demon, of creating the first-person viewpoint that comprises the illusions of intentionality, freedom, and purposefulness. Really, we have no more intentionality, freedom, or purposefulness than a computer—which has no more of these than does a rock.

We have adduced reasons for choosing neither option. The first-person viewpoint won't do nearly as much as Goetz and Taliafero hope, but it is not a universal delusion, either. What is the alternative? How do we conceive of humans as neither ape nor angel?

The alternative, as Aristotle recognized long ago, is that humans are rational animals. We have all of the passions and drives of other animals, but we also have a rational faculty that is uniquely ours. You grossly misunderstand human beings if you neglect or diminish either the animal of the rational aspects. Both must be given their full due. We are animals; in fact, we are primates, closely related to the great apes, so closely related that our evolutionary lineages diverged only five million years ago. We share over 98 percent of our genes with the chimpanzees, our evolutionary cousins to whom we are more closely related than rats are to mice. On the other hand, as Aristotle said, we have a seemingly unique ability to reason and to base our actions on reasons.

Many have regarded human nature as a deep mystery. It does not seem so to me. It seems to me that humans are exactly what you would expect to get if you take a peculiar kind of social, bipedal ape living in the Great Rift Valley of central Africa and endow it with intellect and language. You get a creature with a remarkable capacity to learn, and, thanks to its linguistic aptitude, with a facility for cognitive collaboration, leading to the accumulation and transmission of knowledge. Yet you still have a creature with vehement passions and an apish temperament—and one far more capable of causing trouble by virtue of that superior intellect.

So the general nature of human nature is not too mysterious. Of course, Rosenberg says that "folk psychology" and the first-person viewpoint are comprehensively erroneous pseudoexplanations, and we see their futility when we see how poorly they predict human behavior (211). Does "folk psychology" poorly predict human behavior? Such a claim is so vague and has so many ostensible counterexamples and necessary qualifications that it is hard to know what to make of it. In many ways people seem depressingly predictable.

Of course, we are not good at predicting the human future in any detail. Had you asked people fifty years ago what we would be doing now, in the second decade of the twenty-first century, most would probably have said that we would be touring the solar system, not sitting around staring at computer screens. But all complex things defy detailed, long-term predictions, even when we have quite a good understanding of their causes. For instance, evolutionary theory explains the history of life on planet earth quite cogently and comprehensively, but any predictions about the future course of evolution are speculative since there are just too many imponderables. So, predictive capability is not always the best test of the adequacy or acceptability of explanations. There is, then, no reason to think that neuroscience will consign Shakespeare to the rubbish heap of folk-psychological pseudoexplanations. On the contrary, neuroscience may someday *catch up* with Shakespeare when it comes to understanding human beings, but it still has a long way to go.

Even if we can have our science and our Shakespeare too, then, as we saw in chapter 3, many will still fear that if we fully accept the implications and assumptions of evolutionary theory and neuroscience, including the MTB thesis, we will be pushed toward metaphysical naturalism and atheism. "Intelligent design" activist Professor Phillip E. Johnson argues this (see references for chapter 3). Let's assume, for the sake of argument, that a thoroughly scientific worldview strongly supports metaphysical naturalism. Christian philosopher William Lane Craig holds that a naturalistic worldview, with its concomitant atheism, implies that human life is meaningless:

> If God does not exist, then both man and the universe are inevitably doomed to death. Man, like all biological organisms, must die. With no hope of immortality, man's life leads only to the grave. His life is but a spark in the infinite blackness, a spark that flickers, and dies forever. Compared to the infinite stretch of time, the span of man's life is but an infinitesimal moment; and yet this all the life he will ever know. . . . For though I know now that I exist, that I am alive, I also know that someday I will no longer exist, that I will no longer be, that I will die. This thought is staggering and threatening: to think that the person I call "myself" will cease to exist, that I will be no more! (1994, 57)

The implied premise of this passage is that life must be everlasting to be meaningful, but there is no obvious reason for thinking that. Why not draw the opposite conclusion—that the fact that life is short is our motivation for filling it with meaning? If we are keenly aware of "time's wingéd chariot," we will strive to fill our lives with experiences that, as Aristotle noted, are intrinsically valuable; we will seek *eudaimonia*. We will fill our lives with learning and gaining wisdom, with healthy physical activity, with compassion for the less fortunate, with enjoyment of beauty in nature and art, with love for friends and family, with doing a job well, with fighting against evil and obscurantism, and, yes, with enjoying sex, TV, pizza, and ball games. To live such a life is to flourish as a rational animal. If someone still denies that a life rich with the enjoyment of such goods is meaningful, then he or she must have a deeply dubious notion of *meaning*.

Evolution has made some wonderful things: sauropod dinosaurs longer than two city buses; predators of immense cunning, speed, and strength, from *Velociraptor* to the Bengal tiger; tiny toxic frogs that advertise their lethality with brilliant color; fragile birds that annually migrate thousands of miles over forbidding terrain; penguins that cradle an egg on their feet through the howling darkness of an Antarctic winter; mayflies that hatch in their millions, and have but one day to find a mate; and, at the heart of it all, the double helix of DNA, the archives of organic form, incomprehensibly ancient and potentially immortal.

Arguably, though, the most remarkable of evolution's creations is the human brain, an organ with the potential to save or ruin the world. Our brains are physical objects, subject to the same kinds of causes and laws that govern everything else, and nothing miraculous happens between our ears. Yet we are just beginning to figure out what the brain does and how it does it. The twentieth century was marked by a number of definitive discoveries, for example, subatomic particles, the expansion of the universe, and the structure and function of DNA. Perhaps the twenty-first century will be the century of the brain. Perhaps the main reason we should be suspicious of people who tell us that the brain is "only" this and "only" does that is that we still have so much to learn.

Georg Wilhelm Friedrich Hegel, perhaps the leading German philosopher of the nineteenth century, famously held that philosophy only issues its judgments after the fact, when the real intellectual work has already been done. Actually, the reverse is the case. Philosophers rush in where scientists cannot yet tread, making bold leaps of conjecture and extrapolation and pronouncing on their projections. Nobody can be sure how we will understand human beings a hundred years from now when neuroscience may be much more complete. From what we see so far, though, it looks like Disraeli's "ape or angel" dichotomy is a false dilemma and that Aristotle already had it right over 2,300 years ago: We are neither apes nor angels but rational animals.

FURTHER READINGS FOR CHAPTER EIGHT

Oliver Sacks's books are riveting. It is hard to recommend one over another. If you had time to read only one, I would recommend *An Anthropologist on Mars: Seven Paradoxical Tales* (New York: Vintage Books, 1995). It contains the story of "the last hippie" and six other equally enthralling, and sometimes-disturbing stories. It is hard to read these stories and resist the conclusion that we *are* our brains.

The Internet contains a little gold and much dross. One site that is gold is the Secular Web. It has both a modern and a historical library of secular thought, and it also publishes theistic responses to those critiques. In 2007, the Secular Web hosted the "great debate" between Andrew Melnyk for physicalism and Stewart Goetz and Charles Taliafero against. Melnyk's essay "A Case for Physicalism about the Human Mind" opened the debate. Taliafero and Goetz replied in "Objections to Melnyk's Case for Physicalism." Melnyk replied to these criticisms in "Physicalism and the First-Person Point of View." The page references to Taliafero and Goetz refer to the essay just mentioned. The page numbers for Melnyk refer to his first essay when the parentheses contain "Case" and to the second essay when it says "Reply." I also quote from Goetz's chapter "Human Persons Are Material and Immaterial (Body and Soul)" in *Debating Christian Theism*, edited by J. P. Moreland, Chad Meister, and Khaldoun A. Sweis (Oxford: Oxford University Press, 2013) pp. 261–69. This volume, by the way, is a lively set of debates between Christians and their critics. I was privileged to contribute to this volume as well, on a *very* different topic than the ones we are discussing here. See "Heaven and Hell," pp. 534–45.

The appeal to NDEs is often associated with "new age" types and has a rather-dubious reputation. Two respectable authors, J. P. Moreland and Gary Habermas, support this line of argument in *Immortality: The Other Side of Death* (Nashville, TN: Thomas Nelson, 1992).

As mentioned in the "Further Readings" section for the last chapter, William Lyons, *Matters of the Mind* (New York: Routledge, 2001) is an outstanding book. The quote is from page 25. The dualist *tu quoque* is found, for instance in Sandra Menssen and Thomas D. Sullivan's *The Agnostic Inquirer: Revelation from a Philosophical Standpoint* (Grand Rapids, MI: William B. Eerdmans Publishing, 2007), p. 108. I reviewed this book on the online *Notre Dame Philosophical Reviews* in 2008. Theists often respond to their critics with *tu quoque* arguments. I find that consistently these arguments have rhetorical bark but little logical bite. The reference to Hilary Kornblith is once again to his *Knowledge and Its Place in Nature* (Oxford: Clarendon Press, 2002).

As is clear in the text, I think that there are many deep problems with Alex Rosenberg's *The Atheist's Guide to Reality: Enjoying Life without Illusions* (New York: W. W. Norton, 2011). Nevertheless, I strongly urge that everyone read it. At the end of the last chapter I raised the question of the value of philosophy in a scientific age. Philosophers have a freedom, unmatched by any other intellectual discipline, to say things that are not only false but outrageously so. Sometimes, indeed, it is hard to tell the difference between a philosopher and a high-IQ crackpot. But we live in a culture of astonishing intellectual laziness and complacency, and sometimes it is good to have someone come along and say, in the words of the old Firesign Theater album, *Everything You Know Is*

Wrong! Sometimes it takes radical claims, backed by clever arguments, to motivate us to stop being intellectual couch potatoes and to do some real thinking. Rosenberg's book is very valuable in that regard. It is also written in a very lively and in-your-face style that is refreshing to encounter in a philosophical work.

Owen Flanagan's remarkably readable and insightful book *The Problem of the Soul: Two Visions of Mind and How to Reconcile Them* (New York: Perseus Books Group, 2002) is the antithesis to Rosenberg's manifesto. Flanagan modestly does not claim that dualism has been flatly disproven. How could it be? He does clearly think that the evidence is overwhelmingly in favor of the MTB thesis. If we do accept that thesis, how should we live and how do we think of ourselves? Flanagan says that we must bite the bullet on some of our hopes and aspirations, but that much of the "manifest image" survives, sometimes in altered form. I would put things a bit more strongly. I think that the fact that we think (and feel, imagine, perceive, etc.) with our brains is just a discovery of science and no more dubitable than the fact that organisms have evolved over geological time. Concomitantly, I view the continued efforts of dualists to resist this conclusion as of the same sort as "intelligent-design" advocates who oppose evolution.

Paul Thagard knows more about the brain than Alex Rosenberg and is just as committed to understanding life scientifically, but he reaches very different conclusions. His book *The Brain and the Meaning of Life* (Princeton: Princeton University Press, 2010) is a wonderfully lucid primer on how the brain works and the implications for human life of the advances in neuroscience. Thagard is a most helpful and readable antidote to the exaggerated claims and *non sequiturs* of Rosenberg. Like Flanagan, Thagard concludes that much that has always made life meaningful survives even when we fully accept that the mental is something we do with our brains.

Benjamin Disraeli's "on the side of the angels" pronouncement no doubt strikes modern readers as glib. He was not an intellectual lightweight, however. To be a Jew and to become prime minister in Victorian Britain, you had to have more than glibness going for you. *Wikiquote*, s.v. "Benjamin Disraeli," accessed May 16, 2014, http://en.wikiquote.org/wiki/Benjamin_Disraeli.

The quote from William Lane Craig is from his *Reasonable Faith: Christian Truth and Apologetics*, revised edition (Wheaton, IL: Crossway Books, 1994). In addition to being a distinguished philosopher, Craig is also a professional debater. I have had the honor of debating him twice, once at Prestonwood Baptist Church in Dallas in 1998 and in 2003 at Indiana University at Bloomington. The first debate is available on the Internet, but the second is hard to get. I respond to Craig at greater length on the topic of the meaning of life in my essay "Seven Common Misconceptions about Atheism," which is available at the Secular Web.

INDEX